Reso₁

Water as a Catalyst for Peace

Examining international water allocation policies in different parts of the world, this book suggests that they can be used as a platform to induce cooperation over larger political issues, ultimately settling conflicts. The main premise is that water can and should be used as a catalyst for peace and cooperation rather than conflict.

Evidence is provided to support this claim through detailed case studies from the Middle East and the Lesotho Highlands in Africa. These international cases – including bilateral water treaties and their development and formation process and aftermath – are analyzed to draw conclusions about the outcomes as well as the processes by which these outcomes are achieved. It is demonstrated that the perception of a particular treaty as being equitable and fair is mainly shaped by the negotiation process used to reach certain outcomes, rather than being determined mechanistically by the quantitative allocation of water to each party.

The processes and perceptions leading to international water conflict resolutions are emphasized as key issues in advancing cooperation and robust implementation of international water treaties. The key messages of the book are therefore relevant to the geopolitical and hydro-political aspects of water resources in the context of bilateral and multilateral conflicts, and the transboundary management of water resources, which contributes insights to political ecology, geopolitics, and environmental policy.

Ahmed Abukhater is the Global Director of Product Management for Pitney Bowes, and an Assistant Professor of Geography and Planning at the University at Albany, State University of New York, USA. Previously, he served as Esri's Global Industry Manager for Planning and Community Development and Director of PLACES in California, an instructor at the University of Texas at Austin, and Department Director, Planning & Sustainability Department, Ministry of Planning, Gaza, Palestine. He holds a Ph.D. in Community and Regional Planning from the University of Texas at Austin, with a focus on transboundary water resources management and conflict resolution and hydro-diplomacy, a Master's degree in Urban and Regional Planning from the University of Illinois at Urbana-Champaign and a Bachelor's degree in Architectural Engineering. Throughout his career, Dr. Abukhater has authored numerous publications, served on many governing and advisory boards, and received over 20 prestigious awards for his work.

Earthscan studies in water resource management

Water Management, Food Security and Sustainable Agriculture in Developing Economies
Edited by M. Dinesh Kumar, M.V.K. Sivamohan and Nitin Bassi

Governing International Watercourses
River basin organizations and the sustainable governance of internationally shared rivers and lakes
By Susanne Schmeier

Transferable Groundwater Rights
Integrating hydrogeology, law and economics
By Andreas N. Charalambous

Contemporary Water Governance in the Global South
Scarcity, marketization and participation
Edited by Leila Harris, Jacqueline Goldin and Christopher Sneddon

For more information and to view forthcoming titles in this series, please visit the Routledge website: *http://www.routledge.com/books/series/ECWRM/*

Water as a Catalyst for Peace

Transboundary Water Management
and Conflict Resolution

Ahmed Abukhater

Routledge
Taylor & Francis Group

LONDON AND NEW YORK

earthscan
from Routledge

First published 2013
by Routledge
2 Park Square, Milton Park, Abingdon, Oxon OX14 4RN

Simultaneously published in the USA and Canada
by Routledge
711 Third Avenue, New York, NY 10017

Routledge is an imprint of the Taylor & Francis Group, an informa business

© 2013 Ahmed Abukhater

British Library Cataloguing-in-Publication Data
A catalogue record for this book is available from the British Library

Library of Congress Cataloging-in-Publication Data
Abukhater, Ahmed.
Water as a catalyst for peace : transboundary water management and
conflict resolution / Ahmed Abukhater.
pages cm. — (Earthscan studies in water resource management)
Includes bibliographical references and index.
1. International rivers—Political aspects. 2. Water resources
development—Political aspects. 3. Water resources development—Law
and legislation. 4. Water-supply—Political aspects. 5. Peace-building. 6.
Conflict management. I. Title.
JZ3700.A25 2013
341.4′42—dc23
2013002618

ISBN: 978-0-415-64213-2 (hbk)
ISBN: 978-0-203-08111-2 (ebk)

Typeset in Goudy
by Prepress Projects Ltd, Perth, UK

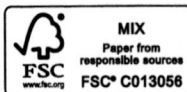

Printed and bound in Great Britain by
TJ International Ltd, Padstow, Cornwall

This book is dedicated to my heroes: my deeply beloved parents *Baha' El-Deen* and *Fatima*, who are still standing in their homeland and yearning for freedom. Despite insurmountable adversity, they set the most poignant example through their patience, courage, and humane character, charting the course for a more hopeful and peaceful destiny.

I also dedicate this work to my wonderful wife *Hilary* and my three beautiful sons, *Baha' El-Deen*, *Yusef*, and *Mustafa*, who came into this world during my journey through this work.

And finally, this work is dedicated in the memory of my late friend and mentor Professor Kent Butler, whose guidance helped kindle the foundation and development of this book . . . he is greatly missed.

"Water breathes life into landscapes. Water binds people together and divides people from one and other. Water enables some people to prosper while others struggle. Water divides rich from poor, the healthy from the ill, the leafy from the dusty. This is especially evident in times and places of aridity. The fair allocation of water lies at the heart of Ahmed Abukhater's *Water as a Catalyst for Peace*. The author views water as a potential 'venue for cooperation' and suggests useful strategies to pursue this course. Furthermore, Dr. Abukhater argues that water can be a catalyst for cooperation, peaceful interactions, and regional stability. As a result, his book offers a clarion framework for those interested in equity, environmental sustainability, and a peaceful planet."

Frederick Steiner,
Dean, School of Architecture and Henry M. Rockwell Chair in Architecture,
University of Texas at Austin, USA

"We tend to think of scarce natural resources as generators of conflict. This book turns that idea on its head and urges us to think of co-riparians as more than just rivals. Abukhater's principles of process equity help us tame the wicked problem of cross-border water allocation."

Paul C. Adams,
Ph.D., Associate Professor and Director of Urban Studies,
University of Texas at Austin, USA

"A good addition to the many existing water negotiation frameworks, especially for those practitioners who desire to link equity, in its myriad forms, into the process of governance. Written through the direct experience of a native Palestinian, the book introduces another transformative approach to conflict resolution that integrates rules of engagement, mechanisms of engagement, and neutral third-party mediation that is worthy of serious consideration by Middle East scholars in water diplomacy."

W. Todd Jarvis,
Director, Institute for Water and Watersheds,
Oregon State University, USA

"It is often thought that water issues are divisive and a likely source of war. Many water experts consider this mistaken, and that water can and should be a source of cooperation. Abukhater's book examines various water disputes and shows that water treaties can indeed lead to cooperation and mutual benefits."

Franklin M. Fisher,
Jane Berkowitz Carlton and Dennis William Carlton Professor of
Microeconomics, Emeritus,
Massachusetts Institute of Technology, USA

Contents

Figures

Tables

Foreword

Ahmed Abukhater is trying to help us come to grips with issues of fairness in transboundary water resource allocation – process fairness, outcome fairness and perceived fairness. He focuses on the Jordan–Israel Water Treaty of 1994 and the Lesotho Highlands Water Project Treaty of 1986. He selected these two cases after considering a larger global sample. He uses his findings to build an argument for a "transformative approach to water conflict resolution." Although water allocation decisions must always be understood in their unique historical, ecological, and political context, he makes a careful effort to generate prescriptive advice that will be useful to water negotiators all over the world.

Abukhater scans his global sample of water conflicts to get a sense of the historical factors and negotiation methods that affect levels of perceived fairness in water allocation disputes. The bilateral nature of the Jordan–Israel agreement, for example, left out key parties. This has led to long-term instability. The secret nature of the negotiations undermined the perceived sense of fairness in the minds of some internal constituencies, particularly on the Jordanian side. Water agreements that fail to take ecological impacts and natural system requirements into account (and include set-asides of water to maintain basic ecosystem services) are unstable. In the Jordan–Israel negotiation this failure has led to continued environmental stresses on water resources. In the Lesotho–South Africa case, the failure to take adequate account of environmental impacts raised ongoing concerns during project design and construction.

The South African–Lesotho negotiations also reveal that successful water allocation negotiations depend on the provision of real benefits to both sides. South Africa got the water it needed while Lesotho got the economic aid it required. Both perceived the result as fair. We do not see the same mutually advantageous outcome in the Israel–Jordan agreement. Israel maintained control over the water it required, but it is not clear that Jordan realized commensurate benefits. (Perhaps Israel's recent commitment to generate 70 percent of its water from desalination by 2040 will produce tangible benefits for the Jordanians.)

As we think about the prerequisites for fairer water agreements, it is important to assess who is invited to the table, what information (including forecasts) they have at their disposal, how negotiations are managed, what form the agreement takes (and how adaptable it is), what the parties can realistically expect to achieve if the negotiations fail, what the prevailing legal standards are for fairly allocating water resources, and what the prospects are for compliance with whatever agreements are reached. In many transboundary water negotiations, as Dr. Abukhater points out, the goal is to head off armed conflict and not necessarily to improve long-term relationships. But, if shared management is approached properly, it can help to transform long-standing conflicts. Although the mechanics of water management ("low politics") can be worked out even if larger political battles ("high politics") are not resolved, a transformative approach to conflict resolution will ultimately have to address these core issues.

When I think about recent research on water security, the key process lessons Dr. Abukhater highlights fit quite nicely. First, perceptions of fairness regarding the allocation of water are crucial to the successful implementation and evolution of water agreements. Second, the involvement of trusted neutrals (such as the World Bank in the Lesotho–South Africa case) may be enormously helpful in keeping treaty negotiations on track and ensuring that equity concerns (both perceived and actual) are addressed. Third, transparency, both in terms of the accountability of the negotiators to the constituencies they represent and the engagement of civil society representatives, is important to perceived equity. Fourth, information sharing (or joint fact finding) can enhance the legitimacy of any negotiated outcome. Finally, to people on the ground – the actual water users – equity has more to do with whether their needs are met than with symbolic victories. Coming away with a right to more water than your neighbor does not necessarily translate into the provision of water when and where you need it. New water-sharing arrangements that meet the needs of all sides simultaneously are much more important.

It is vital to the long-term success of transboundary water management that water professionals and water users give more attention to the issues raised in this book.

Lawrence Susskind
Ford Professor of Urban and Environmental Planning, MIT
Vice Chair, Program on Negotiation at Harvard Law School
March 15, 2013

Executive summary

This book examines international water allocation policies in different parts of the world and suggests that such policies can be used as a platform to induce cooperation over larger political issues, and ultimately to settle conflicts. The main premise is that water can and should be used as a catalyst for peace and cooperation rather than conflict. Evidence is provided to support this claim through detailed case studies from the Middle East and the Lesotho Highlands in Africa. These international cases – including bilateral water treaties, their development and formation process, and aftermath – are analyzed to draw conclusions about outcomes, as well as the processes by which these outcomes are achieved.

The persistence of water conflicts in many arid regions is not simply a matter of water shortages, but rather the lack of equitable agreements that govern the allocation of disputed water resources to mitigate the adverse impacts of hostility and resentment. As such, equity problems are at the heart of many transboundary water disputes. Mindful of the dynamics and implications of inequitable water allocation on inter-state relationships and on the overall stability of a region, this book aims to elicit and develop pragmatic criteria for the equitable distribution of water (*process equity*) as a route to creating equitable outcomes and their perception as such. This book makes the distinction between *process equity* and *outcome equity* and their impact on attaining and sustaining water security, peace, and hydro-stability. These parameters of equitable processes are developed through a review of current literature addressing the issues of water equity in arid regions, coupled with case study analyses, cross-case comparisons, and semi-structured interviews with key water negotiators. These key cases are selected through a systematic screening methodology that analyzes nine pertinent cases. The in-depth analysis and findings are applicable to other international water disputes, particularly in the context of arid regions.

It is demonstrated that the perception of a particular treaty as being equitable and fair is mainly shaped by the negotiation process used to reach certain outcomes, rather than being determined mechanistically by the quantitative allocation of water to each party. The processes and perceptions leading to international water conflict resolutions are emphasized as key issues advancing cooperation and robust implementation of international water treaties. The key messages of the book are therefore relevant to the geopolitical and hydro-political aspects of water resources in the context of bilateral and multilateral conflicts, and the transboundary management of water resources, contributing insights to political ecology, geopolitics, and environmental policy.

Proposing an alternative strategy that views water as a catalyst for peace and cooperation rather than conflict and altercation, this book further advocates the development and adoption of an interdisciplinary, transformative approach to conflict resolution to advance water disputes through plausible and implementable agreements. Aiming to inform the theory and practice of hydro-diplomacy along disputed water resources, this approach encapsulates three key components, including rules of engagement, mechanisms of engagement, and neutral third-party mediation. Water satiety is identified as a major characteristic of equitable water allocation agreements that ensure the level of satisfaction of all stakeholders involved and the extent to which acceptable agreements, durable implementation, and sustainable relationships among co-riparians are attained and maintained.

Acknowledgments

It is my great pleasure and privilege to have the opportunity to write this book, where I am able to stitch together both theoretical and practical ideas for the advancement of transboundary water management and conflict resolution. These ideas were shaped by my own firsthand experience that reinforced the major theme of the book, which I have held so close to my heart throughout my entire life. Being a native Palestinian from the Gaza Strip, raised during the Palestinian *Intifada* (uprising) in a region engulfed in a weighty political power struggle, I became cognizant of the genuine value of water as the sustainer of *life* and *peace*.

I owe special thanks to Tim Hardwick and Ashley Wright at Earthscan, Routledge/Taylor & Francis Group, for the invitation to write this book, and for their continued support. I also would like to express my sincere gratitude to Professor Paul Adams of the University of Texas for his constructive guidance and feedback. With his contribution and tireless effort, this research was brought into fruition.

I greatly appreciate the time and feedback of all interviewees who agreed to participate in this work, particularly Ben Crow, Marwa Daoudy, Munther Haddadin, Ramaswamy Iver, Ainun Nishat, David Phillips, Uri Shamir, Juha Uitto, Melvin Woodhouse, and Mark Zeitoun.

Much of the credit for any merit this work may possess goes to my parents, who planted the seeds of inspiration for the making of this book. I am truly grateful for their support and continuous guidance and insight toward the development and enrichment of this work. I am extremely indebted to their unconditional love and unwavering support.

Gratitude and admiration is expressed to my wonderful and beloved wife, Hilary, for her invaluable support and thoughtful feedback, without which this book would have been incomplete. She has been by my side in all stages of this research, the outcome of which bears her imprint.

Acronyms and abbreviations

ADR	alternative dispute resolution
ADZ	Armistice Demarcation Zone
ANC	African National Congress
AWIRU	African Water Issues Research Unit
BATNA	best alternative to a negotiated agreement
BAR	Basins at Risk
BCM	billion cubic meters
BNP	Bangladesh National Party
CBM	confidence-building measure
CCL	conflict–cooperation level
CFA	Cooperative Framework Agreement
CIA	Central Intelligence Agency
CSIS	Center for Strategic and International Studies
CSRD	Centre for Sustainable Regional Development
DBSA	Development Bank of South Africa
DOP	Declaration of Principles
DWA	Department of Water Affairs
EAP	Environmental Action Plan
EIB	European Investment Bank
EXACT	Executive Action Team
FAO	Food and Agriculture Organization of the United Nations
FRIEND	Flow Regimes from International and Experimental Network Data
GAP	Greater Anatolia Project
GDP	gross domestic product
GBM	Ganges–Brahmaputra–Meghna
GLOWA	Global Change and the Hydrological Cycle
GOLD	General Organization for Land Development
HDR	Human Development Report
IBWC	International Boundary and Water Commission
ILC	International Law Commission
IDRC	International Development Research Center
JFF	joint fact finding

JPTC	Joint Permanent Technical Commission
JRC	Indo-Bangladesh Joint Rivers Commission
JTC	Joint Technical Committee
LHDA	Lesotho Highlands Development Authority
LHWC	Lesotho Highlands Water Commission
LHWP	Lesotho Highlands Water Project
MCM	million cubic meters
MAP	mean annual precipitation
NBI	Nile Basin Initiative
NGO	non-governmental organization
NRLW	Water Development and Management Unit
NWRS	National Water Resources Strategy
ORACOM	Orange-Senqu River Commission
ORP	Orange River Project
OVTS	Orange–Vaal Transfer Scheme
PASSIA	Palestinian Academic Society for the Study of International Affairs
PISGA	Palestinian Self Governing Authority
PLO	Palestine Liberation Organization
PNA	Palestinian National Authority
PWC	Permanent Water Commission
PWV	Pretoria Witwatersrand Vereeniging
SACU	Southern African Customs Union
SADC	African Development Community
STN	Single-Text Negotiation
TCTA	Trans-Caledon Tunnel Authority
TFDD	Transboundary Freshwater Dispute Database
TVA	Tennessee Valley Authority
UN	United Nations
UNDP	United Nations Development Programme
UNEP	United Nations Environment Programme
UNESCO	United Nations Educational, Scientific and Cultural Organization
UNRWA	United Nations Relief and Works Agency
US	United States (of America)
USSR	Union of Soviet Socialist Republics
VNJIS	Vioolsdrift and Noordoewer Joint Irrigation Scheme
WHO	World Health Organization
ZOPA	Zone of Possible Agreement

Introduction

till taught by pain
Men really know not what good water's worth

(Lord Byron)

We cannot live without water. Water is essential for human survival and development. The significance of water as an essential human need has led to the creation of many dramatic and complicated conflicts among different nations throughout human history. Simply stated, nations vie for their fair share of water for different uses, but the intricacy of man-made boundaries and the natural delineation of the concerned parties as upstream or downstream riparians make the issue of international water disputes a formidable and volatile one (Mirumachi, 2007).

Reflecting the growing competition for water in terms of quality and quantity, the issue of water shortage has become a driving force and the focus of many conflicts around the world in general, and more specifically in the context of arid regions. In many arid regions, water availability trends provide evidence of catastrophic water shortages, and water supply remains an unresolved conundrum facing many nations in these regions. For example, in the Palestinian–Israeli conflict, it is one of the major issues, along with Jerusalem and the right of return, that have been deferred to the final future round of negotiations. Consequently, water has been presented as a source of serious disagreement and outright conflict in many parts of the world. This potential obstacle to a lasting peace agreement could be viewed as good news, however, in the sense that water could propitiously be a source of cooperation, rather than a *casus belli* (Fisher, 2006). In arid regions, water is an extremely confined and precious natural resource that has evidently triggered wars in the past; however, this does not mean that it could not possibly become the reason for peace in the future. In support of this argument, the former UN Secretary General Kofi Annan stated that "fierce competition for fresh water may well become a source of conflict and wars in the future," but he later added, "[T]he water problems of our world need not be only a cause of tension; they can also be a catalyst for cooperation. If we work together, a secure and sustainable water future can be ours."[1]

Building on these observations, I contend that the water problem in arid regions is not simply a matter of water shortage, but rather it is related to the lack of equitable water allocation. To that end, Mirumachi (2007: 177) astutely points out that "there is a growing consensus that water scarcity is not the major and sole factor that prompts war." The persisting inequity in its allocation to and across national boundaries fosters *hydro-hostility* (and therefore expropriation of water) instead of *hydro-stability* and multinational cooperation. Far too often, water resource allocations are challenged on grounds of equity. Equity is, not surprisingly, a major concern in water conflict management. To that end, Wolf (1999: 4) states, "application of an 'equitable' water-sharing agreement along the volatile waterways of the world is a prerequisite to hydropolitical stability which, finally, could help propel political forces away from conflict in favor of cooperation." He also adds:

> Not surprisingly, up-stream riparians have advocated that the emphasis between the two principles be on "equitable utilization," since that principle gives the needs of the present the same weight as those of the past. Likewise, down-stream riparians (along with the environmental and development communities) have pushed for emphasis on "no significant harm," effectively the equivalent of the doctrine of historic rights in protecting pre-existing use.
>
> (Wolf, 1999: 4)

Many states have recognized that equity is at the core of such conflicts and a variety of conflict resolution methods have been developed to manage the equitable sharing of transboundary water resources, as well as to create peaceful relationships among riparian states. Nevertheless, severe conflicts over water still exist in which a group of people receive more water than they actually need, while others suffer as a result. As is the case in many arid regions of the world, inequitable water distribution is common and means that certain groups will suffer from lack of water access, whereas other groups may enjoy disproportionate benefits from accessing more than enough water for their needs (Isaac, 1999; Shmueli, 1999). Developing water allocation processes based on principles of equity tends to foster optimal water management, which in turn procures an atmosphere conducive for seeding cooperation and rooting out altercation; however, the path to equitable water management is unclear and even the meaning of "equitable" remains unresolved. Understanding the true meaning and implications of such an elusive concept is crucial in order to successfully manage shared natural resources among multiple users, and, meanwhile, boost environmental justice for all.

In light of the aforementioned conceptualization of the nature and characteristics of environmental conflicts in arid regions, the issue of equitable water allocation emerges as a fundamental concept in international water negotiation and is therefore the focus of this book. Aiming to study the importance of equitable water allocation, I am interested in differentiating between *process equity* and *outcome equity* and their impact on attaining and sustaining water security, peace, and hydro-stability. In particular, their impact on hydro-diplomacy and treaty negotiation will be given particular emphasis in order to decipher the relationship between *process equity* and *perceived equity*: the extent to which an outcome of a water agreement is perceived to be equitable. Toward that end, this book explores whether there are general principles of process equity that can be applied to the theory of water negotiation and conflict resolution. Through detailed case analysis, this book hopes to determine if and how these general principles can be applicable.

In the spirit of objectivity and transparency, I would like to explain why I chose to study the topic of equity in transboundary water resources allocation. This research is motivated by personal experience at the heart of the Palestinian–Israeli conflict, which exhibits many aspects found in disputes over natural resources around the world. The Palestinian–Israeli conflict is very intriguing as it illustrates a situation in which conflicting representations of hydrological resources, coupled with a power imbalance, have created tension, conflict, and injustice. In addition, it is regarded as the *sine qua non* from which many troubling aspects emanate in different parts of the Middle East and beyond. The desire to, ultimately, shed some light on this conflict motivates certain aspects of this study such as the selection of case studies, a third of which are located in the Middle East.

Plan of the book

Topic focus

> Water is the true wealth in a dry land; without it, land is worthless or nearly so. And if you control water, you control the land that depends upon it.
>
> (Wallace Stegner in Gleick, 1993: 9)

Focusing on the topic of equitable water resources allocation, as manifested in binational water treaties, this book examines the interplay of water allocation and equity attainability in the context of arid regions through a multinational, geopolitical, hydropolitical analysis of a few selected cases. The impact of inequitable allotment of scarce water resources on the degree to which water-related conflicts are adequately resolved is given considerable attention. Equity is a vague and disputable concept which is examined in considerable detail in this book. Literature on water negotiations is unclear on what constitutes universal parameters for equitable water allocation and in what situations equity can appear to be refuted. The notion of equity is hard to quantify and can mean different things to different people. Although compelling and desirable, the goal of boosting equitable allocation of water resources seems unattainable when the determinants of the term *equity* are inconclusive. Therefore, there is a need to define an acceptable interpretation of the notion of equity in the context of water resources allocation. Because delineating a clear definition and well-rounded criteria of equity is essential to resolving the main issues surrounding water allocation disputes in arid regions, establishing such parameters of process equity, as distinguished from outcome equity, is also warranted and useful.

This book attempts to characterize this divergence of interpretation and unravel the ambiguity surrounding the concept of water-sharing equity by examining the application (or lack thereof) of equity in sharing transboundary water resources. The idea is to ascertain the extent to which equity in water allocation schemes and international treaties has been attained. The primary objective is to develop criteria for process equity applicable to theories that support water conflict resolution grounded on an understanding of the outcomes of the case studies analyzed and the processes responsible for creating them. In other words, the author is interested in understanding how process equity relates to perceptions of outcome equity, and in turn, developing an improved understanding of the role of process equity as a means for achieving water conflict resolution that can be perceived as satisfactory by the parties involved.

Using the preliminary criteria of equity developed by the author (listed as process equity parameters in Chapter 2, Figure 2.1) based on the most prominent body of literature, this book provides analysis of various case studies and cross-case comparison by investigating any nuanced, unique and instrumental factors that are intrinsic to each selected case. Through an interdisciplinary, intensive case study analysis, this book seeks to identify and elicit the spatial and temporal dynamics of environmental equity patterns and processes in an attempt to develop theoretical criteria for equitable distribution of water (parameters of process equity) in bilateral contexts. The goal is to examine relevant bilateral international water treaties and resolutions reached by different disputing nations over the distribution of shared water resources. This book identifies factors for success, principles, and essential qualities in practical contexts and evaluates lessons learned from these cases to make inferences about appropriate processes of equitable allocation. Ultimately, the aim of this book is to develop a conceptual framework for investigating this issue, monitoring implications, and recommending a transformative approach to conflict resolution, applicable to the context of arid regions, to enhance environmental justice and promote peaceable interaction between humans and nature in a way that mitigates harm to either the environment or each other.

Approach

The book takes a proactive stance in responding to the ramifications of conflicts over water supply in many arid regions around the globe by (1) fostering equity attainability in water allotment, and (2) enhancing peace and prosperity on a multiscalar approach. To do so, the author (based on a holistic cross-regional outlook) uses a comparative multiple case study analysis for the purpose of cross-unit comparison. These cases include the Israeli–Jordanian Water Treaty and the Lesotho Highlands Water Project (LHWP) Treaty between Lesotho and South Africa. These in-depth, major cases, conjoined with nine minor cases covered in less depth, are carefully selected to reflect a variety of geographic contexts and degrees of equity attainment in a range of settings that inform policymaking on various scales and levels, including at national and regional levels. By examining the issue on multiple levels, this book can help policymakers at each level seek and draft more acceptable water agreements by tying local concerns with overall national and international policies.

Methodologically, the book employs a mixed-method approach in order to adequately address and decipher the research questions and provide in-depth analysis and meaningful conclusions. The author utilizes various methodological approaches, including detailed case study analysis, expert panel opinion, semi-structured interviews, and hydrological data analysis in selected cases. The objective of this qualitative approach is to explore how and to what extent *process equity* parameters correlate to cooperative and sustainable agreements over the sharing of water.

Within the overall field of planning, as the book is focused on water resources management and conflict resolution, it engages (and hopes to contribute to the advancement of the theory and practice of) a continuum of various disciplines, ranging from environmental justice to transboundary water negotiation, social science, conflict resolution and particularly alternative dispute resolution (ADR), mediation, political ecology, geopolitics, and studies of international law.

One of the most vexing questions surrounding this topic involves different dimensions of equity that extend across a wide continuum of disciplines including political, environmental, technical, cultural, economic, and social aspects. Therefore, drawing upon a greater number of disciplines and investigating prior and contemporary theoretical frameworks that are applicable to the topic is critical to enhance the applicability of the findings and inferences to a vast spectrum of fields.

Book outline and layout overview

The first chapter outlines the theoretical and epistemological context of the book. It sheds new light on the most prominent literature pertinent to the issue at hand to provide an appropriate theoretical and epistemological background that situates this research within the overall scholarly context. An introduction to the research project and a brief outline of the research agenda, structure, and objectives are provided at the end of the chapter. The second chapter delineates the research questions and hypotheses and lays out the preferred methodological design for the study. In an attempt to develop and summarize a systematic approach to the case selection strategy, the third chapter aims at highlighting major trends and patterns in international water conflicts. It provides a snapshot analysis of nine cases, two of which are addressed more thoroughly in the fourth and fifth chapters. These nine cases were carefully selected to represent a verity of geographic, hydro-political, cultural, environmental, technical, and socioeconomic contexts. Upon an initial examination of these nine cases, two specific cases (the 1994 Israel–Jordan Treaty of Peace and the 1986 LHWP Treaty) will be analyzed in greater detail in the fourth and fifth chapters. The sixth chapter provides a cross-case comparison and analysis of these two cases and further draws significant conclusions and inferences about implications for the theories and practice of water negotiation and equitable solutions. The findings in this chapter will be used to formulate a comprehensive conflict resolution strategy, referred to as a "transformative approach to conflict resolution" that offers guidance to engender equitable hydro-diplomacy and satisfactory agreements, transforming water negotiation from a non-cooperative to cooperative state. At the end of this chapter, a conclusion will be provided to highlight key findings and important nuances. The seventh and last chapter provides concluding remarks and recommendations for future research.

1 Water past and present
Power, conflict, perception, and equity

Injustice anywhere is a threat to justice everywhere.

(Martin Luther King, Jr., 1963)

This chapter serves as a theoretical and epistemological underpinning for the book by outlining the book's research approach and scholarly contribution in light of a critical evaluation of relevant literature. By focusing on what has been written about water in relation to power, conflict, perception, and equity, I will provide an overview and analysis of the different bodies of literature that tackle the issue of water scarcity and conflict resolution in general, and equitable allocation concepts and measurements in particular. Moreover, several popular movements that take a stand on environmental justice and discrimination are introduced and questioned. Although far from an all-encompassing review, this chapter nonetheless provides an anecdotal review and critique of selected research that has attempted to tackle the issues of environmental equity and water allocation. By describing and analyzing relevant previous studies, the research draws conclusions about the effectiveness of each approach introduced by different literature in order to formulate a rigorous research agenda that avoids their shortcomings and drawbacks and further builds on their foundation. In addition, the overall research agenda and scope of this book are delineated briefly and generically in this chapter. More specifically, I will show how this book engages with methods and approaches from the planning and environmental justice disciplines while incorporating concepts and debates from other disciplinary approaches.

Conflicts over transboundary water resources: the nature of the problem

Water conflicts are complicated in nature. This complexity emanates from the very nature of water, which tends to flow through various geopolitical, social, and economic divides irrespective of the actual political and state boundaries. Water conflicts also exemplify the fact that humans compete for their share of water (in terms of its quality and quantity) for survival, consumptive uses, and economic development. When these aspects are conjoined with natural, environmental, ecological, physical, psychological, spatial, temporal, technical, institutional, jurisdictional, and political factors the issue becomes even more complicated. In addition, water problems are related to multiple layers of uncertain variables related to social, natural and political systems that constantly interact and change across various scales, and at different levels (e.g., local, regional, national, and global) in an unpredictable and uncontrollable fashion (Islam and Susskind, 2013). As such, water issues are considered "wicked" problems that cannot be simply resolved based on a solely technological or scientific approach, but rather through many other less objective measures and solutions (Islam and Susskind, 2013).

Because of these complex aspects of transboundary water systems, when countries share parts of a river basin, they tend to either cooperate or clash. Relations depend on the overall political, historical, and socioeconomic atmosphere of the shared river basin in general, and the individual and collective interests and political will of riparian states sharing it. Conflict over water allocation can erupt over essential human water needs as well as economic development needs. As the population of co-riparian states continues to grow, so does the demand for additional water. What complicates the problem is the fact that water is needed for both essential human needs and economic development purposes, yet it is limited in nature, particularly in arid regions. Because of population growth, resource scarcity and access limitation, conflict over the quantity and quality of water access will often occur, creating a situation of hostility that undermines the political and diplomatic relationships between disputing states, which is prone to escalate into military confrontation.

Given this, it is no coincidence that the term "rivals" is derived from the Latin word *rivalis*, which refers to two or more people who share the same river by living on its opposite banks (Dellapenna, 2002; McDermott, 2009). This intriguing fact illustrates the potentially conflictive nature of water if not properly managed through cooperative arrangements among the rivals sharing the river system (Allouche, 2005). Water, as such, can be either an obstacle to achieving peace and regional stability or a reason for achieving a lasting and robust cooperative arrangement among disputing states sharing a river system. Water acted as a deterrent to achieving peace and cooperation in contexts such as the Israeli–Syrian relationship over the Yarmouk, the relationship between China and Thailand apropos the Salween River, and the relationships among the newly independent states of the former Soviet Union over the Aral Sea (Postel and Wolf, 2001). Similarly, there are many other examples that demonstrate the fact that water can be a catalyst for peace and stability, such as the 1960 Indus Waters Treaty between India and Pakistan and the LHWP Treaty between Lesotho and South Africa. In these cases, not only did water act as an incentive to achieving peace, but it also helped in avoiding tensions, building trust and confidence, and fostering mutual social and economic development.

Water in literature: theoretical perspectives

The book examines two controversial issues, in particular, that are pertinent to distributive and operational justice in water allocation: environmental equity and hydro-hegemony. Here I will review early and contemporary methodological controversies surrounding the issues of environmental equity *vis-à-vis* hydro-hegemony. Illuminating a variety of research trends, my summary will be informed by literature which focuses on several aspects of the issue. These can be grouped into the following categories: (1) psychological environment and people's perceptions; (2) water, environmental knowledge, and power (*hydro-hegemony*); (3) water as a "weapon" and "water wars" rhetoric; (4) water equity measurements and principles of international law; and (5) theory of access, game theory, and the commons. These issues and topics will be presented, for the most part, in the context of arid regions including, but not limited to, the Middle East.

Psychological environment and people's perceptions

Much of water research has focused on social and "spatial imaginaries" and perceptions (Wolford, 2004). Differing perceptions of water resources and their relationship to the larger political issues illustrate the conflictive opinions of both optimists and pessimists, which are discussed and explained in an important book by Tony Allan (2001). Water shortage is an impending political issue that impacts local communities in the Middle East but has solutions that can only be found and explained in the global arena (Allan, 2001). Allan, akin to several leading scholars, has astutely explained that representing water conflict solely as a result of water shortage is an oversimplification of the issue at hand. In arid regions, where severe inter-state and intra-state disputes of "high politics" are the norm (Zawahri, 2004), the most important factors determining the way riparian states acquire and use water are motives and perceptions rather than realities. This is because, overall, water use is not simply a matter of local consumption (Dolatyar and Gray, 2000). Riparians' behavior is influenced by their perception of the adversary, which is strongly related to their motives (Lowi, 1995; Rouyer, 2000). A significant body of literature contends that such conflicts connote dramatically diverse and conflicting ideological descents and representations (Lowi, 1995; Dolatyar and Gray, 2000; Rouyer, 2000).

Consequently, riparians' behavior is influenced, and in many cases driven, by their perceptions of water and the *other* (an adversarial geopolitical term often used to describe the opposite side of the discourse). This discursive construction of water resources defines, in one way or another, how each nation negotiates, on what issues, and to what limits (Lowi, 1995). Discourse tends to shape not only their perceptions about each other but also their politics and policies (Dolatyar and Gray, 2000). In other words, people's perceptions and attitudes toward water, and toward each other, determine the way they interact with and consume this valuable resource, on one hand (Dolatyar and Gray, 2000; Lipchin, 2003), and the extent to which they tend to cooperate or altercate, on the other (Lowi, 1995; Rouyer, 2000). The 1996 Ganges Water Treaty between Bangladesh and India is a good example of how people's perception of the other party on both sides can undermine acceptance of a treaty as being fair. In this case, long-standing animosity fueled the overall adversarial perception of the other party, and resulted in skepticism and a lack of acceptance in the terms of the Treaty.

Irrespective of how equitable a treaty is in reality, perceptions can be detrimental to its implementation.

According to Islam and Susskind (2013), water issues are complex in nature, representing various issues related to processes, people, and institutions that continuously and unpredictably interact at various levels and scales (e.g., local, regional, and global). This multidimensional nature of water conflicts necessitates a profound understanding of the interplay between all related issues at these various levels and scales. However, monolithic regional and national policies are often suggested by policymakers as appropriate but they may disregard profound contextual and perceptual factors, which in many cases results in a clash between global and local concerns (Lipchin et al., 2005; De Châtel, 2007). When regional and national polices and solutions suffer from a lack of understanding of the local consumption trends that are strongly related to people's perceptions, these policies often fail to propose acceptable solutions (Dolatyar and Gray, 2000; Turton and Henwood, 2002; Lipchin, 2003). Centralized decision making regarding water resources and allocations in many arid regions usually results in one policy influencing all parties, regardless of differences among localities (Lipchin, 2003; Lipchin et al., 2005; De Châtel, 2007). The key argument made here is that the implementation of any plan, and therefore its practical success, rests solely on the extent to which locals tend to either accept the plan and take a role in its implementation or reject it and disregard its implementation. Having said this, a fundamental understanding of the origin of a water conflict can only be obtained by examining the policy implications at different levels (Dolatyar and Gray, 2000; Lipchin, 2003). Authors who have studied these considerations have suggested that an international water conflict or crisis is most effectively analyzed in a multilevel, multifaceted fashion that regards local, national, and regional perceptions and perspectives of the issue (Lipchin, 2003; De Châtel, 2007).

Psychological and perceptual factors are strongly at play in many contexts involving disputed water resources around the globe. For example, in his work on water politics in Israel and Palestine, Alatout (2000, 2003) demonstrates how a specific historical and political struggle regarding the interpretation of state identity, namely Israel's identity as a Jewish state, influences the construction of water scarcity narratives, which solidifies Israel's control over water resources by imposing and legitimizing certain institutional and techno-political arrangements. Israel, through its political, intellectual, and popular discourses, makes the argument of *needs*, as opposed to *rights*,[1] as a basis not only to justify exclusive control over, and consumption of, international water, but also to provide support for the many *ad hoc* water resolution proposals for the water shortages that the Palestinian communities must face (Shmueli, 1999; Alatout, 2000, 2003). Employing these prominent forms of communicative discourse, Israeli discourses often tend to dissociate the term "occupation" from the debate about water, dismissing the Palestinians and their water needs (Lowi, 1995; Rouyer, 2000). Palestinians, on the other hand, envision their environment and its natural resources to be part of their identity and sense of pride. Although the Palestinians do not reject the idea of allocating water based on actual needs per se, a major issue that remains of great concern to them is the fact that representing the Palestinian water crisis as a matter of absolute *needs* in separation from *rights* is a misrepresentation of their real issue (Rouyer, 2000; Alatout, 2006). As such, the Palestinian perception regarding their water use provides a fundamentally contradictory ideology to that of Israel, asserting that "Israel is using the Palestinian water" (Rouyer, 2000). Palestinians blame Israel not only for attempting to deny their identity by occupying their land and consuming its natural resources, but also for denying their inherent rights (Rouyer, 2000; Alatout, 2006). Saeb Erekat, a high official in the Palestinian National Authority (PNA), reiterated the Palestinian belief that water is part of their occupied land by stating that "the Israelis are stealing our water and it must stop. They must be reminded that they are sitting on our chests; and not by an act of God, but by an act of war" (Rouyer, 2000).

Scholars of "contentious politics" argue that an understanding of space, meaning, and individual and collective experience is imperative in understanding the dynamics of conflict and resistance (Wolford, 2004), as well as in establishing a critical perception of the construction of *self* and *other* (Lowi, 1995; Rouyer, 2000; Shiva, 2002). The perception of the *other* as the adversary triggers sentiments of resentment and hostility, which can be used to promote different forms of resistance. To that end, *othering* can be used as an effective tool to acquire political power and economic leverage, and to inequitably distribute resources among co-riparians, which often impacts the disenfranchised and destitute most severely (Lowi, 1995; Rouyer, 2000; Wolford, 2004).

Water, environmental knowledge, and power (hydro-hegemony)

The mainstream framing of transboundary water conflict resolution often holds that equitable and complete cooperation is hard to attain and, therefore, that promoting cooperation of any sort is a positive sign worthy of encouragement (Selby, 2013). This view has, however, been challenged by recent work related to hydro-hegemony, which is deeply entrenched in many conflicts, particularly the Palestinian–Israeli water conflict, and it tends to reinforce the continued domination and control over water resources (Zeitoun, 2008a; Zeitoun and Allan, 2008). Zeitoun (2008a) provides a framework for the analysis of water-related conflicts based on hydro-hegemony theory. The concept hydro-hegemony reflects several dimensions related to states' power (including military, political, and economic strength), riparian *hydro-strategic* position (upstream riparians have the advantage over downstream riparians) (Wolf, 1994), and exploitation potential (including infrastructure development and technological capacity) (Zeitoun, 2008a).

Hydro-hegemony emerges as a key influential factor in determining the outcome of any agreement, the analysis of *hydro-political* situations, and the perception of oneself and others (Lowi, 1995; Zeitoun, 2008a). For instance, the inextricable interplay between water acquisition and hydro-hegemony is seemingly pronounced in many arid regions' water policies in a number of ways. First, within the course of the ongoing cycle of violence and military intervention, the more powerful states take advantage of their military superiority to maximize their share of water with little or no regard for others. The results are clear; a group of people (i.e., the most powerful state) receives more water than their basic needs, whereas many others (the disenfranchised and destitute) suffer because of that (Isaac, 1999; Shmueli, 1999). Of course, "needs" are defined differently depending on a community's ability to process and export goods, access foreign markets, leverage capital to fund intensive agriculture, and so forth.

Second, cooperation offers little benefit to the most powerful state, which tends to be less interested in negotiation (Wolf, 1994) since it often means relinquishing its most favorable position (Zeitoun, 2008a). Lowi (1995) makes the argument that the dominant state, usually the least needy state, often tends to avoid cooperation and has more inclination toward resolving water issues by altercation. Conversely, because of its very limited and weak BATNA (best alternative to a negotiated agreement), needier and less powerful riparian states tend to seek cooperative arrangements, despite the larger political conflicts (Lowi, 1995; Wolf, 1995). In such cases, if the dominant regional power agrees to cooperate, their cooperation will be confined to specific cases where the dominant riparian will receive the most gain. Thus, such cooperative arrangements may not yield a successful and lasting cooperation and cannot be viewed as a potential avenue for a prosperous and politically stable region (Lowi, 1995). This is because these agreements, which tend to be mostly based on coercion, tend also to be flimsy and prone to disintegrate.

Third, powerful states tend not only to extract considerably more water than their basic needs, but also to manipulate scientific facts about water use to underplay or conceal detrimental environmental ramifications (Alatout, 2006). This is because the production of scientific data and environmental narratives regarding the use of water and its environmental consequences is influenced by power imbalances, as well as the hegemonic structure (Alatout, 2006). This combined use of power and information is referred to as "games of the state," where the state's power and manipulation of data are used to build consensus over a singular form of environmental "truth" (Blaikie and Muldavin, 2004). By so doing, these powerful states gain control of natural resources and exclude any contradicting evidence as being inaccurate or biased. Evidently, as Walker (2006) puts it, such narratives may display a "flawed and power-laden" interpretation of hydrological reality.

A good example that illustrates this point can be found in the relationship between India and Bangladesh. In this context, precise data about water quantity and quality is often distorted or withheld on the basis of confidentiality and secrecy. In many other cases, such as the Tigris–Euphrates river dispute between Turkey and Syria, the former state, being the most powerful, is generally uninterested in developing regional regimes to monitor water resources and share data (Zawahri, 2006). Even after the two countries were able to reach an agreement in 2001, the role of hydro-hegemony became apparent in controlling aspects related to data acquisition and dissemination. Another prevalent example that illustrates this issue is provided in the work of Alatout (2006). According to this research, Palestinian and Israeli environmental narratives display an enriched tradition in theories of hydro-hegemony and environmental knowledge. The divergence in perception of power between the Palestinians and Israelis is shaped by the history of the conflict and driven by the ideology of occupation and its territorial policies, and the lack of Palestinian sovereignty and ability to form comprehensive environmental policies (Alatout, 2006). This divergence, according to Alatout (2006), determines the nature of resistance and the extent to which cooperation is attained. The relationship between Jordan and Israel, akin to the relationship between Lesotho and South Africa, provides an example of how cooperation is attained notwithstanding the detrimental impact of power structure imbalance. In spite of the overall political conflict, the two states were able to achieve an agreement in 1994 apropos of the allocation of shared water resources. A great deal of emphasis is placed on maintaining national security and the commitment to peace and eliminating hostility in the region (Treaty of Peace, 1994: Article 4). Despite the many obstacles that it encounters, this water Treaty still provides hope that an equitable agreement is feasible and warranted, regardless of rivalry and the perils of hydro-hegemony.

Water as a "weapon" and "water wars" rhetoric

Water has frequently been utilized as a powerful weapon by riparian states to direct negotiations with others. A brief perusal of work by Zawahri (2004, 2006) shows that the degree to which the water weapon is used varies along a continuum from stable cooperation, unstable cooperation, to conflict. According to Zawahri (2004), in the case of stable cooperation, the military is not used and the water weapon is minimally utilized, whereas unstable cooperation involves heavy use of water as a weapon. Conflict occurs when riparian states are unable to resolve their water disputes and resort to military means to manage their interdependent relationships (Zawahri, 2004).

There are numerous examples around the world that show how water can be used as a weapon to influence other riparians or achieve political gain. The conflict between India and Pakistan over the Indus River represents a good illustration of this point. Prior to achieving the 1960 Indus Waters Treaty, India was successful in utilizing water as a weapon to exert pressure on Pakistan by interrupting the water flow to Pakistan in 1948 (Bhatti, 1999). As this action threatened its agricultural production capabilities, Pakistan was forced to provide payment for water released by India (Priscoli and Wolf, 2009). A similar example emerges from the Tigris–Euphrates river basin when Turkey threatened a blockage of water to Syria because of its support of the Kurdistan Workers' Party, as will be discussed in detail in Chapter 3 (Berman and Wihbey, 1999). The Nile case provides another good example of how these factors work. In this region, water has become a source of power, which has always been employed to the highest extent to influence the decision-making process, and therefore direct the conflict in favor of one side or the other. The relationship between Egypt, the more powerful state, and Sudan, the weaker state, provides an example of unstable cooperation in which water and military advantage are utilized together.

According to Gleick (1992), states that utilize the water weapon (by restricting water flow to other states for political and military ends) may succeed in attaining and maintaining uninterrupted access to water resources, but their action tends to invoke violence. According to Lowi (1995), states engaged in such a protracted conflict are unable to achieve progress on resolving their disputes by simply focusing on technical matters of mutual concern without parallel efforts to resolve political issues. Opting to utilize water as a weapon to influence and pressure other riparian states tends to exacerbate problems, deepen political divides and threaten regional stability (Gleick, 1992). Hence, political instability can both be caused by, and exacerbate, international water disputes (Wolf, 1998).

The concept of "water wars," as suggested by Starr (1991), has increasingly become an important element of the political rhetoric, and is well suited to the Palestinian–Israeli case, where disproportional allocation of water has emerged as a source of conflict and an obstacle to peace and prosperity. Under this argument, lack of water and the need to secure adequate water supply can become a matter of national security and can be prioritized in political and military objectives (Mirumachi, 2007). In 1967, and prior to the Six-Day War between Israel and other Arab countries, the Israeli Prime Minister, Levi Eshkol, stated that "water is a question of survival for Israel," and that "Israel will use all means necessary to secure that the water continues to flow" (Biliouri, 1997). Commenting on the same event, the Crown Prince of Jordan stated that the war of 1967 "was brought on very largely over water related matters," and he predicted that, without an international water agreement in the Middle East by 2000, "countries in the region will be forced into conflict" (Irani, 1991). These statements and others are frequently referenced in support of the "water wars" rhetoric. Because of dramatic population growth and the staggering technological advancement of the twentieth century that made great volumes of water extraction possible (Grover, 2007), many scholars have gone even further to suggest that the next war will be fought over water. Gleik (1993), for example, explains that water in these arid contexts can become an instrument for war and can easily escalate to a contentious issue. To that end, Dr. Ismail Serageldin, former vice president of the World Bank, remarked in 1995 that "the wars of the next century will be about water."[2]

However, there is a growing literature challenging the prediction of a future war over water. Supporters of the water war paradigm affirm that water evidently caused wars in the past and will most likely be the reason for a looming water war in the future. On the other hand, opponents of this paradigm assert that the water war rhetoric is misleading and irrelevant at best (Wolf *et al.*, 2006; Gleick, 2009). Notwithstanding their points of contention, both opponents and proponents of this argument agree that the lack of a continuous clean freshwater supply, exacerbated by a growing water demand, could cause instability, which might result in conflict and political impasse (Wolf, 1998). Many scholars, such as Fisher and colleagues (2005), Wolf (1996), and Haddadin (2002a, 2002b), argue that allocation of water resources should be viewed not as a catalyst for war, but rather as an initiative for diplomacy and potential conflict resolution. Those who adopt this point of view promote an alternative idea that water can serve as a catalyst for peace and cooperation (Fisher *et al.*, 2005; Wolf *et al.*, 2006).

A good example that illustrates the impact of water in fostering cooperation emerges from the LHWP Treaty between Lesotho and South Africa. In this case, water was the catalyst for cooperation as these two states were able to find common ground and capitalize on the shared water resources to reach a satisfactory agreement that catered to both of their respective needs. Water, in this case, was delivered to South Africa for economic development while financial assistance was provided to Lesotho to allow it to build its economy (Priscoli and Wolf, 2009). The idea here is that water acted as an incentive for peace and as an incentive for disputing states to enter into negotiation for the purpose of settling their disputes.

Water equity measurements and principles of international law

> Rivers have a perverse habit of wandering across borders . . . and nation states have a perverse habit of treating whatever portion of them flows within their borders as a national resource at their sovereign disposal.
>
> (John Waterbury, 1979, *Hydropolitics of the Nile Valley*)

International water treaties literature generally identifies two types of equity: *distributive* and *procedural* justice.[3] Central to scholarly research are the relationship between them and their role in collective action (Cohen, 1987). Rawls (1971), reflecting on distributive social justice and elements of social contract, contends that negotiating parties are neither naturally altruistic nor purely egotistic. Rather, they have interests they seek to achieve through cooperation on mutually acceptable grounds (Rawls, 1971). Based on the notion of *justice as fairness*, Rawls suggests that these mutual grounds include a model of a fair choice situation, rather than utilitarianism or libertarianism (Rawls and Kelly, 2001). In his account of procedural justice, two elements of justice surface: the equal liberty principle and the difference principle. To Rawls, procedural justice entails three accounts: *perfect procedural justice*, with independent criteria of fairness and a procedure that guarantees fair outcomes, *imperfect procedural justice*, with independent criteria of fairness but no method that guarantees fair outcomes, and *pure procedural justice*, with neither (Rawls, 1971). The work of Tyler and Huo (2002) also highlights the notion of process-based, or procedural, justice as a key component of legitimacy and acceptability. To that end, they point out that people's response to conflict resolution decisions made by a third party is greatly determined by their view and assessment of the fairness of the procedures based on which these decisions are made (Tyler and Huo, 2002).

The veil of ignorance concept is particularly intriguing, as it ties inequalities to people's and states' awareness of their innate differences in abilities and assets. According to Rawls, this awareness may lead to efforts to preserve one's advantage when developing general principles of justice. Based on the notion of the veil of ignorance, people and states should not be able to determine what outcome would benefit them. Therefore, the impact of these innate differences on the decision-making process can be mitigated or eliminated in order to produce an outcome that can be characterized as equitable (Rawls, 1971). Nozick (1974) provided an alternative normative theory of distribution, emphasizing three principles: justice in acquisition, justice in transfer, and just rectification (Cohen, 1987). Later, Rawls refers to two types of justice, *objective* and *subjective* (Rawls and Kelly, 2001), both of which coincide with the outcome equity parameters introduced in this book: *commensurable* and *perceived* respectively. However, only distributive justice has been directly highlighted in the context of theory of negotiation and resources allocation.

An important strand of recent literature, which consists of analysis of international law, emphasizes the need to incorporate equitable water allocation measures and criteria into international water agreements (Wolf, 1999; Kuttab and Ishaq, 2000; Rouyer, 2000; Giordano and Wolf, 2001; Selby, 2003). This venue, which is increasingly gaining acceptance (Kuttab and Ishaq, 2000), is crucial to the foundation of this research because many adversarial riparian parties involved in transboundary water disputes often ground their arguments in the negotiation over water allocation on principles of international law, which often attempt to offer general guidelines and elements for distributive justice. By attempting to address the issue of sovereignty over regional water resources, international law poses the question of who has the right to what resources (Kuttab and Ishaq, 2000). The major international laws deemed applicable to many international water disputes[4] are the Hague Regulations of 1907, the Fourth Geneva Convention of 1949, the Helsinki Rules of 1966, and the United Nations (UN, 1997) *Convention on the Law of the Non-navigational Uses of Watercourses* (Wolf, 1999). Recognizing that water equity has been an under-theorized domain of research, a handful of scholars have begun to critically examine the practice of international law. Among the first to try to fill this lacuna were Giordano and Wolf (2001).

International water treaties literature indicates that common water allocation practice is based on either hydrographical or chronological grounds (Giordano and Wolf, 2001). First, the hydrographical argument is pertinent to the origin of a river or aquifer and how much of its boundary falls within each riparian state (Lipper, 1967; Eckstein, 1995). This includes "the doctrine of absolute territorial sovereignty" and the principle of "territorial integrity," which are opposite to each other. The doctrine of absolute territorial sovereignty argument is often made by upstream riparians who claim, based on the Harmon Doctrine,[5] absolute sovereignty over water falling within their boundary (Eckstein, 2002; Spiegel, 2005; Upreti, 2006).

The application of the Harmon Doctrine can best be seen in the US legal position in the dispute with Mexico over the sharing of the Rio Grande river in 1896 (Upreti, 2006). This claim was also made by India in 1948 to justify its interruption of the Indus River flow from India to Pakistan, and most recently by China in 1997 (Bhatti, 1999; Upreti, 2006). Conversely, the principle of territorial integrity, usually advocated by the downstream state, promotes the argument that upstream states cannot alter the flow of a shared river. Many examples that illustrate this principle include allegations by downstream states such as Argentina,

Egypt, Spain, Bangladesh, Syria, and Iraq against upstream riparian states (Upreti, 2006). For instance, Egypt objects any water diversion plans by Ethiopia or Sudan that would alter the flow of the Nile (McKinney, 2008). Similarly, Indian leaders have expressed objections to several irrigation and hydropower project proposals by Nepal on the basis of the principle of territorial integrity (Shrestha and Singh, 1996; Upreti, 2006).

Second, the chronological argument, referred to as "prior appropriation," reflects the historical use of water: that is, first in time, first in right (Wolf, 1996). Downstream riparians often rely on this argument to support their claim of historic rights to water. For example, Egyptian leaders make the claim of historic right over the Nile waters and have threatened Sudan and Ethiopia with military action to deter them from developing water projects that might impede the flow of the river (Kliot, 1994). Egypt also proposed to divert the waters of the Nile to the Sinai despite Ethiopian criticism, and Egypt furtherwarned Ethopia that "we [Egypt] do not need permission from Ethiopia or the Soviet Union to divert our Nile water. If Ethiopia takes any action to block the Nile waters, there will be no alternative for us but to use force" (Wolf, 1994: 30).

Wolf (1999) further classified the legal principles of international law into two main categories: (1) extreme principles, such as "the doctrine of absolute territorial sovereignty," referred to as "the Harmon Doctrine," "the doctrine of absolute riverian integrity," and "prior appropriation doctrine" and (2) moderated principles, such as "the doctrine of limited territorial sovereignty," "the Helsinki Rules of 1966," and "the 1997 UN Convention on the Non-navigational Uses of Watercourses." Extreme legal principles[6] address issues pertinent to sovereignty, absolute river integrity, and historic rights, whereas moderated principles focus on issues of equitable use and the obligation not to cause significant harm (Giordano and Wolf, 2001).

The Helsinki Rules of 1966, which put emphasis on the right to "beneficial use" rather than the right to water *per se* (Housen-Couriel, 1994), provided 11 factors pertinent to hydrographic and sociopolitical criteria (Wolf, 1999). Those factors, it was suggested, should be considered as a whole to resolve water allocation conflicts. The two main principles introduced in the 1997 UN Convention are "equitable and reasonable utilization," set forth in Article 5, and "no significant harm" to other watercourse states, set forth in Article 7 (Diabes-Murad, 2004). The UN Convention further gives "special regards" to the "requirement of vital human needs" (Wolf, 1999). Although the concept and guidelines for "reasonable and equitable" sharing of common waterways were introduced in the Helsinki Rules of 1966, as well as the 1997 UN Convention, no clear definition of this concept was provided (Caponera, 1985; Kuttab and Ishaq, 2000; Rouyer, 2000). This makes it easy for disputing parties to interpret the law differently according to their own perceptions and desires, which tends to generate a great deal of conflict (Kuttab and Ishaq, 2000).

Giordano and Wolf (2001) identified various measures of water-sharing equity introduced in international law, including rights-based, needs-based and economic-based measures, all of which evolved over time. According to Kally

(1993), rights-based measures introduced in international law give priority to both "existing" and "historical" uses of water. Generally, three different types of water conflicts are suggested by water conflict and negotiation literature. These are conflict between existing uses, conflict between existing and new uses, and conflict over future uses (Vinogradov et al., 2003). First, conflict between existing uses occurs when the demand for water exceeds the amount of water available for extraction. The result is a situation, such as in the Aral Sea basin, where the shared water resource is subject to over-exploitation and downstream riparians receive a declining share. The law in this situation is clear regarding causing "no significant harm" to other riparians sharing the river system (Diabes-Murad, 2004). Second, conflict between existing and new uses occurs when new planned activities interfere with existing uses. In this situation, many scholars, such as Wolf (1999), assert that trends of past agreements show a pattern of underscoring the significance of historical (existing uses) and geographical considerations in international water disputes. Based on that, Rouyer (2000) points out that the Palestinian historical rights to water were always protected during Ottoman, British, Egyptian and Jordanian rule and further argues that only under the Israeli occupation have these rights been summarily denied. Rouyer (2000) also asserts that the heart and soul of the Arab legal claim to water resources of the Gaza Strip and the West Bank rests on the principle of "belligerent occupation." Conversely, refusing the term "occupation," Israel has denied the applicability and relevance of these laws (Rouyer, 2000).

Third, conflict over future uses occurs when the planned activities are not certain. In this situation, the law does not allow states to allocate water for uncertain future uses (Vinogradov et al., 2003). The inability to address uncertain future needs seems to be extremely problematic, as a key component of what could be involved in water conflicts may be related to future uses. In particular, many conflicts are present in situations in which one or both riparians are undergoing rapid changes in population or land use, such as urbanization and industrialization, or some other development that would affect its water "needs" according to certain interpretations of "needs." This is at the core of the Israeli–Palestinian disagreement, in which the Israelis often blame the Palestinians for failing to control their population, whereas the Palestinians blame the Israelis for using more water per capita in order to support agricultural exports and industries that are part of a much more developed and externally subsidized economy. In these situations, the law is not reliable.

Highlighting the inadequacy of international law in addressing the question of water-sharing equity, a great deal of literature concludes that the applicability and effectiveness of these conflicting doctrines in many arid regions' disputed water resources is questionable. Given this, many assert that international law is an area of great controversy (Wolf, 1995, 1996; Rouyer, 2000). A significant body of literature indicates that international water law does not have teeth in the sense that it does not provide a clear enforcement mechanism to ensure the applicability of its rules (Rouyer, 2000). In general, international law has intended to provide general rules to allow room for flexibility and adaptability for states to more easily find solutions for various circumstances. However, in many cases around the world, this flexibility arguably has caused more disagreements among disputing parties over the meaning and interpretation of the laws, as each party seeks to justify and legitimize their positions (Rouyer, 2000). These international laws, in addition to relying on conventional resolution methods of position-based negotiation (Rouyer, 2000), have in fact been focused on *dividing* the resources of the region, namely land and water, rather than *sharing* them (Scheumann, 1998). However, international law is effective in the sense that it supports a shift from rights-based equity (based on country-specific allegations of territorial sovereignty) to needs-based equity[7] (based on principles that generally allocate water based on consumption trends, regardless of the country in question) (Wolf, 1999; Mohamed-Katerere and van der Zaag, 2003).

Given the ambiguity of the equity-based principles of the law, namely "equitable and reasonable utilization" and "no significant harm," international law has also been considered vague regarding the meaning of these concepts (Caponera, 1985; Rouyer, 2000; Diabes-Murad, 2004). In short, international law is necessary, yet not sufficient by itself. Literature addressing the questions of water equity measurements comes to a tidy conclusion in admitting the absence of such measures in international law, and advocating for additional research that tackles this crucial issue (Mohamed-Katerere and van der Zaag, 2003).

Theory of access, game theory, and the commons

"Theory of access" literature provides a different, yet unique, take on how natural resource *access* is contingent upon power (Leach *et al.*, 1999; Ribot and Peluso, 2003; Robbins, 2004; Korf and Funfgeld, 2005). According to Peet and Watts (2004), those with greatest access to power are in the best position to control access to water resources. As Ribot and Peluso (2003, italics in original) astutely note, "'bundles' and 'webs' of powers . . . configure resource *access*." This arrangement of unequal *access* to resources emanates from, and is strongly linked to, the question of hydro-hegemony and the unequal distribution of power. This argument, made in a considerable body of literature, contends that power imbalance tends to monopolize *access* to water by restricting others from accessing water in order to harness the greatest benefit (Mohamed-Katerere and van der Zaag, 2003). Ribot and Peluso (2003) argue that this imbalance of garnering benefit from natural resources is due to the difference between *access* and *property*, where the former incorporates the notion of "bundle of powers," while the latter incorporates the notion of "bundle of rights." The recognition of how *access*, the ability to benefit, rather than mere *rights* and entitlement, determines who derives benefits from what resources at what time, helps develop a better understanding of the dynamic processes and relationships of access (Ribot and Peluso, 2003). Korf and Funfgeld (2005) illustrate these dynamic processes and relationships of access in the context of political economy and the political geographies of war. They refer in their discussion of access to the "spatial dynamics of military control," which tend to define and confine the access of common-pool resource users and their livelihood. These frameworks presented by Ribot and Peluso (2003) and Korf and Funfgeld (2005) help in understanding how powerful nations gain benefit from certain resources while others do not.

These mechanisms, although itroduced in relation to gaining benefit, are also delineated in relation to exclusion (Mitchell, 2002; Robbins, 2004; Korf and Funfgeld, 2005; Schwartz, 2006). According to Schwartz (2006), who examines a case study of two Western-supported initiatives in national park management in Latvia in the late 1990s, the perception of the local population as ignorant "rapacious natives" personifies the impact of these mechanisms of exclusion at work to restrict their access to natural resources. To that end, gaining, controlling, and maintaining access is not just a function of obtaining and allocating utility alone, but also a function of exclusion and prevention of others from doing the same, which entails a great deal of competition and sometimes conflict, clash, or even war, according to Korf and Funfgeld (2005). Many scholars, including Schwartz (2006), Mitchell (2002), and Robbins (2004) explore these relationships in the context of human sustenance and survival, rather than as a measure of imbalance between local interests and pristine natural resource conditions. They argue that this perception of the local population and the Western view of "untouched nature" exemplify a basic misconception of nature in the context of the well-established and rich cultural landscapes of the local population. It nonetheless explains how certain schemes are justified at the expense of these local populations (Schwartz, 2006). Following the same scheme, similar arguments have been made by many scholars, akin to Zeitoun (2008a) and Lowi (1995), who assert that access to water is strongly linked to the degree of hydro-hegemony that each riparian state has achieved. Zeitoun (2008a) contends that understanding how the role of hydro-hegemony as a dominant underlying factor in determining who gets access to water, regardless of claims of right and entitlement, helps to explain how Israel for example, being the most powerful co-riparian, receives the highest amounts of water in the region. Equitable ubiquitous *access*, efficiency, and sustainability were emphasized nevertheless as a prerequisite for fruitful cooperation by various scholars of transboundary water conflicts (Turton and Henwood, 2002; Ribot and Peluso, 2003).

Access to natural resources is also introduced in relation to property rights and common pool resources. In his examination of the concept of private property (and its relevance to deriving utility) and his assertion of the detrimental impact of the freedom of access to common pool resources, Hardin (1968) illustrated how the rational calculation of utility can cause individual decisions to lead to abominable environmental harm, referred to as the "tragedy of the commons" syndrome. Interestingly, the notion of the collective action problem and its impact on irrigation systems in the United States was introduced by Katharine Coman (1911) almost a half century before Hardin's description of the concept of common pool resources (Islam and Susskind, 2013). In the context of the Middle East, as is the case in many arid regions around the world, each state tries to understand the game and to rationally enhance its access to water with no regard for the other states' interests (Rogers *et al.*, 1978; Zawahri, 2004). A considerable number of scholars on this subject view and portray water access in arid regions in light of game theory, where each player reacts based on enlightened self-interest and rational behavior to "win" (Hardin, 1968; Wolf, 1995). When there is no control in place to regulate the use of common resources shared by more than one party, it is expected that the faster one uses the resource the less of it the other parties can obtain. What prevents this mad dash to consume is the establishment of a meaningful definition of "fair division" of the resource.

The difficulty of attaining "fair division" of disputed resources becomes more severe when the parties claim sovereignty and entitlement (Brams and Taylor, 1996). A procedure of fair division requires the agreement on criteria for a fair division and the selection of valid procedural rules to follow (Brams and Taylor, 1996). Fair division criteria, introduced by Brams and Taylor, include a proportional or simple fair division, envy-free division, exact division, efficient or optimal division, and a practically grounded equitable division. However, the issue of water allocation goes beyond the "division" of water per se to include sharing of the resources, including proportional distribution of costs and benefits among all parties involved (Glassner and Fahrer, 2004).

Under the assumption that riparian states have, at some level, a shared interest in cooperation over water access, the behavior and strategy chosen by each player – whether to work unilaterally (defect) or to cooperate – depends mainly on the geopolitical relationships among them, which is a function of water availability, level of hostility, distribution of power, and legal and constitutional settings (LeMarquand, 1977; Falkenmark, 1989; Wolf, 1995), as well as religion and cultural affiliations and divides. Within hostile settings, however, the prospect of cooperation is threatened by the "prisoner's dilemma," whereby if all states defect the commons are destroyed, and if one cooperates while the other defects, the "sucker's payoff" involves substantial losses to those who cooperate; but if all are able to cooperate, then significant losses are avoided (Wolf, 1995; Zawahri, 2004). Having said this, unless general rules of the game are agreed upon and put in effect, each player competes to divert the greatest amount of water and prevent others from doing the same.

Interestingly, this body of literature offers a new understanding of the concept of freedom and resolution opportunities on the basis of necessity (Hardin, 1968), which is synonymous to the notion of needs-based allocation, rather than allegations of rights and entitlement (Wolf, 1999).[8] It suggests the need to rely on interest-based, rather than right-based negotiation strategies over water allocation (Wolf, 1999). This is because, unlike rights, needs are commensurable; that is, they are easy to quantify and therefore negotiate over (Wolf, 1999). Water needs can be defined and quantified on the basis of irrigable land or on catering for the population or the requirements of specific projects (Wolf, 1999). The notion of necessity is also introduced in the context of social and economic interdependence.[9] This social and economic interdependence is described in terms of equitable distribution of the common surplus (Gibson-Graham, 2006). However, Wolf (1995: 116) argues that the tragedy of the commons and game theory literature "has not been yet developed to the point where it can adequately model complex international decision-making."

Book contextualization: research approach and scholarly contribution

This research is motivated by (first) the lack of research that focuses on equity attainability rather than power imbalance and (second) the inadequacy of literature in theorizing and tackling the issue of water allocation equity in a meaningful way (Wolf, 2003). This created major gaps, unexplored areas, and points of contention, on which further studies are warranted. Much of the literature, which tackles the issue of water scarcity in arid regions, echoing traditional conflict resolution methods, focuses on a vast spectrum of key issues related to power structure imbalance (hydro-hegemony), ideologies, and position-based negotiation. Since these factors act as deterrents to achieving hydro-stability, the resulting outcome so far has reflected a state of altercation, *hydro-hostility*, and in turn, potential clashes ranging from minor policy disagreements to outright violence. The literature further suggests that the problem is so sophisticated that a single approach is not sufficient (Lipchin, 2003). Although it is important to understand the impact of power structure on water acquisition and allocation plans, recognizing hydro-hegemony alone does not offer adequate resolutions to such a sophisticated problem. Rather, there is a pressing need to approach the issue from another angle that, instead of limiting its scope to only diagnosing the root cause of the problem, provides a multidisciplinary approach that viably extends its scope to finding workable resolutions. Evidently, these potential resolutions lie in a second alternative, proposed in this book, by focusing on pacifiers, such as shared values, equity attainability, and interest-based negotiation, which will be introduced and discussed in detail later in the book.

In a nutshell, this topic is rich, yet under-investigated. The lack of attention to the question of *hydro-equity* in recent literature on bilateral conflicts in arid regions is indicative of the reality that inequitable allocation of water has become a taken for granted fact in the current state of many water conflicts around the world. In addition, the notion of equity has not been adequately theorized, even though it is frequently mentioned superficially, and often recommended as a meaningful future approach for further study (Wolf, 2003). Further, the vast majority of literature focuses on outcome-based measures of equity (whether commensurable or perceived), with only cursory attention given to the processes that create those outcomes and patterns. It is therefore warranted to analyze these processes and factors (procedural justice) in order to understand outcome equity (or distributive justice). This way, the path to an equitable solution can be charted.

The greatest advantage of this approach is its unique approach toward attaining hydro-equity as a way to combat or cope with hydro-hegemony. The literature provides evidence of how often powerful states have practiced coercion, injustice, obduracy, and doctrinal manipulation to enhance their share and deprive others and, in turn, wreak havoc on other disenfranchised and marginalized groups. Unmasking and decoding such dynamics of inequity and the processes responsible for creating them falls under the broad rubric of environmental justice, a growing field of scholarship to which this book will make new contributions. The literature also provides evidence that water is becoming a greater source of cooperation rather than altercation. Cognizant of this important fact, the book attempts to develop an understanding of how achieving cooperation over water resources can result in achieving cooperation over other complicated political issues. In broad terms, it is hoped that findings from this book will provide new knowledge needed for equitable water allocation and will reveal new insights and understandings that will support meaningful conflict resolution processes to mitigate environmental injustice and hydro-hostility, and in turn promote regional hydro-stability. Because, in part, of the lack of such studies and the fascination of this topic, this book will contribute to filling this gap in the literature, and to the advancement of the theoretical aspects and practical applications relevant to this field. These overlooked aspects of the issue being investigated offer great venues for this book and future studies to decipher.

2 Research questions, hypotheses, and methodological design

This chapter aims to provide a detailed description of the research methodology, including key questions, hypotheses, and methodological design, adopted for this book. Furthermore, this part presents a methodical account of (and justification for) employing a detailed case study analysis and cross-case comparison for a robust methodological design. The research methodology explained in this chapter will be applied in the case selection and analysis parts of the book, which will be guided by the logic of the model developed and presented in this chapter.

Research questions and hypotheses

To serve the purpose of the research outlined in the previous chapter, the book, in essence, seeks to explicate three key questions:

Q1: How does equity matter in the process of moving from a non-cooperative to a cooperative state in bilateral water allocation negotiations, treaties, and other arenas of international law?
Q2: What are the key predictors (defined as *process equity*) that explain equitable outcomes in bilateral water allocation treaties, and what are their policy implications?
Q3: What are the lessons and parallels that can be extrapolated to benefit the theory and practice of equitable concepts in water negotiation in general?

Operationalization of the concept of equity, which is seemingly context specific, presents a definitional challenge. To that end, it is necessary to formulate concrete yet flexible and adaptable parameters of process equity (parameters of treaty formation), which are a moving target themselves, to be able to evaluate the impact of a policy-development approach, or lack thereof. Furthermore, there is a set of hypotheses layered into the main questions that will be explained and then examined in the research. These assumptions include:

H1: Water allocation influences inter- and intra-state conflict settings.
H2: The long-standing nature of political conflict between riparian states serves as a disincentive to cooperation.

H3: Equity is a desirable goal for water allocation that has favorable impacts on the degree to which states choose to either cooperate or clash.

H4: Riparian states' tendency or reluctance to cooperate is steered by their perception of what constitutes equity.

H5: Water policy is a potential venue for future conflict resolution and reconciliation.

H6: Attaining mutually recognized equity is a prerequisite for cooperation, promoting hydro-stability, and deterring hydro-hostility.

H7: Following key procedural components of equity enhances chances of attaining equitable outcomes and therefore of *perceived equity*.

To address such a profound research problem and examine such perplexing questions, the research seeks to explore the following logic diagram shown in Figure 2.1. The logic diagram proposed by this research is composed of three main parts: parameters of

1 process equity;
2 outcome equity; and
3 control factors, representing the context of each individual case.

This logic diagram will guide the selection process of cases and the overall analysis focus of these cases.

First, parameters of *process equity* are informed by theory as a compilation of preliminary hydro-equity criteria. They are also informed and verified by consultation with the most widely cited scholars and experts. These parameters, representing components of *process equity*, incorporate three aspects:

1 negotiation process or treaty formation;
2 planning analysis; and
3 structure of the agreement.

Second, parameters of *outcome equity* include *commensurable* (operational) *equity* to reflect the shift from *rights* to *needs* made by the literature, and *perceived equity* (level of satisfaction or satiation) to reflect the shift from *needs* to *satiety* proposed by this research. Based on a review of the literature, all of the elements listed in the diagram are assumed to contribute to positive correlations between *process equity* and *outcome equity*. Third, control factors include the contextual component of the model. This component represents elements of cultural hydrology and incorporates control factors, such as arid regions, bilateral conflicts, cultural variations, history of aggression and hydro-hostility, power structure imbalance, and geographic variations, which will be qualitatively described to contextualize each individual case. These control factors will be used to select the nine cases and will be maintained constantly across all cases to allow for variation in only the *process equity* parameters and the corresponding *outcome equity* factors.

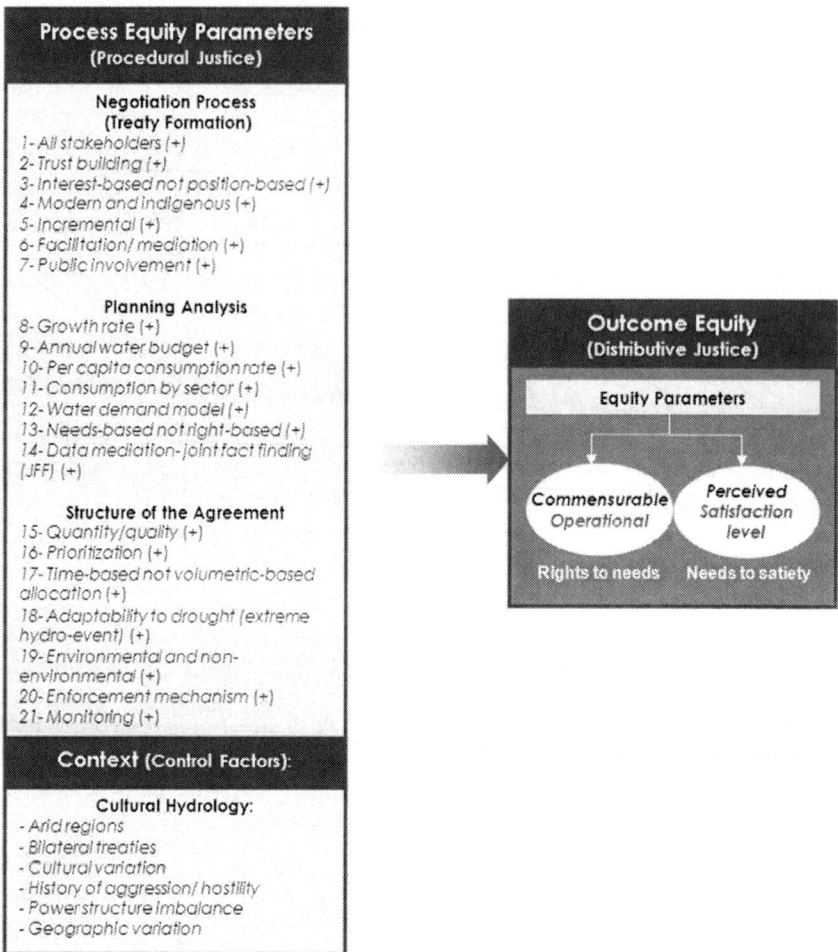

Process Equity Parameters
(Procedural Justice)

Negotiation Process
(Treaty Formation)
1- *All stakeholders (+)*
2- *Trust building (+)*
3- *Interest-based not position-based (+)*
4- *Modern and indigenous (+)*
5- *Incremental (+)*
6- *Facilitation/ mediation (+)*
7- *Public involvement (+)*

Planning Analysis
8- *Growth rate (+)*
9- *Annual water budget (+)*
10- *Per capita consumption rate (+)*
11- *Consumption by sector (+)*
12- *Water demand model (+)*
13- *Needs-based not right-based (+)*
14- *Data mediation- joint fact finding (JFF) (+)*

Structure of the Agreement
15- *Quantity/quality (+)*
16- *Prioritization (+)*
17- *Time-based not volumetric-based allocation (+)*
18- *Adaptability to drought (extreme hydro-event) (+)*
19- *Environmental and non-environmental (+)*
20- *Enforcement mechanism (+)*
21- *Monitoring (+)*

Context (Control Factors):

Cultural Hydrology:
- *Arid regions*
- *Bilateral treaties*
- *Cultural variation*
- *History of aggression/ hostility*
- *Power structure imbalance*
- *Geographic variation*

Outcome Equity
(Distributive Justice)

Equity Parameters

Commensurable Operational

Perceived Satisfaction level

Rights to needs   Needs to satiety

Figure 2.1 Research logic model.

A framework for analyzing key international water treaties

Given the complexity of the water issues that this book attempts to tackle, the research relies on a variety of methods representing a multidesign approach. In essence, the book employs a mixed-method approach that includes the following:

1   literature review;
2   water experts' opinions;
3   content analysis of secondary literature and initial screening of nine ratified international water allocation treaties (snapshot case analysis);
4   comparative case study analysis of detailed cases and pattern matching; and
5   in-depth explanatory one-on-one interviews in each of these detailed cases.

A more detailed description of the research methodological designs that the book employs is outlined in Appendix A.

The objective of the in-depth analysis of key cases is to develop a richer and deeper contextual and anecdotal understanding of the circumstantial evidence and lessons. Focusing on two key cases better served the research objectives and helped answer the aforementioned research questions. By making inferences pertinent to significant parameters of *process equity*, this also helped in developing an understanding of the theoretical framework of the issue at hand. With this in mind, case presentations and analyses were guided by the methodological components and central tenets of this thesis (parameters of *process equity*). The overall framework for the case analysis was primarily based on the logic diagram presented earlier in this chapter. The structure of the two cases was based on the chronological order of events as told in the context of storytelling. Within this organization, the logic model was used to track and highlight each parameter related to process equity as it surfaced in the analysis.

The application and explanation of the methodological design, as manifested in the analysis of the two case studies, reflected the overall goals and tenets of the research and aimed at providing a stronger link between research methodology and analysis. This analysis agenda was used consistently in both cases, but more so in the cross-case comparison chapter, where the logic model was used to not only structure the analysis but also guide its key elements and consequent discussion. In this regard, a cross-case analysis of procedural equity components was guided by the logic model and was broken into three elements. This included the negotiation process (treaty formation), planning analysis, and structure of the agreement. Thus, this conceptual frame provided in the logic model tied the research methodological design to the analysis and implementation and allowed for a careful exploration of this framework in both case studies.

Background information regarding the basin spatial hydrogeomorphology, including (1) the geographic, demographic, and hydro-meteorological context and (2) cultural hydrology, will be introduced prior to the analysis of each treaty to develop a well-informed research base. The framework for analyzing key international water treaties adopted in this research was grounded in seven elements:

1 the intent of the treaty;
2 process equity and treaty formation;
3 content analysis;
4 perceived equity;
5 analysis of the outcome resulting from the treaty;
6 exploration of its environmental consequences; and
7 overall assessment of the treaty.

Each of the key treaties will be analyzed based on the extent to which they conform to the overall parameters of equity outlined in the logic model, and on how well they support hydro-stability for both parties involved in these bilateral diplomatic efforts. Of particular interest was expanding the spectrum of water- and non-water-related issues to finding common places of differential value to trade across. The goal was to provide a comprehensive analysis of each treaty and their major implications for the theories of water allocation negotiation. By drawing lessons and parallels that help break logjams created by intransigent position-based negotiations, the resultant agreements are more likely not only to be fair from a distributive justice standpoint, but also to be regarded as fair from a perceptual standpoint (Innes and Booher, 1999). It will be shown that the two cases differ in how water was allocated and in the negotiation process by which outcomes were reached and implemented. Although the two cases are similar in regard to the existence of a high level of hostility at the time the states entered into negotiation, the Lesotho Highlands Water Project (LHWP) Treaty was successful in focusing on the benefits of water rather than on volumetric water allocation itself. Furthermore, unlike the Israel–Jordan Treaty, the LHWP Treaty brought all key stakeholders together in a multinational effort to reach a comprehensive allocation and management of the basin water resources. The case study comparison will also show how important it is that negotiators consider extreme hydro-events, proactive public involvement, and third-party mediation, for example to ensure success in acceptance of outcome, implementation, and therefore hydro-stability. These and other lessons learned from this analysis (outlined in later chapters) will set the stage for the development of a set of critical elements that are deemed essential in transforming water negotiation from a non-cooperative to a cooperative state, which is outlined in detail later in this book and referred to as a "transformative approach to conflict resolution."

3 Case selection and first-cut analysis

This chapter provides a systematic approach for selecting ratified bilateral, international water treaties from which in-depth cases will be selected and studied. To set the stage for a sound selection approach, the chapter provides, in a broad sense, a review of international water treaties in an attempt to highlight major trends and patterns that are pertinent to the assessment and understanding of processes and outcomes. This provides grounds for the selection of an array of cases, nine in all, which are scanned, documented, and analyzed. In this scanning assessment of cases, the chapter seeks to elicit a modicum of additional background data and information on the range of equity conditions, from high to low, in both "perceived" and "outcome" conditions. In this way, scanning these nine treaties will provide a global outlook of the issues and forces surrounding international water conflicts, negotiation, and allocation, and a characterization of the spectrum of equity-related factors and considerations, whereas the subsequent analysis of two detailed key case studies will allow for a richer and deeper investigation.

Overview of trends in international water treaties

Variously known as an international agreement, protocol, covenant, convention, or accord, the term "treaty" is used here to mean an agreement signed by international players, namely states and international organizations, under international law (United Nations, 1999). By and large, there are two types of treaties: bilateral treaties, involving only two parties, and multilateral treaties, involving more than two parties. According to the United Nations, the term "convention" refers to formal multinational treaties, although it has been used to describe bilateral treaties. The term "agreement" often refers to less formal instruments involving a narrower range of topics than treaties, such as economic, cultural, scientific, and technological cooperation (United Nations, 1999; Towfique, 2002). "Treaty" and "agreement" are used interchangeably in many contexts, including in this book.

Components of effective treaties, as outlined in Hirji and Grey (1998), incorporate efforts on national and international levels. National efforts aim at leveling the playing field, whereas international efforts aim at moving focus from dialog to action. The following are the key components of effective treaty negotiation identified by Hirji and Grey (1998):

- *Building the national capacity and identifying national priorities* involves assessment of national water resources and effective water management strategy informed by a locally driven participatory process. Another successful objective of building the national capacity entails addressing the asymmetry in acquiring, analyzing, and interpreting basin-wide data and information available for co-riparian countries in the river basin to enhance their ability to conduct meaningful negotiation and an informed decision-making process.
- *Developing simultaneous dialog between basin co-riparians on multiple tracks* by fostering information dissemination, trust building, and partnership, and recognizing and emphasizing common goals and interests.
- *Focusing on achievable goals and avoiding complex matters* by incrementally developing agreements over less controversial issues. This means finding common ground for negotiation, and enhancing institutional development and regional joint management efforts of shared water resources, rather than tackling issues of great contention, such as rights and entitlement, at the onset of the negotiation. Once a certain degree of trust and common ground is reached, parties can then tackle more complex issues.
- *Recognizing that resolving water-related disputes is a slow and laborious process* that takes a considerable amount of time, effort, persistence, and patience. Given this, negotiation needs to be rejuvenated and sustained, reasons for tension or resentment need to be removed, and trust needs to be re-established, whenever progress seems to lag or come to a halt.
- *Viewing water-related negotiations as a venue for cooperation and resolving more troublesome issues among disputing countries,* in order to enhance parties' ability to cooperate over water-related matters as a first step, laying the groundwork for more cooperation on other issues of high politics in the future.
- *Seeking cooperation on other venues and projects of common interest and benefit to both disputing parties,* whether related to the shared basins or outside the shared water resources in question. This tends to create a win–win situation and promote an atmosphere conducive to cooperation.

Because these criteria emphasize procedural aspects of negotiation, such as national capacity building, dialog, viewing water as a venue for cooperation, common interests, and achievable goals, they represent key process equity parameters. These process equity parameters were previously listed in the diagram in Figure 2.1.

International water treaties are the most commonly utilized water-related conflict resolution tools to resolve disputes over shared international waterways. This is because they are viewed as economically efficient and strategically and hydropolitically sound strategies. Although the reality of many water treaties might suggest otherwise, treaties are considered a preferred option to resolving conflicts over water resources. Although the territories of an estimated 145 countries are located in international basins, the number of countries sharing international watercourses is significantly higher (Wolf, 2002). It is estimated that over 245 river basins are shared and peacefully managed by two or more riparian states in spite of the lack of agreed-upon international law governing the non-navigational uses of international waterways prior to the 1997 UN Convention on the respective matter (Salman *et al.*, 1998).[1] According to Wolf (1998; 2002), approximately 263 international river basins are identified as potential sites for future conflict, comprising more than a half of the earth's land area, 60 percent of its freshwater flow, and supporting 40 percent of its population (Salman *et al.*, 1998; Wolf, 2002). These basins are associated with a comparable number of international watercourses (IWC), governed by a similar number of treaties, and representing both international and transnational rivers (Upreti, 2006).

As illustrated in Figure 3.1, international rivers (alternatively referred to as *boundary* or *contiguous rivers*) are those rivers that run more than 10 km along the border (Gleditsch *et al.*, 2007), demarcating international boundaries between two or more riparian states. Transnational rivers (alternatively referred to as *successive rivers*) are those that flow across international boundaries from one riparian state to another, creating a spatial configuration of an upstream/downstream relationship (Gleditsch *et al.*, 2007). *Mixed rivers* are those that combine the attributes of both of the above (Zawahri, 2004; Allouche, 2005). Depending on the type of river and the nature of the relationship between riparian states, sharing water resources can create different types of problems and dispute intensity, with therefore differing potential for cooperative resolutions. Generally, it is argued that the upstream/downstream relationship (*transnational river*) situation often results in the most intense and serious types of conflicts (Gleditsch *et al.*, 2007). In the

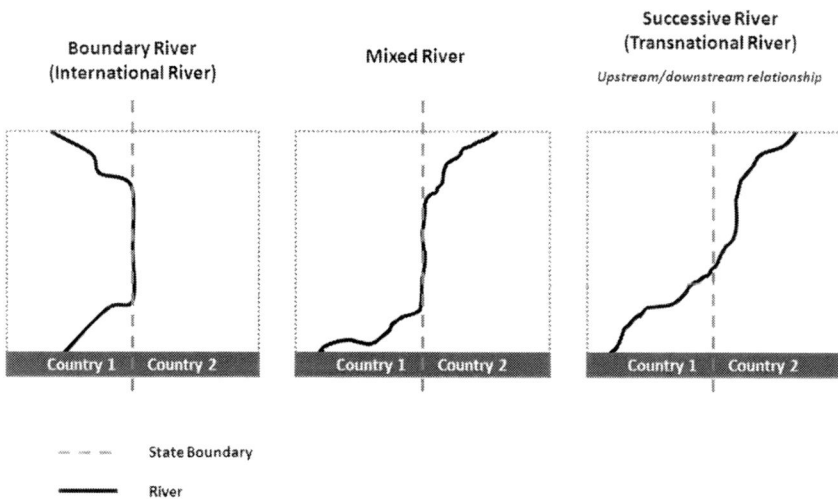

Figure 3.1 Types of rivers and co-riparian relations. Source: adapted from Gleditsch *et al.* (2007).

boundary river situation, country A tends to extract water for its own benefit disregarding country B, which will retaliate by diverting water from its own side of the boundary. In both cases, conflict related to the common calamity and to other navigational issues will ensue. Trans-border pollution problems are also common in the boundary river situation (Gleditsch *et al.*, 2007). However, because of a zero-sum situation provoked by the upstream/downstream relationship situation and the lack of trust, the likelihood for cooperation is slimmer than in the boundary river situation (Gleditsch *et al.*, 2007).

Although the exact number of these international basins is not precisely specified or agreed upon, and is subject to modification due partially to changes in international boundary configuration, Figure 3.2 shows that the largest number of these river basins – about 69 basins – are located in Europe, followed by 59 basins located in Africa, 57 in Asia, 40 in North America, and 38 in South America (Wolf, 2002). In many documented cases, water, arguably, has acted as a source of conflict and an obstacle to achieving peace, but in many other cases water acted as a reason and catalyst for peace and cooperation, prompting the use of the term hydro-stability. Understanding these geopolitical and hydro-political dynamics of water-related conflict realities (as well as the complex physical, political, and human interactions within international river basins) can provide adequate explanation of how conflict occurs and why, under what circumstances treaties are negotiated and signed, and who wins and loses and with what mechanisms.

A multitude of bilateral and multilateral treaties have taken place both at the regional and international levels throughout human history to manage conflict and enhance cooperation over shared water resources (Salman *et al.*, 1998).[3] It is estimated that between the years of 805 and 1984 over 3,600 treaties have been negotiated and signed to resolve conflict and tension over international waters (UNFAO, 1978; 1984). Approximately 300 treaties have been signed since 1814 pertinent to numerous water issues, including the non-navigational aspects of water management, flood control, hydropower projects, and the allotment for consumptive or non-consumptive uses of international water resources (Hamner and Wolf, 1998). An analysis of these treaties reveals that approximately 145 treaties could be directly related to water issues *per se*, 86 percent of which were bilateral whereas 14 percent were multilateral (Wolf, 1998).

Although many bilateral treaties occurred in watersheds shared by only two co-riparians, many (such as the Jordanian–Israeli water treaty of 1994) also occurred in watersheds shared by more than two parties. An explanation for this phenomenon is found in the negotiation theory literature, which holds that these bilateral treaties occur in part because of the difficulty entailed in conducting multinational negotiations that integrate all of the involved and impacted parties (Zartman, 1991). Table 3.1 provides a statistical summary of these 145 treaties documented and described in more detail in Wolf (1997) and in Hamner and Wolf (1998).

Over 1,800 water events (representing incidents of either conflict or cooperation that occurred within an international river basin) were studied in considerable detail by Wolf and colleagues (2003) in an attempt to assess indicators for

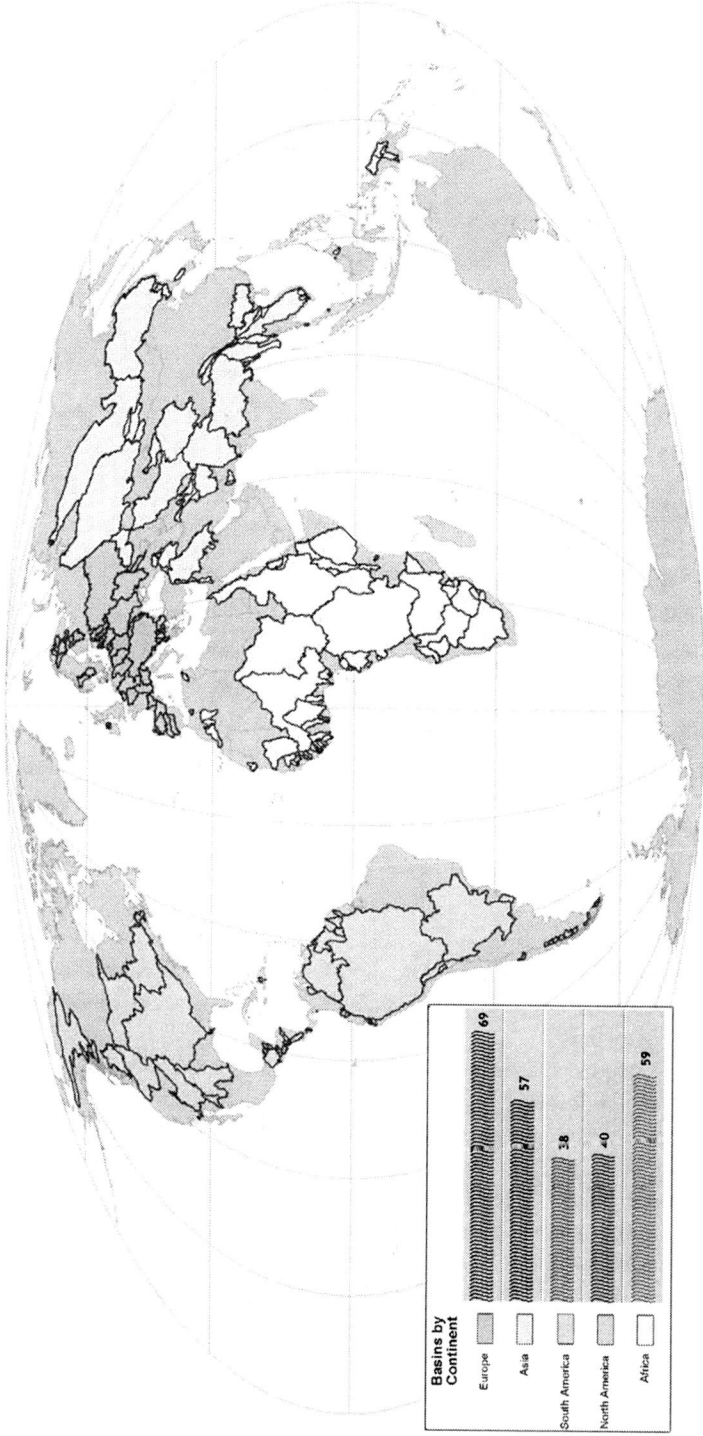

Figure 3.2 International river basins, as delineated by the Transboundary Freshwater Dispute Database project, Oregon State University, 2000. Data source: Atlas of International Freshwater Agreements, Wolf (2002).[2]

Basins by Continent

- Europe 69
- Asia 57
- South America 38
- North America 40
- Africa 59

Table 3.1 Treaty statistics summary sheet

Signatories		*Enforcement*	
Bilateral	124/145 (86%)	Council	26/145 (18%)
Multilateral	21/145 (14%)	Force	2/145 (1%)
Principal focus		Economic	1/145 (<1%)
Water supply	53/145 (37%)	None/not available	116/145 (80%)
Hydropower	57/145 (39%)	*Unequal power relationship*	
Flood control	13/145 (9%)	Yes	52/145 (36%)
Industrial uses	9/145 (6%)	No/unclear	93/145 (64%)
Navigation	6/145 (4%)	*Information sharing*	
Pollution	6/145 (4%)	Yes	93/145 (64%)
Fishing	1/145 (<1%)	No/not available	52/145 (36%)
Monitoring		*Water allocation*	
Provided	78/145 (54%)	Equal portions	15/145 (10%)
No/not available	67/145 (46%)	Complex but clear	39/145 (27%)
Conflict resolution		Unclear	14/145 (10%)
Council	43/145 (30%)	None/not available	77/145 (53%)
Other governmental unit	9/145 (6%)	*Non-water linkages*	
United Nations/third party	14/145 (10%)	Money	44/145 (30%)
None/not available	79/145 (54%)	Land	6/145 (4%)
		Political concessions	2/145 (1%)
		Other linkages	10/145 (7%)

Source: Hamner and Wolf (1998).

identifying basins at risk for the purpose of enhancing preventive diplomacy and ameliorating potential conflicts (Wolf, 2004). Basins at risk are identified and illustrated in Figure 3.3.

   This study included events of more than 231 river basins occurring over a period of 50 years (between 1948 and 1999), where water appears to have been the driving force of the events, excluding those where water was trivial to the dispute (Grover, 2007). It utilized the *Basins at Risk (BAR) Scale* (shown in Figure 3.4), categorizing water-related events based on the degree and intensity of conflict and cooperation that each of these events exhibited. Included were 13 categories of key issues, including water quality and quantity, hydropower, hydroelectricity, navigation, fishing, flood control/relief, general economic development, joint management of the river basin, irrigation, infrastructure of development projects (such as dams, canals and storage facilities), technical cooperation, and shared border and territorial issues (Towfique, 2002). The results of this study are represented in Figure 3.4, which shows the number of events and the corresponding

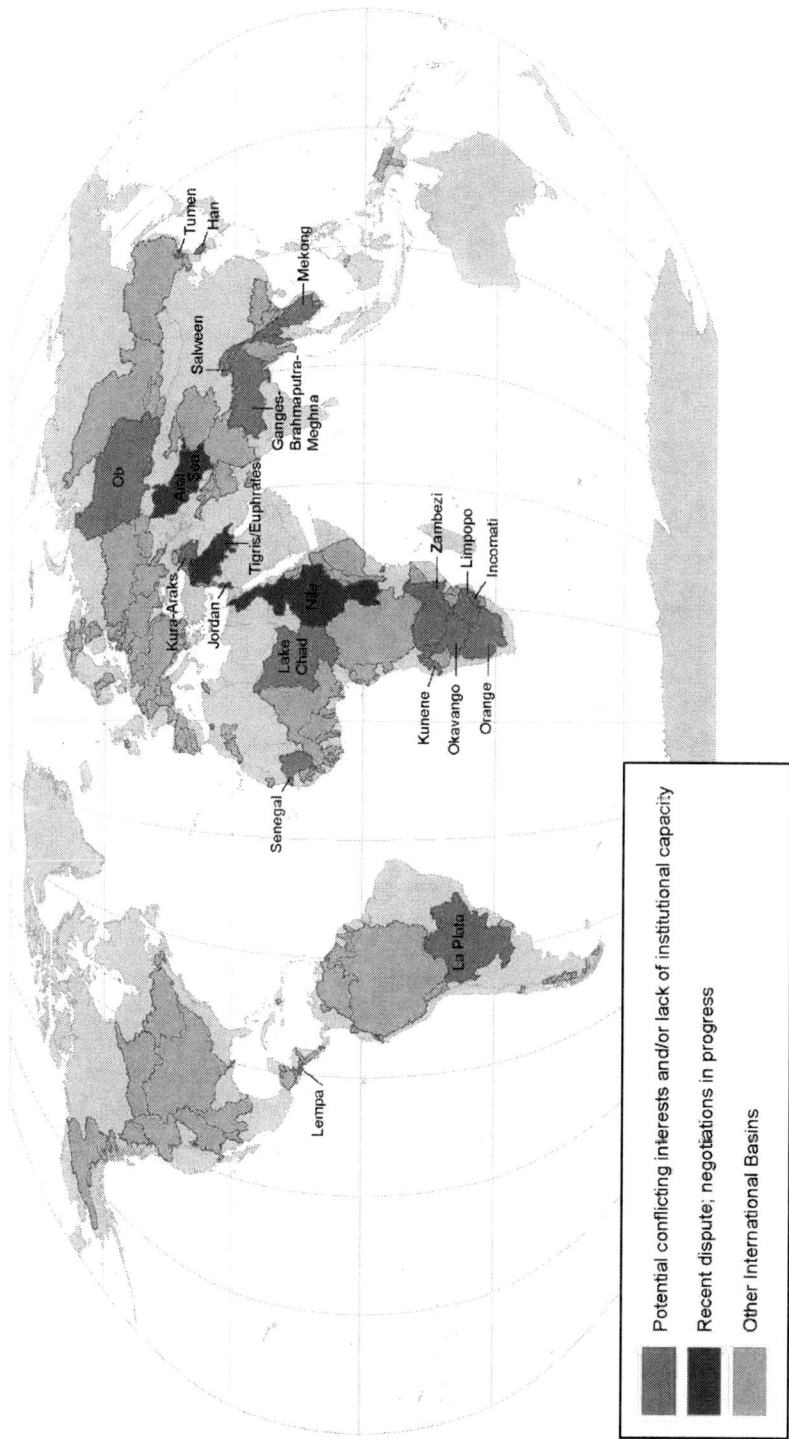

Legend:

- Potential conflicting interests and/or lack of institutional capacity
- Recent dispute; negotiations in progress
- Other International Basins

Figure 3.3 Basins at risk. Source: *Wolf et al.* (2003).[4]

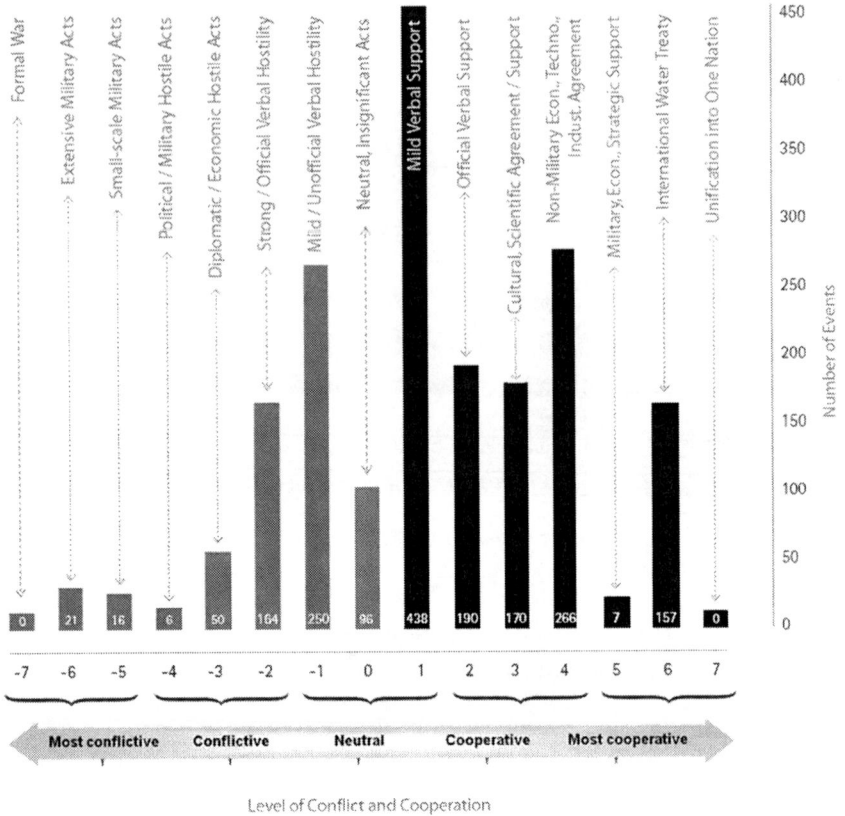

Figure 3.4 Number of events by bar scale, showing the spectrum of experience in international water basins. Source: adapted from Wolf *et al.* (2003).

conflict–cooperation level (CCL), ranging from most cooperative to most conflictive events.

Water quantity and infrastructure, being closely and inextricably related to one another, were the two main issues comprising about 64 percent of the total water events (Wolf, 2006). On the other hand, water quality-related events seemed to constitute a minute percentage, totaling only 6 percent. The study reveals that conflict occurs more often over water quantity and infrastructure, which is more likely to involve an armed component. Conversely, cooperation occurs more often over a wider range of water-related issues, involving water quantity, infrastructure, joint management, hydropower, water quality, technical cooperation, flood control and relief, and others (Wolf *et al.*, 2003).

Moreover, during the 50 years covered by the study, there was no indication of any war fought over water (Grover, 2007) and only 37 severe disputes involving violence occurred over water (30 of which occurred between Israel and one of its neighbors) (Wolf, 2002). It is conclusively noted that cooperation over shared water resources appeared to be the norm, comprising more than three-quarters of

Distribution of cooperative
events by issue area

Distribution of conflictive
events by issue area

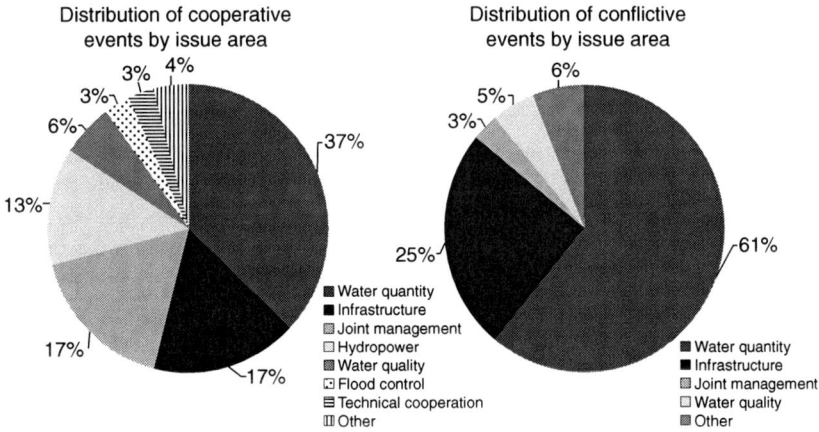

Figure 3.5 Distribution of cooperative and conflictive events by issue area. Sources: adapted from Wolf *et al.* (2003) and Wolf (2004, 2006).

water-related interactions during the last century (Grover, 2007). Conversely, a quarter of these water-related interactions involved some form of aggression and violence, which was confined to the intra-state and sub-national levels, namely between tribes and indigenous populations, rather than the inter-state level (Grover, 2007). In addition, as shown in Figure 3.5, most of these conflicts were confined to issues of water quantities and infrastructure. This number of violent incidents seems very trivial when compared with the number of treaties (the most cooperative types of events) that were negotiated and officiated during the same period, totaling 157 treaties (Wolf, 2006). The study also suggests that there is a noticeable trend in recent years indicating a growing interest in water-related issues (Wolf *et al.*, 2003).

Collectively, there have been about 507 conflictive events compared with 1,228 cooperative events, suggesting that water-related events are shifting toward more cooperative trends rather than violent clashes (Wolf, 2006). The study's main conclusion contradicts common predictions that the future will bring an increasing number of conflicts over shared water supplies. It sent shockwaves into the community of researchers espousing the "water war" paradigm. It concludes that water is increasingly becoming a "unifying agent" and a source of cooperation rather than conflict. Conflict, which seems to be the anomaly, is perceived as being economically inefficient, strategically unsound, and hydrographically unviable (Wolf, 2002; 2003; 2006). It is critical to understand, however, that history demonstrates that wars over water did occur among many nations in the past and that this will probably be the case in the future if water conflicts are not proactively managed (Abukhater, 2006). Water can and should be used as a unifying factor for bringing riparians to the negotiating table. The genuine value of water lies in its potential ability to promote hydro-political stability. This approach echoes the conclusions of this book that conflict remains an anomaly, that nations tend to gravitate more toward cooperative arrangements, and that water, after all, can and should become a catalyst for peace and stability.

First-cut selection and scanning

The goal of this section is to identify and briefly scan nine international water allocation cases. Reflecting the book's primary concern of equity, case selection is focused on ratified bilateral, international water allocation treaties. The main objective of the case selection is to provide a number of cases which share common characteristics identified below and are representative of arid region contexts. A number of international water treaties that address the issue of water quantity and quality as well as the parties involved were compiled based on various sources. To be able to select adequate cases that are not only relevant to the scope of this book, but also representative of arid region cases in different contexts around the world, a number of methods were utilized. By relying on a systematic review of the available published treaty collections, online water treaty databases, water expert panel opinion surveys and phone interviews, and other secondary sources of data, the first-cut selection will help in identifying and reviewing candidate cases, two of which will be selected for more in-depth analysis. To ensure that they are constant across all cases, the control factors listed in the logic model (as shown in Figure 2.1 in the previous chapter) helped in selecting compatible cases that, although exhibiting similar attributes, may differ to various degrees with regard to equity attainability. In this way, we were able to triangulate both commensurable equity and perceived equity as parts of the case selection exercise.

The case selection process involved the selection of at least one case to represent each of the categories identified in the following criteria, including perceived equity, level of implementation, and geographic location. For any case to be selected, it had to match all three criteria listed in Table 3.2 combined. As such, negating one criterion was enough to exclude a case.

Because this research is focused on the contextual understanding of water treaties in arid regions reflecting a variety of geographic contexts, the selection

Table 3.2 Case selection criteria

The nine cases will be selected based on the following criteria:

- **Degree of implementation/level of monitoring:**
  We are interested in full and partial implementation. Data on this criterion can be obtained from expert opinion and other secondary sources

- **Perceived equity – level of satisfaction or satiation:**
  We are interested in high and low perceived equity. This can be obtained and triangulated from expert opinion and other secondary sources

- **Geographic location:**
  African case (AF)
  Asian case (AS)
  Eurasian case (E)
  Middle Eastern case (M)
  5- Western case (W)

Compatible cells = $2 \times 2 \times 5 = 20$ cells

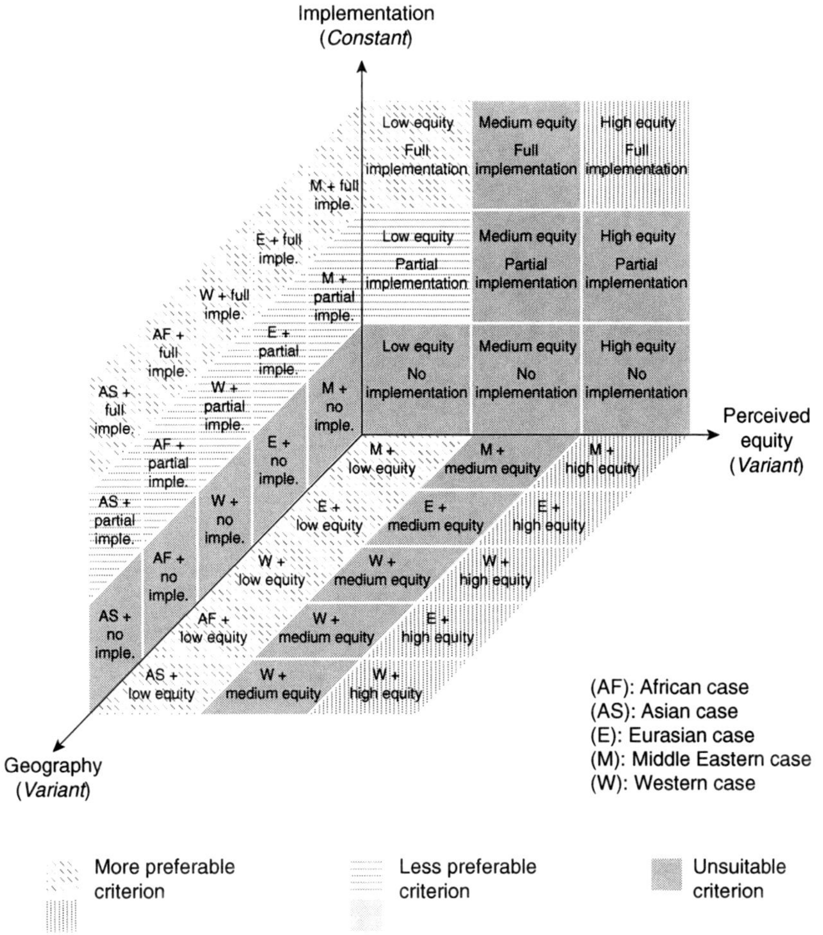

Figure 3.6 Case selection criteria.

criteria for case studies utilized the three-dimensional model illustrated in Figure 3.6. This systematic model encompasses the selection criteria listed in Table 3.2, including the degree of implementation, which was kept constant across all cases, perceived equity, and geographic location, both of which were variant. The goal was to find cases that not only comply with these criteria but also indicate full or partial implementation, which shows that the treaty is currently in effect. To that end, this hermeneutic model shows the range of cases that are deemed acceptable for the first-cut case selection. These cases can be selected based on this three-dimensional model so long as they coincide with any of the 20 compatible cells shown in Figure 3.7.

For any case to be considered for the selection process, it needed to be suggested by water experts and verified by secondary data sources. For this particular purpose, expert panels and a review of secondary sources were conducted. This

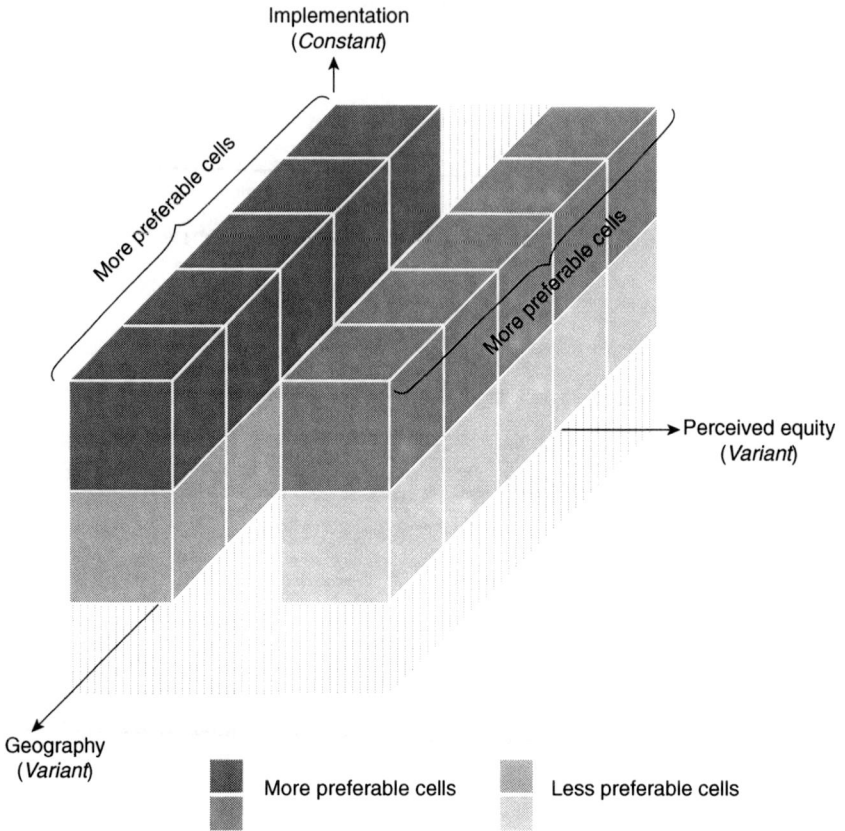

Figure 3.7 Compatible cells for case selection.

was achieved by exploring and collecting the different attitudes and perceptions of water experts toward the current and previous water allocation schemes introduced by these international treaties and conflict resolution plans concerned with water allotment. A number of international water treaty experts were surveyed to determine a shortlist of comparable treaties.[5] Based on this survey, a list of the most-cited treaties was compiled. Cases that were the most referenced were considered, whereas other less-referenced cases were discarded. These most-cited cases were verified against the three-dimensional model and only those cases that complied with the selection criteria were considered.

These results were further validated using a variety of online water databases, such as the Transboundary Freshwater Dispute Database (TFDD), the International Freshwater Treaties Database, and International River Basins of the World. These online tabular and spatial water treaty databases provided a systematic examination and verification of these treaties. By reviewing these secondary sources of data, the contextual and circumstantial details of each treaty were clarified. In addition, based on these databases, texts of these treaties were initially screened to evaluate the compatibility of each treaty and to validate their selection. As a result, the initial list of treaties was then narrowed down to nine, reflecting a triangulation process that incorporated complementary methods of inquiry. These nine treaties are shown in Table 3.3.

First-cut case analysis (snapshot case analysis)

This section seeks to provide a case scanning and documentation assessment of these nine bilateral treaties. By presenting a short description of each of these cases, this segment outlines (1) the case background by introducing the general characteristics of the river basin and the parties involved in the conflict; (2) a synoptic history of water dispute issues and hydro-diplomacy, including the history of the conflict and the key issues under dispute, previous diplomatic efforts and treaty formation; and (3) treaty outcome, which traces and analyzes the water allocation scheme provisioned by the water agreement. The goal of this first-cut analysis of the nine cases is to provide a global outlook on issues surrounding international water diplomacy and conflict resolution trends and to elicit a better understanding of water-related issues on a global scale. It will also provide ample theoretical and epistemological grounds for the selection of the detailed cases to be analyzed in the following chapters.

The Nile river basin: Agreement between the Government of the United Arab Republic of Egypt and the Government of Sudan, November 8, 1959

Background

Egypt and Sudan primarily depend on the Nile (*al-nahr al-khalid*, the "eternal river") for survival and socioeconomic development. Egypt is heavily reliant on the Nile as its only major source of water, although the source of the river does not originate within its boundaries and Egypt is thus the downstream riparian. This poses a great threat to its water security, as the river is considered to be the "life of Egypt." The Nile plays host to about 150 million people, as it drains an area equivalent to about 10 percent of the African continent, exceeding 2.9 million km^2 and stretching to about 6,500 km in length (Salame and Zaag, 2010). Ten nations share the Nile river basin, as it stretches across the territories of Burundi, the Democratic Republic of the Congo, Egypt, Ethiopia, Eritrea, Kenya, Rwanda, Sudan, Tanzania, and Uganda (Nicol, 2003). Compared with these other states sharing the river basin, Egypt is the most powerful, both economically and militarily (McKinney, 2008). Because of its large size, the extreme aridity of the region, and the multicultural and heterogeneous nature of its riparians, the Nile offers a striking case where water distribution is challenged by various layers of added complexity to an already complex situation (Nicole, 2003).

Table 3.3 Initially selected cases

	Date	Treaty basin	Signatories	Conflict issue	Geographic context	Compatibility (perceived equity)
1	November 8, 1959	Nile	Sudan, Egypt	Water flow	Africa	High
2	March 29, 1946	Tigris–Euphrates–Shatt al-Arab	Iraq, Turkey		Eurasia	Low
3	August 23, 2001	Tigris–Euphrates–Shatt al-Arab	Syria, Turkey	Dams, water flow	Eurasia	Low
4	September 19, 1960	Indus	India, Pakistan	Border issues, water flow	Asia	High
5	October 24, 1986	Orange/Senqu (LHWP)	Lesotho, South Africa	Hydropower/hydroelectricity	Africa	High
6	September 14, 1992	Orange/Senqu	Namibia, South Africa	Water flow	Africa	Moderate
7	October 26, 1994	Jordan, Yarmouk, Araba/Arava groundwater	Israel, Jordan	Water flow	Middle East	Low
8	December 12, 1996	Ganges–Brahmaputra–Meghna (GBM)	Bangladesh, India	Siltation, flooding	Asia	Low
9	November 14, 1944	Rio Grande/Bravo and Colorado	US, Mexico	Water flow, quality, and salinization	Western	Very low

Synoptic history of water dispute issues and hydro-diplomacy

The water dispute along the Nile started with the upstream states' increasing interest in river development in the wake of their independence from colonial rule (McKinney, 2008). Leaders in Egypt, the downstream and relatively most dependent riparian on the Nile, perceived this as a great threat to their water security, which triggered hostility and contentious water conflict, particularly with Sudan (Wolf and Newton, 2008). The major issue of contention between the two countries is centered around the question of sovereignty vis-à-vis historic water rights, reflecting the classic tension and divergence between the claims of upstream and downstream riparians (Priscoli and Wolf, 2009). In Egypt in 1952, while the Aswan High Dam project was under way – a large water storage project with a projected capacity of 156 billion cubic meters (BCM) per year – hostility between the two countries intensified (Wolf and Newton, 2008). This hostility ended in November 1958 after a military coup that gave rise to a new friendly government in Sudan that was willing to negotiate with Egypt (Waterbury, 2002). A year later, in 1959, the two governments were able to sign the 1959 Nile Waters Agreement, to which we now turn (McKinney, 2008; Wolf and Newton, 2008). This agreement was in response to Sudan's objection to the 1929 treaty, which had been concluded under the British colonial regime in Egypt and Sudan (Upreti, 2006).

Treaty outcome

The Treaty aimed to plan for integrated watershed development and to achieve equitable sharing of the water of the Nile between Egypt and Sudan (Priscoli and Wolf, 2009). The Treaty largely advocated the principle of prior appropriation and safeguarded Egypt's existing and historic water use, although Egypt does not contribute to the water of the river (Upreti, 2006). This was a direct result of Egypt's powerful position in the basin. Under this Treaty, Sudan consented to the construction of the Aswan High Dam, whereas Egypt, in return, consented to mitigate the harmful impacts of the construction of the dam and to regulate water flow[6] (Haftendorn, 2000). Based on an estimated average flow of 84 BCM/ year, as measured at the Aswan High Dam, and evaporation loss of 10 BCM/ year, the Treaty allocated 55.5 BCM/year to Egypt and 18.5 BCM/year to Sudan (Mekonnen, 1999). The two countries agreed that any additional water acquisition, whether from natural flow or river development, would be split evenly between them. Although the Treaty did not consider the water needs of other co-riparians, it required the establishment of the Permanent Joint Technical Committee to address the other co-riparian states' water claims and to determine a water allocation scheme in case of exceptionally low flows (Wolf and Newton, 2008). For Egypt, the ability to build the Aswan High Dam, which allowed it to capture the annual flood in one place within its territory, constituted an essentially symbolic solution to its water insecurity. However, the principle of equitable utilization was largely ignored by the makers of the Treaty (Upreti, 2006). Although some may perceive it as satisfactory to both countries, the Treaty failed to involve and equitably address the needs of all riparian countries sharing the basin, besides its evidently detrimental environmental impacts (Mekonnen, 1999). Moreover, the water allocation scheme provisioned by the Treaty represented a case of inequitable division, with the most powerful state in the basin (Egypt) benefitting from its advantageous position. Nevertheless, the Treaty implementation has been successful, as the water allocations stipulated by it have been adhered to until today.

Since then, the basin has witnessed several attempts to resolve conflict and to establish cooperative relationships among its riparians. These attempts tackled the issue from various angles, including socioeconomic, political, and knowledge and capacity building, the most recent of which is the multinational FRIEND initiative of 1996 (Salame and Zaag, 2010). This initiative aimed at establishing cooperation and trust among the various parties through the collection and sharing of critical basin data for better regional management of its resources. Although Egypt has traditionally been the most powerful state in the basin, and therefore the most influential in negotiating and tilting the 1959 Treaty with Sudan in its favor, evidence suggests that over the past decade the other upstream riparians have been successful in utilizing their hydro-strategic positions and collective relationship as a bargaining tool to influence Egypt and establish a more balanced cooperative relationship (Cascao and Zeiton, 2010). The Cooperative Framework Agreement (CFA) was established in 1997, which was the first tipping point for the recent change in the basin's hydro-political dynamics (Cascao and Zeiton, 2010). The Nile Basin Initiative (NBI), a multinational cooperative partnership among the 10 co-riparians, was established in May of 1999, to collaboratively manage the development of the Nile water resources (Abdalla and Eldaw, 2002; Nicol, 2003). The objective of the NBI is to "achieve sustainable socioeconomic development through the equitable utilization of, and benefit from the common Nile Basin water resources" (NBI, 2006). These recent initiatives provided a successful example of a more balanced relationship among co-riparians along the Nile and addressed the various issues surrounding its water conflict.

The Tigris–Euphrates river basin

Two water-related agreements are included in this analysis:

(A) Treaty of Friendship and Good Neighbourly Relations between Iraq and Turkey, signed at Ankara, March 29, 1946;
(B) Joint Communiqué between the Republic of Turkey and the Arab Republic of Syria, August 23, 2001.

Background

The Tigris–Euphrates river basin comprises three key countries, including Iraq (the downstream riparian), Syria (the middle riparian), and Turkey (the upstream and most powerful riparian in the basin) (Elhance, 1999). Iraq and Syria are the most dependent countries on the flow of the Euphrates and the Tigris, whereas

Turkey is the least dependent on the rivers for its water needs (Haftendorn, 2000). Both rivers originate in Turkey's south-eastern highlands and flow downstream into Syria and Iraq, where they merge into the Shatt al-Arab waterway as they eventually disappear into the Arabian Gulf (Shungur, 2005).

Synoptic history of water dispute issues and hydro-diplomacy

Prior to the 1960s, the relationship between the three co-riparian states was relatively harmonious (Kibaroglu and Unver, 2000). On March 29, 1946, Turkey and Iraq signed the Treaty of Friendship and Good Neighbourly Relations, and six annexed protocols. Through this Treaty, the two countries were able to set forth a framework for managing the sharing and utilization of the rivers and solidi-fied friendly relationships between them. It was not until the 1960s that hydro-political tension increased, as the three countries announced their plans for water development projects to unilaterally utilize the waters of the Tigris–Euphrates river system for human consumption, energy, and irrigation purposes (Towfique, 2002). As these plans were merely concerned with augmenting water supply, they were uncoordinated and seemed to ignore the demand-side management, both on the local and regional, basin-wide levels. Not only did these major develop-ment projects – particularly the Greater Anatolia Project (GAP) of Turkey and the Euphrates Valley Project of Syria – reduce water flow to Syria and Iraq and increase the tension over the regional water resources, but they also threatened to compromise political relations between the countries for many years (Lowi, 1991). In this regard, water allocation and states' relations are intimately inter-twined with and impacted by political issues. For example, Turkey blamed Syria for its support of the Kurdistan Workers' Party (best known as the PKK Kurdish rebels) and, as a result, further threatened a blockage of water to Syria[7] (Berman and Wihbey, 1999). Although regarded as international rivers by Syria and Iraq, the Euphrates and Tigris rivers are viewed by Turkey as Turkish waters subject to unlimited Turkish utilization, as the need arose (Haftendorn, 2000; El-Fadel *et al.*, 2002). This view was best expressed by Turkey's former prime minister, Süleyman Demirel, who stated that "neither Syria nor Iraq can lay claim to Turkey's rivers any more than Ankara could claim their oil. We have a right to do anything we like. The water resources are Turkey's, the oil resources are theirs. We don't say we share their oil resources, and they can't say they share our water resources" (quoted in Zawahri, 2006: 1046).

Meetings and talks between the two countries have been taking place since the mid-1960s until today. These meetings attempted to negotiate an equitable allocation of the Euphrates' water resources among the co-riparian states (Priscoli and Wolf, 2009). The escalated tension between Syria and Turkey in late 1998, and the three-year drought from 1999 to 2001, challenged the countries to resort to diplomacy. The Joint Technical Committee (JTC) was established in 1980 to govern the utilization of these shared transboundary watercourses and coordinate the three states' efforts to collect and share data and monitor the sharing of the river basin waters[8] (Kibaroglu, 2002; Zawahri, 2004). However, the asymmetrical relationship of interdependence among the co-riparian states, namely Turkey being the upstream riparian and least dependent on the river system, created disinterest in developing such regional water regimes and undermined the role of the JTC to adequately and collaboratively govern the management of the regional water allocation (Zawahri, 2004; Shungur, 2005). In October 1998, Turkey and Syria broke new ground by signing the Adana Agreement, according to which Syria agreed to ban the PKK rebels from its territories (Ilter, 2000). Later in August 2001, the two countries once again were able to agree on a protocol of cooperation regarding Turkey's GAP project and Syria's General Organization for Land Development (GOLD) (Priscoli and Wolf, 2009).

Treaty outcome

First, the 1946 Treaty between Turkey and Iraq is a remarkable legal text as it was the first diplomatic attempt to collaboratively manage and monitor the river basin, namely its annexed Protocol 1, which addressed issues related to the regulation of the Tigris–Euphrates waters and their tributaries (TFDD, 2008). While providing a framework for addressing each country's interests along the river system, this protocol called for constructing hydrological infrastructures (including dams and storage facilities) and meteorological observation stations along the rivers in Turkey, from which both countries would benefit (Kibaroglu, 2002; Zawahri, 2006). The Treaty, however, failed to address the regional concerns of all involved parties as it excluded Syria. In addition, the Treaty did not call for the establishment of a joint trilateral water regime to govern the river basin as a whole. Cold political relations still dominate the two countries' diplomatic realm as no progress has been achieved to address the interests and concerns of Iraq's water security, which largely remains unsatisfied until today (Towfique, 2002).

Second, although the Adana Agreement of 1998 helped in opening the channels of communication and diplomatic dialog between Syria and Turkey, the 2001 protocol of cooperation provided a point of departure from the unilateral nature of the river development plans. By securing the mutual consent of the GAP and GOLD projects, the protocol encompassed collaborative training programs, joint projects, exchange programs, and partnerships (TFDD, 2008). However, tension still lingers as this protocol of cooperation did not yield a substantial long-term agreement to govern the water allocation of their shared rivers (Haftendorn, 2000). As populations continue to grow and the demand for development increases the demand for water, the issue of water sharing of the Tigris–Euphrates among the three co-riparians remains unresolved. These diplomatic strides were unsuccessful in convincing Turkey in 2003 to agree to a final accord for the allocation of shared waters with Syria and Iraq (United Press International, 2003). Not only could reaching an agreeable settlement provide mutual economic advantages, it also could achieve a reasonable and practical solution to the Kurdish problem (Haftendorn, 2000). Many experts believe that the recent change in the political landscape of the region, in the wake of the United States' incursion into Iraq, might provide the platform for a more productive future dialog among the three co-riparians that could potentially resolve their distributional conflict over the sharing of the Tigris–Euphrates water basin (Wolf and Newton, 2008). However, despite their previous disagreements, Syria and Iraq have formed a united front in facing Turkey, whereas Turkey allied with Israel in terms of its military and economic relations, which formed a new power parallelogram that could reshape the dynamics of future regional hydro-politics (Haftendorn, 2000). This relationship has recently deteriorated in the wake of Israel's military actions and the continuous blockade of the Gaza Strip.

The Indus River basin: Agreement between Pakistan and India on West Pakistan–India border disputes, signed in New Delhi, January 11, 1960

Background

Originating in the Himalayan Mountains in the north and meandering into the dry alluvial plains of Sindh in the south, the Indus River basin stretches across a large area that penetrates many territories, including China, India (and Indian controlled Kashmir), Afghanistan, and Pakistan, before it empties into the Arabian Sea (McKinney, 2008). Extending to more than 3,000 km in length, with five major tributaries fed by glacial melt, the Indus River plays an important role in supporting the agricultural sectors of both India and Pakistan (Wolf and Newton, 2010a). In this basin, India and Pakistan are the two major powerhouses that compete over accessing the river water resources. Although Pakistan occupies more than half of the river basin's area and houses much of its extensive canal systems,[9] the headwaters of the river and its tributaries are located within Indian territories, which provides an advantageous hydro-strategic position for India as the upstream riparian (Bhatti, 1999).

Synoptic history of water dispute issues and hydro-diplomacy

India and Pakistan have a long-standing animosity and conflict over the waters of the Indus for irrigation and hydropower purposes, fueled by political tension and unresolved population and territorial issues (Wolf and Newton, 2010a). This conflict, which is reflective of the classic upstream versus downstream riparians' argument, dates back to 1947 with the partitioning of colonial India that caused the waters of the river to be split between the two new rival states: India and Pakistan (McKinney, 2008). The water dispute between the two countries is tied to a wider political and cultural conflict, namely the control over disputed Kashmir and its upper reaches of the river. In the wake of independence from Britain in 1947, the countries' boundaries were delineated with little regard to water resources, causing Pakistan to become dependent on India for its water supply (Wolf and Newton, 2010a). With the expiration of the "Standstill Agreement" of 1947, and in the absence of cooperation, tension between the two countries mounted, leading India to take advantage of its strong hydro-strategic position by interrupting the water flow to Pakistan on April 1, 1948 (Bhatti, 1999). This threatened Pakistan's water security, which predominantly depends on the river for its agricultural production. An interim agreement was established in 1948, according to which Pakistan shouldered the burden of annual royalty payments to India in return for India's agreeing to release water to Pakistan (Priscoli and Wolf, 2009). As this temporary arrangement was insufficient to provide a permanent and acceptable agreement, negotiations between the two riparians were carried out with the involvement of the World Bank as a third-party mediator in 1952 (McKinney, 2008). The idea was to find a practical solution to the water problem through negotiation over technical and engineering aspects rather than pure politics (Wolf and Newton, 2010a). These negotiations that lasted for more than a decade culminated on September 19, 1960, with the signing of a water treaty between India and Pakistan, known as the Indus Waters Treaty, which provided a permanent solution to their water dispute (Bhatti, 1999; Shungur, 2005). With this Treaty, the two co-riparians were able to reach agreement on independently managing their water resources (Wolf and Newton, 2010a).

Treaty outcome

This was a historical treaty that aimed at attaining equitable management and utilization of the river's water and existing canal systems between the two countries. The Treaty, which outlined the technical aspects and financial arrangements pertaining to the shared water resources, established the right of Pakistan to use the three western rivers – namely the Indus, the Jehlum, and the Chenab – and the right of India to use the three eastern rivers – namely the Ravi, the Beas and the Satluj (Mechlem, 2002; Shungur, 2005). This allocation scheme was developed by the World Bank and based on a reconciliation of the two separate proposals submitted by the two riparians (Bhatti, 1999). In this case, the role of the World Bank as a third-party mediator was critical in requesting and comparing the two countries' plans to find common ground for negotiation (Wolf and Newton, 2010a). In addition, the Treaty put forth a 10-year transition period, during which Pakistan would be able to build multipurpose dams, canals and barrages while continuing to receive uninterrupted water from India. This period also allowed the latter to pay its financial obligation toward the building of these hydrological infrastructures.

The Treaty established the Permanent Indus Commission to oversee implementation and ensure future cooperation, data exchange and developed mechanisms for ongoing conflict resolution. To manage future disputes, the two parties agreed to utilize "neutral experts," impartial third-party mediation, and a Court of Arbitration consecutively (Wolf and Newton, 2008). The financial involvement of the World Bank provided sufficient incentive for the two countries to achieve a breakthrough in their negotiations and a pragmatic resolution that addressed their respective hydrological concerns, apart from heated political debates (Wolf, 2001). The fact that the negotiation focused on technical issues and was led by senior engineers on both sides helped in managing and neutralizing the impact of the technical complexity of the shared water system and expediting the negotiation (Wolf and Newton, 2010a).

Although it did not consider a basin-wide development and management plan, the Treaty was very successful in managing the individual utilization and operation of the river system and was adhered to even during the most contentious periods, namely the wars of 1965 and 1971, except for a few occasions of operation-related disagreements that were resolved through bilateral negotiations (Roy, 1997; Klare, 2001; Shungur, 2005). This aspect of the Treaty is critical to understand. The fact that both countries were interested in issues related to sovereignty and control over their individual resources, rather than integrated water management, made it possible for them to reach an agreement that was acceptable to both irrespective of the wider political dispute (Wolf and Newton, 2010a). As the two countries continued to be satisfied with its terms, the Treaty has been valid since 1960 when it was first officiated, notwithstanding the current tension between them. As such, it is the consensus of scholars that the Indus Waters Treaty emerges as one of the most successful water agreements, as it conclusively provided ample evidence of how water scarcity can be used as an incentive for cooperation rather than conflict (McKinney, 2008; Upreti, 2006).

The Orange/Senqu River basin

Two water-related agreements are included in this analysis:

(A) Treaty on the Lesotho Highlands Water Project between the Government of the Kingdom of Lesotho and the Government of the Republic of South Africa, signed at Maseru, October 24, 1986;

(B) Agreement between the Government of the Republic of Namibia and the Government of the Republic of South Africa on the Establishment of a Permanent Water Commission, signed at Noordoewer, September 14, 1992.

Background

The Orange River originates in the range of Lesotho's Maluti Mountains, 3,000 m above sea level, where rainfall is significant and exceeds evaporation. The Senqu in Lesotho, the Vaal in South Africa, and the Fish in Namibia are the main tributaries of the Orange River, which eventually terminates in the Atlantic Ocean (Heyns, 2004). Being the most developed transboundary river basin in the region, the Orange River basin houses four states, including South Africa, Namibia, Botswana, and Lesotho, all of which vie for their fair access of the river waters (Hughes *et al.*, 2010). The Vaal, in particular, is the most urbanized catchment in the region, housing more than 48 percent of South Africa's population (Malzbender and Kranz, 2010). Within this geographic context, Lesotho, which is completely surrounded by South Africa, is a less-developed country with limited economic development capacity compared with South Africa's more competitive economy. Lesotho is the least economically affluent riparian, whereas Botswana,

Namibia, and South Africa are the most economically powerful states in the basin, yet suffering from acute water shortage (Earle *et al.*, 2005). Future economic development in Namibia and South Africa is contingent upon their level of utilization of the Orange River's water resources (Earle *et al.*, 2005). Although the water needs of South Africa are similar to those of Namibia, the latter does not have significant industrial development, unlike South Africa, which has a relatively significant industrial development basis. On the other hand, South Africa is located downstream of Lesotho, which is the upstream riparian and is also economically the most dependent on South Africa's economy.

Synoptic history of water dispute issues and hydro-diplomacy

On the whole, the river plays a pivotal role in the economic development of these four co-riparians, including the industrial, livestock, and agricultural sectors. Water is abundantly available and underutilized in Lesotho but much is needed in South Africa for its industrial development (Wolf and Newton, 2010b). Although the idea of transferring water from Lesotho to South Africa's industrial hub had been under consideration since the 1950s, the project was handicapped by payment-related disagreements coupled with the lack of cooperative efforts to manage the river system, which was largely because of the souring of relations in the wake of the apartheid system (Priscoli and Wolf, 2009). Intermittent diplomatic efforts continued between the two countries to cement a plausible, acceptable, and lasting agreement. A plan was proposed by South Africa in 1966, which resulted in the establishment of the Joint Water Committee in 1978 (Wolf and Newton, 2010b). Notwithstanding the larger political turmoil, the two countries were able to agree on a final feasibility study and negotiate a treaty on October 24, 1986, that led to the establishment and implementation of the Lesotho Highlands Water Project (LHWP), the largest and most ambitious water transfer scheme in South Africa.

The dispute between Namibia and South Africa regarding the lower reaches of the Orange River reflect territorial sovereignty and international boundary issues (Ashton, 2000). Collaborative river management has been a concern for the state officials in both Namibia and South Africa, namely regarding the portion of the river that delineates their shared borders (Heyns *et al.*, 2008). With the independence of Namibia in 1990, negotiation efforts between the two states continued resulting in the 1992 Agreement (Heyns, 2004) and the establishment of the Permanent Water Commission (PWC) and the Vioolsdrift and Noordoewer Joint Irrigation Scheme (VNJIS) (Heyns *et al.*, 2008).

Treaty outcome

The 1986 Treaty facilitated the establishment of a five-phase project (LHWP) and provided four protocols covering details pertaining to the technical, economic, and political arrangements (Earle *et al.*, 2005; Wolf and Newton, 2008).

It also provisioned the establishment of a Permanent Joint Technical Commission to oversee the development and operation of the project (Heyns, 2004). Under this agreement, South Africa agreed to finance the construction of infrastructure to transfer water to the Vaal River basin as well as payments for the transferred water, while Lesotho agreed to finance the construction of the hydropower component of the project and received aid from various donors, including the World Bank (Wolf and Newton, 2010b). With the LHWP under way, Lesotho, a state poor in natural resources and economic potential, is able to generate hydropower for its internal consumption and gain a considerable amount of foreign exchange derived from water royalties, whereas South Africa, a state in need of water, is able to transfer water to its industrial heartland (Wyk, 1998; Priscoli and Wolf, 2009). Under the auspices of benefit sharing, the exchange of water for royalties provided South Africa with the water it needed for its industrial development and Lesotho with the economic incentives it needed for nation building (Whann, 1995). Despite its success in swapping water for economic development, the Treaty is criticized on the basis of its focus on the feasibility of the project, which caused detrimental impacts as the social and environmental aspects were overlooked (Wolf and Newton, 2010b). However, despite the political events that ensued after the signing of the Treaty, namely the fall of the apartheid regime, the Treaty is still being adhered to by both countries and the LHWP is being implemented with its second phase, due for completion in 2017 (Wolf and Newton, 2010b).

The 1992 Agreement between Namibia and South Africa also provided a platform for future cooperation and the establishment of a Permanent Water Commission (PWC) to advise the parties on matters related to the management and development of the Lower Orange River, where it forms the border between the two countries. The PWC completed and is managing the work of the Lower Orange River Management Study (LORMS) (Heyns *et al.*, 2008). In particular, because of its vulnerable location downstream and the uncertainty surrounding its future water quantity, quality, and flood risk, coupled with the lack of its water storage facilities, Namibia has been dependent on the regulated flow of water from South Africa (Conley and Niekerk, 2000). These water storage facilities for irrigation, mining, and power generation can only be possible through a cooperative arrangement with South Africa (Heyns, 2004). As such, this bilateral accord served the interests of both states and set the stage for future socioeconomic development (Heyns, 2004). Although the Orange basin embraces four countries, these treaties involve only two states (Wyk, 1998). This situation calls for an integrated, basin-wide management with the involvement and collaboration of all riparian states to enhance equitable and reasonable utilization of shared water resources. The establishment of the Orange-Senqu River Commission (ORACOM) in 2000 constituted a step in this direction (Heyns, 2004; Swain, 2004). These bilateral agreements, however, are proven resilient as they have been adhered to with no significant changes since their signing, despite the drastic changes in the political landscape since (Priscoli and Wolf, 2009).

The Jordan River basin: Treaty of Peace between the State of Israel and the Hashemite Kingdom of Jordan, October 26, 1994

Background

The Jordan River originates in the north from a series of tributaries in Lebanon and Syria and terminates in the Dead Sea in the south. Along its upper course, the river meets its main tributaries, the Dan, the Banias (Hermon), and the Snir (Hasbani), before entering into Lake Tiberias. As it exits the Lake, the lower portion of the river flows southwards, where it meets the Yarmouk, one of its largest tributaries (Haddadin, 2006). As the river meanders toward the lowest point on earth, about 400 meters below sea level, it becomes more and more polluted and its flow increasingly diminishes as a result of extensive water diversion, failing to quench the regional thirst for freshwater. The river basin stretches across the territories of five highly contentious nations: Israel, Jordan, Lebanon, Palestine, and Syria (McKinney, 2008). Although access to the river water is disputed among these nations, Israel controls and utilizes most of the river system for its own benefit. Israel has emerged as the regional hegemonic power, which has led to its advantageous hydro-strategic position relative to the other co-riparian states. Despite its relatively small size, the Jordan River exerts considerable influence on the Arab–Israeli dispute in general, and this has resulted in armed conflicts in the past as well as opportunities for peace in the Middle East.

Synoptic history of water dispute issues and hydro-diplomacy

The Middle East is a semi-arid region with confined water resources and a rising demand for water, to cater to an exponentially growing population and to agricultural and industrial development (Swain, 2004). This water scarcity affects the Middle Eastern political arena as much as it affects the lives of its inhabitants. The relationship between the hydrological resources used and geopolitical representations has always been intimately linked whenever these resources are potentially available from either side of a political border (Abukhater, 2010a). The geopolitical aspect of the conflict resides at the heart of the water scarcity crisis, as water and politics are intimately intermingled, causing pernicious tension among the riparian states. The conflict over water resources dates back to the late nineteenth century with the Jewish immigration to Palestine, the establishment of Israel in 1948 and the consequent displacement of Palestinian refugees, and the 1967 Six-Day War, which resulted in the occupation of Palestinian and Syrian territories continuing to date. The status quo water utilization personifies a situation of inequitable water allocation that fuels resentment and hostility.

Armed conflict as well as opportunities for cooperation over water-related matters have ensued in this region throughout its recent history. In the early 1950s, plans to utilize the waters of the river were unilaterally proposed by several states, notably Israel and Jordan, which sparked clashes between Israel and several of its neighbors (Priscoli and Wolf, 2009). To resolve the water issue along the Jordan

River, several plans were proposed, including the Main Plan, the Cotton Plan, the Arab Plan, and the Unified (Johnston) Plan, none of which were officially ratified (Haddadin, 2002a). As tension continued to mount, intermittent diplomatic attempts, which continued through the 1970s and 1980s, failed to reach a plausible resolution creating a situation of "talk-fight" (Gurr *et al.*, 2001; Priscoli and Wolf, 2009). It was not until the Madrid Peace Process on October 31, 1991, which resulted in two complementary bilateral and multilateral negotiation tracks, that the two countries were able to enter into serious negotiation that aimed at achieving "a just, lasting and comprehensive peace settlement" on the basis of the "land for peace formula," outlined by the UN Security Council Resolutions 242 and 338 (Israel Ministry of Foreign Affairs, 1995). These diplomatic efforts continued, notwithstanding the political challenges surrounding the conflict, leading to the signing of the Israel–Jordan Peace Treaty on October 26, 1994, which became a testimony of how water can be used to deflate conflict and promote a vision of hydro-diplomacy and peace.

Treaty outcome

The Treaty provided details about volumetric water allocation from the Jordan River, the Yarmouk, and Wadi Araba groundwater, and provisioned the establishment of a Joint Water Committee (JWC) to monitor implementation and ensure future cooperation between the two parties (Haddadin, 2002a). Although the Jordanian negotiators assert that the Treaty was negotiated based on the deal offered by the Unified (Johnston) Plan of 1955, the water agreement portion of the Treaty did not support this claim (Elmusa, 1995). Jordan was offered much less water than was originally allocated to it by the Plan. These troubling aspects, related to the volumetric outcome of the Treaty and the fact that it ignored other important regional players, namely the Palestinians and Syria, caused a widespread skepticism of the Treaty's fairness among the public and water experts around the world (Zeitoun, 2008a). Even 19 years after the signing of the Treaty, the Palestinian water issue remains an unresolved conundrum threatening the regional hydro-stability of the Middle East. With the exclusion of the Palestinians and Syria from the negotiation and scope of the Treaty, the issue that remains of a far-reaching importance is the ability to achieve a basin-wide management with multilateral involvement of all interested and impacted parties.

The Ganges–Brahmaputra–Meghna (GBM) river basin: Treaty between the Government of the Republic of India and the Government of the People's Republic of Bangladesh on Sharing of the Ganga/Ganges Waters at Farakka, signed on December 12, 1996

Background

The Ganga, or Ganges, rises into the lofty Himalayan Mountains in Nepal and Tibet, courses through India and joins in Bangladesh before terminating in the Bay of Bengal to the east (Roy, 1997). The headwaters of the river are located in India and Nepal. Four co-riparians share access to, and are extremely dependent upon, the river basin water, including China, Nepal, India, and Bangladesh (Mirza et al., 2004). India is the upstream and most dominant regional power compared with Bangladesh, which is the downstream riparian and the one that is more vulnerable to natural disasters, flooding, and droughts because of its location and topography (Uitto, 2008). In addition to its socioeconomic importance to the four co-riparian countries, the Ganges has a particularly significant religious importance to many Hindus (Swain, 2004).

Synoptic history of water dispute issues and hydro-diplomacy

Despite the abundance of water in the South Asian subcontinent, the issue of water disputes in this part of the world in general (and the Ganga Waters Dispute in particular) is the byproduct of enormous population growth, over-dependence on supply-side water management, lack of demand-side water management, inefficient irrigation practice, the drastic seasonal fluctuation of rainfall (which causes both droughts and flooding), and political and territorial strife. As India and Bangladesh share 54 rivers and streams, the issue of water sharing is of great concern to state-level politics in both countries (Nishat, 2008). The issue first arose in 1951 when East Pakistan protested against the proposed Farakka Barrage to divert water from the Ganges to the Hooghly River (Nishat, 2001; Rahaman, 2006). In 1971, Bangladesh earned its independence and, in 1972, the Indo-Bangladesh Joint Rivers Commission (JRC) was established (Nishat and Pasha, 2001; Priscoli and Wolf, 2009). Water is needed downstream in Bangladesh for agriculture, fisheries, navigation, industries, and domestic purposes during the dry season (January to May) (Nishat, 2001; Mirza et al., 2004). Enough water was flowing into Bangladesh until 1975, when India pursued its plans to unilaterally divert the river's waters at Farakka for its own irrigation,[10] which resulted in a significant reduction in the water supply reaching Bangladesh, causing detrimental environmental and socioeconomic impacts and, in turn, the souring of the relationship between the two countries[11] (Roy, 1997). The construction of the Farakka Barrage (known as the "Farakka Issue") occupies the heart and soul of the Indo-Bangladeshi water conflict as it, on the one hand, caused a great deal of hatred and distrust but, on the other, was the catalyst for negotiation (Ramaswamy, 2008). As their previous negotiation attempts failed in reaching a long-term settlement,[12] the exchange of high-level visits, combined with "facilitatory" efforts at non-official levels, created a window of opportunity and helped the two parties capitalize on good relations (Ramaswamy, 2008). On December 12, 1996, the prime ministers of India and Bangladesh officiated the Ganges River Water Sharing Treaty, which set the stage for an integrated water management of the Ganges–Brahmaputra–Meghna (GBM) region (Nishat, 2001).

Treaty outcome

The 1996 Ganges River Water Sharing Treaty is a long-term agreement that governs the sharing of the river waters at Farakka for a period of 30 years (subject to renewal by mutual consent) and finding a long-term solution for flow augmentation during the lean season (Nishat, 2001; Datta, 2005; Priscoli and Wolf, 2009). Under the auspices of equity, fairness, and no harm to either party, the Treaty aimed at achieving mutual benefit for the peoples of the two countries with respect to flood management, irrigation, river basin development and generation of hydropower (Nishat and Pasha, 2001; TFDD, 2008). According to Annex I, the water available at Farakka is more or less equally shared between the two countries, with three 10-day periods of a guaranteed minimum amount of 35,000 cusecs for both of them (Nishat, 2001; Ramaswamy, 2003). Relying on supply-side water allocation criteria (Uitto, 2008), the Treaty utilized a time-based water allocation mechanism combined with percentages of the flow during the dry season. The treaty-making stage can best be characterized as "position-based" negotiation focusing on rights rather than needs (Crow, 2008), which hindered the outcome of the Treaty (Uitto, 2008). Within this process, India was reluctant to involve Nepal in the negotiation process in fear of losing its favorable position (Nishat, 2008).

The Joint Committee was formed to monitor implementation and operation. A contingency planning clause was included in the Treaty stipulating that an emergency situation will be declared in the case of the river's water level dropping below a certain threshold (50,000 cusecs), at which point the two countries will convene to collaboratively find ways to augment the flows and cope with the drought. In addition, in cases of disagreement, the Treaty set forth conflict resolution mechanisms by referring matters to the Joint Committee, the Joint Rivers Commission, and then to the two governments consecutively (Iver, 2003; Chimni, 2005).

The Treaty suffers from a number of shortcomings. First, the Treaty did not consider water quality and the health of the ecosystem (Datta, 2005). Second, although there was a "guarantee clause" in the 1977 Agreement and a "burden-sharing" formula in the 1985 documents, India refused to include such clauses in the 1996 Treaty (Ramaswamy, 2008). Third, by entering into a bilateral negotiation with Bangladesh *only*, and not with Nepal, India ruled out Bangladesh's proposal for augmentation from within the Ganges system by storing its monsoon flows behind seven high dams in Nepal (Haftendorn, 2000; Nishat, 2001). The bilateral nature of this agreement, akin to the one between India and Nepal, provided more advantage to India (Haftendorn, 2000). Fourth, the lack of third-party involvement in the negotiation of the Treaty solidified the Indian position, being the stronger of the two riparians, by avoiding the formation of any coalition between Bangladesh and Nepal during the negotiation. Many experts argue that excluding non-water-related issues and focusing efforts on sharing water helped in facilitating a better negotiation atmosphere. However, by considering other non-water linkages, namely economic incentives, the parties would have reached a more satisfactory formula to regulate equitable sharing.

Although the outcome of the Treaty might seemingly appear equitable, public perception of the Treaty on both sides is anything but equitable. Lack of public dialog and engagement resulted in widespread public skepticism about the outcome of the Treaty, which was a political rather than a technocratic settlement. The high secrecy of the negotiation process and the confidentiality of related data, which were not shared openly with the public, caused many people to view the Treaty as favorable to the most powerful state, India. The government of Bangladesh was facing particularly severe criticism in the first year of the Treaty's implementation (1997), as the water at Farakka dropped below what had been estimated and Bangladesh did not receive its share as identified in the Treaty[13] (Swain, 2004). Although the Bangladesh National Party (BNP) was in opposition to the Treaty, they did not seek any revision of its terms when they came to power in 2001 (Ramaswamy, 2008). The Treaty improved diplomatic relations between the two countries and created a political momentum and mutual trust, which even the new anti-India government wishes to maintain. Although the Treaty has been operating smoothly since its ratification, many experts believe that it has not yet been put to the test during extreme hydro-events, as the amount of rainfall has improved since its ratification (Swain, 2004). However, the fact that it is still in place today, there is less concern about the Treaty as the sense of grievance and hostility surrounding the water issue seems to have drastically subsided since its ratification (Crow, 2008).

Rio Grande/Rio Bravo and Colorado river basins: Treaty between the United States of America and Mexico Relating to the Utilization of the Waters of the Colorado and Tijuana Rivers and of the Rio Grande, signed at Washington on February 3, 1944

Background

With their headwaters emanating from Colorado, both the Colorado and the Rio Grande, known in Mexico as the Rio Bravo, flow through a semi-arid region southward and eastward toward the Gulf of California and the Gulf of Mexico respectively, supplying water for agricultural, hydropower, municipal, and industrial purposes. A significant portion of the Rio Grande water, about two-thirds, comes from the Rio Conchos. These rivers delineate parts of the international boundary between the United States, the more powerful and economically developed riparian, and Mexico. The United States is the upstream riparian in the Colorado River and the downstream riparian in the Lower Rio Grande. The two river basins were negotiated as part of one agreement, which allows a higher degree of flexibility between the two basins (Fischhendler and Feitelson, 2004).

Synoptic history of water dispute issues and hydro-diplomacy

The water dispute between the United States and Mexico is a result of water scarcity, rapid population growth, extensive industrialization, aquifer depletion, and water quantity and quality issues, including pollution and environmental degradation and salinity. As such, the border region between the United States and Mexico has witnessed severe distributional conflict fueled by absolute water shortage, resembling many other semi-arid regions of the world. In the border region, the two countries exhibit vastly different governmental structures, levels of development, and future planning goals. In 1895, Mexico expressed concerns regarding the United States increasing its utilization of the Rio Grande, which impacted the amount of water it received. The United States' position reflected the Harmon Doctrine, which gave protection to its utilization of the water found within its territories. In 1906 and 1928, the two countries rectified the water issues that had brought them to conflict over the Rio Grande and the Colorado respectively (Haftendorn, 2000). However, long-term transboundary water management that would ideally involve various levels of government, including local, state, and federal government on both sides of the border, has been the key concern for both countries since the onset of the twentieth century (CSIS, 2003). With this in mind, the two countries entered into negotiations that continued for 40 years before reaching a long-term settlement (Fischhendler and Feitelson, 2004). To govern the sharing of the water resources of the border region, the United States and Mexico signed a treaty, which remains the main legal instrument for the management of the utilization of the Colorado and Tijuana rivers and the Lower Rio Grande. The Treaty was signed in Washington on February 3rd, 1944.

The Treaty produced a bilateral legal framework for the management and regulation of transboundary water and established a long-standing regime of border cooperation (Lascurain, 2002; Upreti, 2006). However, the Treaty did not directly address the sharing of groundwater resources or water quality issues (Haftendorn, 2000). To that end, the Colorado River salinity crisis of 1961–73, over which Mexico sought to sue the United States (Wolf, 1997), was the impetus for considering groundwater management and allocation as a viable option to cope with increasing water demand and alleviate pressure on surface water resources. Although the amount of groundwater extraction for both countries is extremely limited along the border, the only agreement between them that governs groundwater is Minute 242 of 1973 (Priscoli and Wolf, 2009), which is challenged by the lack of collaboration between federal and state governments, water quality related issues, and the ambiguity of its terms regarding implementation.

Treaty outcome

Under this Treaty, the United States and Mexico are obligated to trade off water from the Colorado and the Rio Grande respectively (Fischhendler and Feitelson, 2004). It has proven successful in providing adequate mechanisms for delimiting the water rights of the two countries, while providing a resilience mechanism for the modification of its terms without the need to be entirely renegotiated in the future through a "minute" process outlined by Article 25 of the Treaty (CSIS, 2003). The Treaty promoted the notion of equitable utilization between the two parties, notwithstanding the United States' rigid position in negotiating with Mexico (Upreti, 2006). However, public perception of the Treaty as inequitable in terms of volumetric water allocation, access, and water cost, is one reason for the occasionally inharmonious relations between the United States and Mexico (Lascurain, 2001; 2002). Much less water flows into Mexico than was projected by the Treaty because of the exaggeration in estimating available in-stream flow and the extensive use of water in the United States (Haftendorn, 2000). In accord-ance with the Treaty, the International Boundary and Water Commission (IBWC) was established to manage the shared water resources (Priscoli and Wolf, 2009). Until the early 1960s, the focus of the Commission was confined to surface water. However, the management of surface water resources is minuscule when com-pared with the allocation of groundwater resources, which was not adequately addressed by the Treaty, although acknowledged as a vital element, and caused tremendous disagreements over 23 sites along the shared border region (Priscoli and Wolf, 2009). Although the importance of groundwater resources is also ampli-fied in the 1973 Addendum to the 1944 Treaty, the issue of groundwater resources still remains unresolved, as it puzzles water experts on both sides of the border. The Treaty is generally adhered to by the two countries, except during the drought of 1992–1997, which resulted in Mexico's inability to deliver the volume of water stipulated by the Treaty and the consequent political controversy (Fischhendler and Feitelson, 2004).

Recap and selection of the two detailed cases

Good case analysis entails a thorough and rigorous case selection protocol. Generally, the criteria for selecting cases vary greatly depending on the research scope and purpose. Although case selection is a tricky and difficult exercise, a well-selected case is more likely to succeed in providing a relevant analysis than is a case selected without accounting for the type and purpose of the analysis to be conducted and the many nuances involved in each case.

Case study analysis is primarily oriented toward theory building or theory test-ing. Only one or the other of these protocols has been advocated on conflicting grounds. However, a combination of both protocols is touted here for its value in advancing this substantive area of interest, which requires both generating and testing theories.

In essence, the case selection strategy is employed to identify "crucial" cases (Eckstein, 1975) that (1) have the potential to confirm or refute current theories in the field, and (2) provide new insight and substantial anecdotal evidence suf-ficient to generate meaningful conclusions and new theoretical foundations to account for the cases' findings (Yin, 1994). "In a crucial case it must be extremely difficult, or clearly petulant, to dismiss any finding contrary to theory as simply 'deviant' (due to chance, or the operation of unconsidered factors, or whatever)"

(Eckstein, 1975: 157). Because it is pivotal that these crucial cases are representative of a presumably large class of relevant cases, a considerable degree of design and rigor in case selection is required to make a choice. When cases are selected "with a minimum of design or rigor," there is little that characterizes them,

> except the researcher's hopes and intentions, and results can only turn up by good fortune – which the bright will seize and the dull miss, but which the researcher can do nothing to induce. The alternatives are to use at least a modicum of design and rigor in research and not to choose just any case on any grounds but a special sort of case.
>
> (Eckstein, 1975: 145)

These "special sort[s] of cases" are selected in this research to capitalize on the opportunity to benefit from more formalized and systematic procedures of case selection, for the pursuit and creation of credible knowledge by learning from contextual details that otherwise would be overlooked by a purely statistical study.

Two comprehensive cases were selected to reflect a wider range of issues and variability. In principle, these relevant cases were carefully identified based on the previous screening process of the initial nine cases, which reflect a variety of categories and geopolitical contexts, to be able to examine, test, and validate casual associations. These two cases vary with regard to both process equity and outcome equity parameters. Using this method can help in selecting comparable cases that, although different in terms of the process equity parameters, share common features, important issues, and phenomena with each other and the context of the target arid regions. These similarities allowed us to control and look for the impact of exogenous factors and account for, and focus on, the variation in outcome equity parameters in correspondence to the variation in process equity parameters across various cases. It should be noted that, because cases were selected based on the variation of outcome equity parameters, the book thoroughly investigates two detailed cases, whereby one exhibits high outcome and perceived equity and the other exhibits low outcome and perceived equity respectively. This variance of outcome equity parameters allowed for more variables to be examined, hopefully rooting out any possible allegations of selection bias often associated with comparative case study analysis (Geddes, 1990). Furthermore, this variance helps in substituting for the small number of cases and the comparatively large number of factors.

This selection approach incorporated various sources and considerable validation. The main goal of this step was to ensure impartiality and objectivity in selecting cases for the study and relying on a variety of data sources served this purpose. Selection of the detailed cases took into account both outcome and perceived equity considerations. To select the two cases for more detailed case study analysis, we considered a number of factors that comprise not only outcome equity but also "climate of equity." The latter is a reflection of the negotiation process and refers to how equitable a treaty is perceived to be, notwithstanding its commensurable outcome. For this purpose, we utilized a comprehensive approach that relied on a variety of data sources, including practical considerations, experts' perceptions of how equitable treaties were, the dates of ratification and the accessibility of the makers of the treaty (prior to or post-1980), acceptability (level of satisfaction) of the treaty, and a systematic review of the available literature on international water law and treaties (refer to Figure 3.8). Triangulating the results obtained from these different sources helped in identifying the most meaningful and relevant cases.

First, based on experts' responses, five treaties were ranked as low in terms of perceived equity, three others were ranked as high, whereas one treaty was identified as moderate. The five treaties ranked as low equity are the 2001 Joint Communiqué between Syria and Turkey, the 1946 Treaty between Iraq and Turkey, the 1996 Ganges Water Treaty between Bangladesh and India, the 1994 Treaty between Israel and Jordan, and 1944 Treaty between the United States and Mexico. Conversely, the other three treaties ranked as high equity are the 1959 Treaty between Egypt and Sudan, the 1960 Indus Waters Treaty between India and Pakistan, and the 1986 Lesotho Highlands Water Project Treaty between Lesotho and South Africa. The 1992 Treaty between Namibia and South Africa is identified as moderate equity. Not surprisingly, the Jordanian–Israeli Treaty of 1994 was ranked as one of the lowest in terms of its perceived equity. This is because the Treaty offers an example of how lingering, unresolved political issues, namely the Palestinian issue, greatly impact both its *outcome* and *perception* (Mirumachi and Allan, 2007). Fortuitously, and contrary to my own expectation, being familiar with its outcome,[14] the 1959 Treaty between Egypt and Sudan was ranked as one of the highest in terms of equity attainability. This

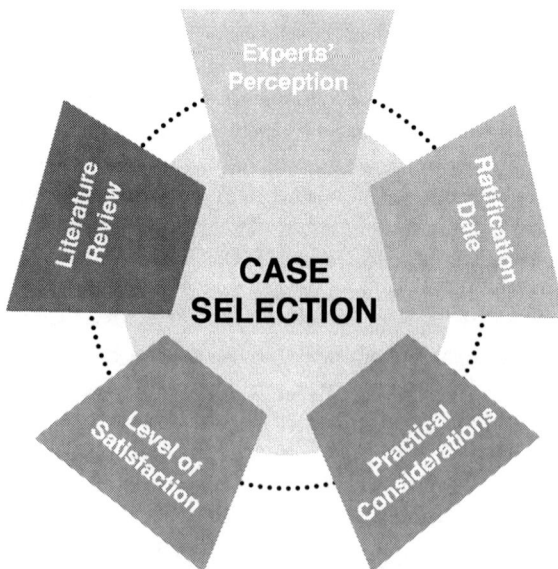

Figure 3.8 A comprehensive approach to case selection.

book, however, recognizes and emphasizes the importance of process equity as a significant determinant in whether treaties are perceived to be either equitable or inequitable, irrespective of their commensurable outcome (which, in this case, clearly favored Egypt).

Second, practical considerations were also incorporated and considered in the selection of the two cases. These concern the questions of how old a treaty is and how long it has been in effect. In that sense, treaties that were too old were excluded because of the difficulty in finding information, evidence, and stakeholders who participated in negotiating the treaty. Similarly, treaties concluded too recently were also excluded, because they have not withstood the test of time in terms of implementation, and therefore it is premature to evaluate their success, or there might be little credible literature written about them. With this in mind, treaties such as the 2001 Joint Communiqué between Syria and Turkey were excluded because of their recent date of ratification and the relatively limited amount of information available. By the same token, treaties, such as the 1959 Nile Treaty between Egypt and Sudan, the Indus Waters Treaty of 1960 between India and Pakistan, the United States–Mexico Treaty of 1944, and the 1946 Treaty between Iraq and Turkey were excluded because of their older dates of ratification and the difficulty in finding negotiators to interview.

Third, according to Lautze and Giordano (2006), it is believed that "equity agreements" are substantially different from "non-equity agreements." "Equity agreements" here refer to those agreements that contain the term "equity" or its derivatives, or agreements that make reference to the 1966 Helsinki Rules or the 1997 Convention on Non-Navigational Uses of International Watercourses (Lautze and Giordano, 2006). Lautze and Giordano, (2006) found conclusive evidence that there is a strong correlation between the inclusion of equity language in transboundary water laws, which for the most part occurred after 1980, and more equitable outcomes. As such, water treaties prior to 1980 are most likely to be inequitable because of the lack of equity language, as opposed to water treaties signed after 1980, which tend to include equity language. In light of this conclusion, treaties prior to 1980 were deemed to be less equitable when compared with those treaties ratified after 1980. Based on this indicator, the Lesotho Highlands Water Project Treaty of 1986, which was ranked the highest in terms of perceived equity, was one of the selected cases. The LHWP makes specific reference to water equity by stating in its Preamble that both Lesotho and South Africa should consider "the mutual benefit . . . to be derived from the . . . equitable sharing of the water resources of the Senqu/Orange River and its effluents . . ." (LHWP Treaty, 1986: 1). Parties involved in the 1986 LHWP Treaty were able to collaboratively capitalize on their common interests by focusing on the broader benefit of water use rather than on the division of shared water *per se* (Wolf, 2002). To that end, in return for diverting water for the use of the Gauteng province, South Africa provides financial support for a hydroelectric/water diversion facility (Wolf, 2002). Not surprisingly, the 1986 Treaty was positively viewed by the experts who ranked it. Similarly, the United States–Mexico Treaty of 1944, which was ranked low in terms of perceived equity, was not considered further for case analysis. This choice is consistent with Lautze and Giordano's criterion, given its date of signature prior to 1980.

However, the Indus Waters Treaty of 1960 between India and Pakistan, the Jordanian–Israeli Treaty of 1994, and the 1996 Ganges River Water Sharing Treaty between Bangladesh and India appeared to be inconsistent with this indicator. It should be noted that, although signed after 1980, the Jordanian–Israeli Treaty of 1994 does not make direct reference to equity attainability in its text. As such, it did not fully comply with this finding, which made it yet more interesting to explore. The Indus Waters Treaty of 1960 was ranked high on equity, despite its relatively old date prior to 1980, and the 1996 Ganges River Water Sharing Treaty between Bangladesh and India, although it was signed after 1980, was ranked as being very low on perceived equity by water experts. Avoiding selecting cases in the same geographic location suggested including either the 1996 Ganges Water Treaty between Bangladesh and India or the Indus Waters Treaty of 1960, given that both treaties are located in Asia. However, the background information provided by the preceding first-cut analysis suggests that the India–Bangladesh Treaty is high on outcome equity, despite its low ranking on perception, which means that it does not fit very well into the selection matrix shown in Table 3.4. Similarly, the inconstancy between its high ranking by experts and its low actual outcome in terms of equity attainability, along with its old signatory date, also make the Indus River Treaty of 1960 inconsistent with the selection criteria as well.

Fourth, acceptability of a particular treaty is a strong indicator of its equitable allocation of water. Acceptability can be measured by how successful the treaty implementation is in reality. Equitable treaties tend to sustain implementation and indicate no conflicts or disagreements between different co-riparians. Based on this indicator, the LHWP Treaty seemed to be, for the most part, acceptable to all parties involved in the conflict, and therefore appeared to be the most equitable since it maintained implementation, as the parties did not encounter major clashes with each other. Conversely, there seemed to be a number of disagreements and points of contention that necessitated a number of consequent amendments to the United States–Mexico Treaty of 1944, which was an indicator of lack of acceptability. Similarly, the 1996 Ganges Water Treaty between Bangladesh and India would seem to lack acceptability in light of the fact that both countries have disagreed on the mechanism of implementation on many occasions, although the Treaty maintained steady implementation. Consequently, this indicator also provided results consistent with those obtained from previous indicators.

Fifth, the research relied on a rigorous and systematic review of literature and databases to verify and ensure the accuracy and consistency of the results, including the Transboundary Freshwater Dispute Database (TFDD),[15] which is comprised of both tabular and spatial components. Tabular databases included the Middle East Water Collection,[16] International Freshwater Treaties Database,[17] International Water Events Database,[18] International River Basins of the World (International River Basin Register),[19] and Water Conflict and Cooperation

Table 3.4 Primary selection criteria for two detailed cases

High *outcome equity* ✓ **High *perceived equity*** Lesotho–South Africa (LHWP) **Date**: October 24, 1986 **Basin**: Orange-Senqu (LHWP) **Signatories**: Lesotho, South Africa **Conflict issue:** Hydropower/Hydroelectricity **Geographic context:** Africa **Compatibility:** High *perceived* and *outcome equity*	Low *outcome equity* High *perceived equity*
High *outcome equity* Low *perceived equity*	**Low *outcome equity*** ✓ **Low *perceived equity*** Israel–Jordan water treaty, 1994 **Date:** October 26, 1994 **Basin:** Jordan **Signatories:** Israel, Jordan **Conflict issue:** Water flow **Geographic context:** Middle East **Compatibility:** Low *perceived* and *outcome equity*

Bibliography.[20] Spatial databases included the Transboundary Freshwater Dispute Database,[21] the Atlas of International Freshwater Agreements,[22] and the Maps and Images Gallery.[23]

In a nutshell, triangulating all of these methods would suggest that the Lesotho Highlands Water Project of 1986 (LHWP) is high on attaining equity in water allocation because it is believed to have provided the most conducive "climate of equity" (both in terms of outcome and perception), whereas the Jordanian–Israeli treaty of 1994 emerges as low on equity because it is believed to be lacking this "climate of equity" (both in terms of outcome and perception). These conclusions are summarized in Table 3.4.

These two cases fit the selection criteria and are reflective of divergent geographic contexts. They also take into account theoretical and empirical considerations, as noted in Chapter 2. They provide a very good basis for an interesting and compelling analysis, as they seem to cover the spectrum of variation needed for such analysis and treaty formation. In this way, we will investigate in the next two chapters the two extreme ends of the spectrum that these two cases represent. We will analyze two complementary aspects of treaty formation (negotiation and implementation): the focus in the Israel–Jordan Water Treaty is more on treaty negotiation, whereas in the Lesotho–South Africa Treaty it is on treaty implementation. However, both aspects (negotiation and implementation) will be discussed in both cases in various levels of detail. These two cases have the potential to generate lessons and parallels that can feed into and inform theories of negotiation and equitable water allocation. As well as providing additional cultural richness and variation, the selection of these cases from two continents has the potential to offer new understanding when considering future cases. The goal of the analysis of these two cases provided in the subsequent chapters is to establish a clear delineation of process equity parameters that have significant influence on the outcome and perception of the treaty.

The scanning of the nine cases provided a general understanding of global trends and common characteristics in international water conflict, negotiation, and hydro-diplomacy. The analysis elicited a set of characteristics that were commonly shared across all cases. First, although some of the nations had a history of conflict, water itself was not the reason for any of the nations to enter into military conflict. Rather, unilateral development of the river basin for the benefit of one riparian is usually the main cause of the rift in relationships and can further complicate the situation, resulting in a more severe state of dispute. Second, conflicts over transboundary water resources can be exacerbated by a set of factors, including growing population and the consequent rise in water demand, coupled with the scarcity of water in semi-arid regions, the regional competition over transboundary water resources that is fueled by conflicts of a broader nature, and the fact that water is utilized for both economic and consumptive purposes. These factors collectively resulted in many cases in a situation of severe competition and hostility. Third, perhaps one of the most intriguing findings that emerged from this analysis is the fact that conflict over water resources is triggered by competition over shared water resources for domestic and municipal uses rather than industrial or other economic purposes. Fourth, water is often used as a "weapon" in many situations by the more powerful riparian to exert pressure on the weaker riparian or impact the outcome for its own political and diplomatic gain. Fifth, the weaker riparian often tends to seek a cooperative agreement in accordance with the principles of international water law since this body of law is often viewed as a mechanism for protecting the weaker riparian. Sixth, the involvement of a neutral and powerful third party as a peace broker can have positive implications on the negotiation and implementation processes and can result in a more satisfactory agreement. Seventh, the availability of economic incentives and capacity building programs, focusing on technical issues, and the separation from the larger political climate can in some cases expedite the negotiation process and may yield practical results.

The scope and approach taken in this book, however, has its own limitations that need to be highlighted. First, focusing on bilateral treaties, this approach is mainly relevant to binational negotiations and applies only in limited ways to multilateral water negotiations. Second, the scanning of these cases resulted in the selection and identification of two cases, one of which (the LHWP Treaty) was ranked high in terms of both outcome and perceived equity, whereas the other (the Israel–Jordan Treaty) was ranked low in terms of both outcome and perceived equity. This indicates that these particular cases may be exceptional and not entirely representative of another context. For example, the LHWP Treaty is exceptional in the sense that it represents a situation where both riparians have an interest in pursuing negotiation for differential, yet compatible water needs. These asymmetrical water needs, coupled with the unique political relationships and non-water-related issues, provided a common platform for negotiations and trading that are not common in many other cases. Notwithstanding this fact, this particular case was selected because of its potential to illuminate best practices in terms of treaty formation (both in negotiation and implementation). With this in mind, the purpose of selecting the two cases for more detailed analysis is to learn more about key aspects and important nuances related to process equity in relation to outcomes in the context of bilateral water treaties. The real value of studying these two cases is the substantive anecdotal evidence that they provide and the opportunity to critically understand contextual nuances that make these cases unique.

4   The Israel–Jordan Peace Treaty of 1994

> Water has been a source over so many years of erosion of confidence, of tension, of human rights abuses . . . That must stop if we are going to be able to develop a climate for peace.
> (Queen Noor of Jordan, http://www.doonething.org/quotes/water-quotes-4.htm)

This chapter provides an analysis of the Israel–Jordan Peace Treaty of 1994, which emerged as a unique case because of its importance and geographic context. Of particular importance are Article 6 and Annex II (included in the Appendix of the 1994 Peace Treaty), which address water-related matters. In this chapter, the 1994 Peace Treaty will be referred to as the "Treaty," whereas Annex II of the Treaty ("Water Related Matters") will be referred to as the "water agreement." Putting emphasis on the significance of water-sharing equity, this chapter provides a brief background to the nature of the water crisis between Israel and Jordan, focusing on the lack of equity in water allocation as a profound element of the water conundrum that permeates through almost every aspect of these diplomatic efforts. A critical part of the analysis of the Treaty provided in this chapter connects it to other related matters that impacted the Treaty, its outcome and implementation, making reference to issues related to the Palestinians and to Syria, as they appear relevant to the Israeli–Jordanian relations, negotiations, and treaty formation. Using the logic model as a guiding principle for the analysis, this chapter provides an assessment of the 1994 Israel–Jordan Peace Treaty in order to draw lessons and conclusions that will be useful in future water negotiations.

The allocation of shared water resources, land division and international borders, normalization, and achieving viable and sustainable peace in the region are all key issues that the two countries need to address through water negotiation. By highlighting critical details about the root causes and nature of the water crisis, this chapter aims to provide a diagnostic account of the problem and detailed narration of essential procedural and outcome aspects. As will be explained in this chapter, trust building measures prior to negotiation, the relationship between "high" and "low" politics, lack of public involvement, environmental consequences, neutral third-party involvement, and the interplay between water, land, border, and the environment surfaced as key issues in determining how equitable the agreement is to both sides and the extent to which it is perceived as such. Focusing on the negotiation process and its relation to perceptions of the outcome of the water agreement, this chapter asserts the importance of water in creating and sustaining cooperation.

Two major sections are introduced in this chapter: the background information related to current and future water demand and supply figures is introduced in the first section, and the water equity analysis part of the case is introduced in the second part. Throughout the analysis, several examples are used to illustrate the severity of the situation, particularly when it comes to the complex regional setting, the relationship, and the history of aggression and hostility. Along with concluding remarks, an outcome-based equitable water allocation plan is provided for comparison purposes and to unravel the acute lack of equity in this particular treaty. Nonetheless, the Treaty provides essential lessons for future negotiation and hydro-diplomacy worthy of discovery.

Background: basin spatial hydrogeomorphology

Compared with other rivers around the world, the Jordan River is considered to be a small stream (Lowi, 1995), but for the co-riparians sharing the river system it is very critical (Zawahri, 2009). This fact becomes evident when considering the total annual discharge of the Jordan River, estimated at an average of 1,500 million cubic meters (MCM) (Elmusa, 1995) and varying from 1,200 MCM to 1,800 MCM, which, compared with other rivers in the region, is very insignificant (Hiniker, 1999; Haddadin, 2003). This annual flow of the Jordan constitutes only 1.5 percent of the annual flow of the Nile and 4.3 percent of the Euphrates (Lowi, 1995; Haddadin, 2006). In spite of its relatively insignificant size, the Jordan River is deemed "the lifeline of the Middle East" and has remained a nettlesome issue for the past century (Elmusa, 1995; and McKinney, 2008). The insignificant size of the river, combined with its great importance in shaping the regional hydropolitics, makes things even more complicated as the co-riparians have to share a very small pie between them (Zawahri, 2009).

Reflecting the same irony of its size *vis-à-vis* importance, the legacy of the Jordan River in the Middle East peace process is monumental. The demise of the Jordan River was the catalyst for peace and cooperation between Israel and Jordan. In October 1994, the two countries, realizing the importance of rendering an equitable water-sharing plan, preserving this natural resource and extending peaceful relations, were able to defy all odds by signing an agreement to govern their behavior with regard to the shared regional water resources and each other, given the long historical animosity and tension (Haddadin, 2002b).

By examining the balance between population growth and water requirements, this section presents the basin spatial hydrogeomorphology with respect to the geographical, climatic, and hydrological conditions, the spatial patterns of current and previous water utilization trends and dynamics, conflict and cooperation over water, and other human-caused events that influence the basin water characteristics and inter-basin relations among riparian nations. It further looks into the future requirements of water, as it provides population projections and associated future water demand. This section is important to the case analysis as it shows current and future population growth and associated water requirements for both nations. The goal is to provide a factual basis for the analysis and some measure of the distribution of water access and related diplomatic discourse. This helps in understanding water stress imposed by population growth and evaluating available water resources to meet this demand. It also provides a brief history of the conflict and previous water agreements and negotiation attempts to give sufficient context and background information essential to the understanding of the history of the conflict and the various diplomatic efforts that ensued.

Geographic, demographic, and hydro-meteorological context

Geography

The region is composed of seven distinct physiographic zones. These physiographic zones include the Jordan Rift Valley, the Jordan Highland, the Jordan Plateau, the South Jordan Desert, the Mountain Belt, the Negev, and the Coastal Plain (EXACT, 2005a). The three major water features in the region are the Dead Sea (a terminal lake), the Sea of Galilee (also known as Lake Tiberias or Kinneret) and the Jordan River. As a result of the climatic and hydrologic characteristics of the region, two major projects were constructed to transfer water from the north into areas lacking water in the south. These projects were the Israeli National Water Carrier and the King Abdallah Canal (which was constructed on a smaller scale by Jordan). The Israeli National Water Carrier, which was constructed in the 1950s, and was in operation in the early 1960s, to transfer water from Lake Tiberias, the most rainy region in the north, to the Negev Desert, the arid areas in the south (Shamir and Haddadin, 2003; Beyth, 2006). The King Abdallah Canal project (also known as East Ghor Canal) transfers water from the Yarmouk to residents of the Jordan Valley (Water Care, 2004).

Bordering five countries and territories, the catchment area of the Jordan River basin extends across a total area of 18,300 km², with a total length of 228 km (Shamir and Haddadin, 2003; Fischhendler, 2008b). Demarcating a strategic location, the basin includes five particularly contentious co-riparian states that claim right of access to its water, two of which are dependent upon its water as their primary water supply (Wolf, 2001). These include Israel, the Palestinian West Bank, Jordan, Syria, and Lebanon (see Figure 4.1).

Table 4.1 illustrates the geographic configuration of the Dead Sea international basin. The largest land area of the basin, about 20,600 km² (more than 48 percent of the total basin area), is located in Jordan, followed by 9,100 km² (about 21.3 percent of the total basin area) located in Israel, with the smallest land area of the basin, 1,500 km² (representing only 1.3 percent), falling in the occupied Golan Heights (Wolf, 2002).

DEMOGRAPHIC CHARACTERISTICS

Collectively, the population of the five co-riparian counties exceeds 37 million people, of whom 18.5 percent are located in the Jordan River basin (Phillips et al., 2007b). As shown in Table 4.1, the Jordan River basin houses approximately 7 million people, of whom 3 million are located in Jordan, 2 million in Syria, 1.4 million in Palestine, 0.3 million in Lebanon, and 0.28 million in Israel, which represents the smallest demographic segment of the basin (Phillips et al., 2007b).

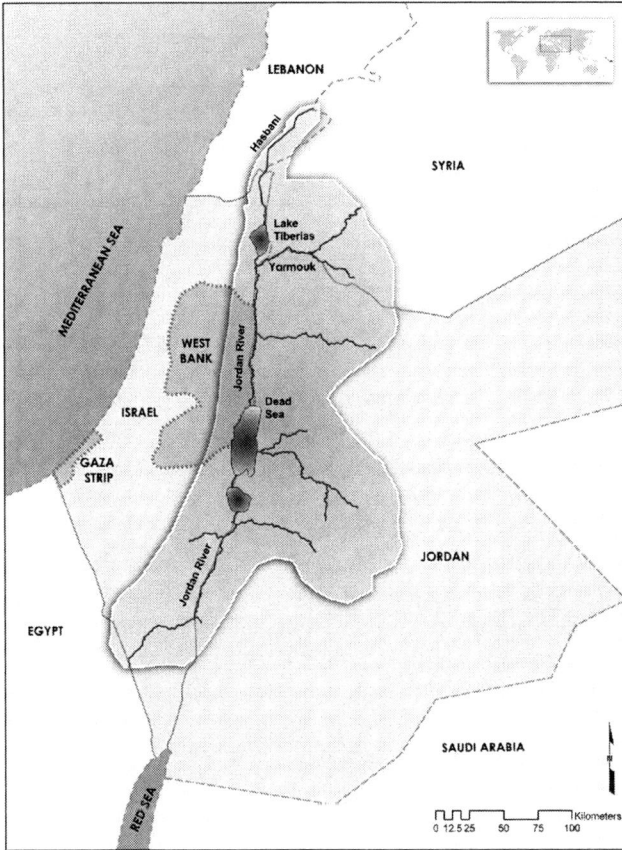

Figure 4.1 A map of the Jordan River, Dead Sea, and Wadi Araba catchment area.[1]

Table 4.1 Jordan River basin

	Lebanon	Syria	Israel	Palestine[2]	Jordan	Totals
Total population (millions)	3.8	18.0	6.7	3.6	5.6	37.7
Population within the basin (millions)	0.3	2.0	0.28	1.4	3.0	6.98
Percentage of total population within the basin	7.9	11.1	4.2	38.9	53.6	18.5
Area of basin in country (km²)	600	6,400	9,100	5,900	20,600	42,600
% of total area	1.33	14.95	21.26	13.79	48.13	100

Total area: 42,600 km² (Dead Sea international drainage)
Data sources: Giordano and Wolf (2001); Phillips *et al.* (2007b); and FAO-Aquastat (2008).

Table 4.2 Population forecast

Year	Israel	Jordan	Palestine
2008	6.7	5.6	3.6
2025	9.3	7.6	7.4
2050	11.2	9.6	11.9

Data sources: Hummel (2006); Phillips *et al.* (2007b); and Dinar (2009).

Syria houses the largest population in the region, about 18 million people, whereas Lebanon houses the smallest, about 3.8 million people (McKinney, 2008). With an average population density of about 837 people per square kilometer, the current population in Israel is approximately 6.7 million people, 92 percent of which is urban (Hummel, 2006). The demographic, ethnic composition of Israel predominantly comprises the Jewish population, about 76.4 percent, and the non-Jewish population, about 23.6 percent of the total population (CIA World Factbook, 2008). Table 4.2 illustrates the current and projected population estimates of Jordan, Israel, and Palestine. According to the Population Reference Bureau, 2009, the 2008 natural population growth in Israel is estimated at about 1.6 percent, which means that Israel's population is projected to reach about 9.3 million people in 2025 and 11.2 million people in 2050 (Dinar, 2009). Israel's population increases as a result of immigration, which is hard to predict precisely. With an average population density of about 163 people per square kilometer, Jordan's current population is approaching 5.6 million people, 79 percent of which is urban (Haddadin, 2006; Hummel, 2006; and Phillips *et al.*, 2007b). Jordan's demographic composition consists of Jordanians (East Bankers), Palestinian refugees from pre-1948 Palestine, Syrians, and people of other origins (Haddadin, 2006). It is estimated that, based on the current natural population growth rate of 2.4 percent, the Jordanian population will increase to 7.6 million people in 2025 and 9.6 million people in 2050 (Hummel, 2006; Dinar, 2009; Population Reference Bureau, 2009).

Hydro-meteorological setting

The Middle East is a semi-arid region that exhibits a Mediterranean climate with long, hot, dry summers and short, cool, rainy winters (FAO-Aquastat, 2008). The climate of the Jordan River basin is identified by its location between the subtropical aridity of Egypt and the subtropical humidity of the Levant (FAO-Aquastat, 2008). The three main water resources in the region are precipitation, groundwater and the Jordan River basin (Haddadin, 2006). Precipitation from rainwater and snowfall is one of the main sources of water, both blue and green water.[3] Receiving more than 300 mm of rainfall annually, about one-third of the region is cultivatable. Because of the availability of rainwater, there are substantially large areas of rain-fed agriculture in the region. Although the rainy season is between October and April, the amount of rain fluctuates from one season to another and from one area to another, depending on various topographical factors. Rainfall is unevenly distributed, intensifying sharply as one moves westwards and northwards. Reflecting this general trend, the amount of rain reaches a nadir (less than 100 mm) in the Jordan Valley and south of the region.

The amount of annual rainfall that the Jordan Valley area receives can be as low as 100 mm and as high as 700 mm (PNIC, 1999). The western side receives between 500 and 600 mm, whereas the eastern side receives an amount of rain between 100 and 450 mm. Precipitation of more than 500 mm is mainly concentrated in the high elevations of the Ajloun and Balqa mountains (Salameh, 1997; Haddadin, 2006). As is the case in other parts of the region, rainfall in Israel and Jordan is limited to winter and declines from north to south and from west to east. The average annual rainfall in Israel varies from less than 100 mm in the south to 1,128 mm in the north. Jordan receives an average annual rainfall of 82 mm, which varies from 50 mm in the eastern and southern desert regions to 650 mm in the northern highlands (Haddadin, 2006; FAO-Aquastat, 2008). It is estimated that more than 93 percent of Jordanian land receives precipitation below 200 mm (EXACT, 2005b; FAO-Aquastat, 2008). Climatic conditions impact not only the amount and distribution of precipitation but also the potential for evaporation. Varying depending on location and accounting for more than 80 percent of precipitation, evaporation rates range from between 122 and 168 cm per year in Israel and the Occupied Territories (Goldreich, 2003) to between 160 and 400 cm per year in Jordan (Haddadin, 2006).

Accounting for more than 75 percent of the total freshwater consumption, groundwater is deemed to be the major source of renewable freshwater for the region. The amount of surface water is extremely limited in the region because of low rainfall rates and high rates of evaporation (Water Care, 2004). The principal source of surface water in the region is the Jordan River and its main tributary the Yarmouk. Originating at Mount Hermon (or *Jabal al-Sheikh*) and terminating at the Dead Sea, the Jordan River is an international watercourse whose historic predevelopment annual discharge fluctuates from 1,200 MCM to 1,800 MCM (Hiniker, 1999; Haddadin, 2003; 2006; and Zawahri, 2009).

The river is composed of two main segments, the Upper Jordan River and the Lower Jordan River, which are divided by Lake Tiberias. The North Springs and Mount Hermon, which receive a high rate of precipitation (as high as 1,300 mm), are the main sources of the Upper Jordan River (PNIC, 1999). The three principal tributaries of the Upper Jordan River, Dan, Banias (Hermon), and Snir (Hasbani), meet where the Jordan River begins to meander southward, gathering runoff water from the Upper Galilee and continuing its flow through Lake Tiberias (Haddadin, 2006). As it exits the lake, forming its lower portion, the Jordan River then meets the Yarmouk, one of its largest tributaries, collecting additional flows from a series of streams until it drains into the Dead Sea. Although the distance between Lake Tiberias and the Dead Sea is 100 km, the actual length of the Lower Jordan is 192 km because of its winding nature (Water Care, 2004).

In the past, until the early 1960s, the overall volume of water flow from the

Table 4.3 Water resources extraction rates (MCM/year)

Water resources	Annual recharge (safe yield)	Total extraction
Western	362	372
Northeastern	150	150
Eastern	172	144
Greater coastal aquifer	250	260
Gaza	55	120
Jordan River	1,100	1,334–1,340
Wastewater	450	450

Data sources: PASSIA (2002); and Princeton University, Woodrow Wilson School of Public and International Affairs (2006).

Jordan River to the Dead Sea was estimated at an average rate of 1,300 MCM/year (Beyth, 2006) and approximately 1,150 MCM/year at a later date (Water Care, 2004). Today, however, this amount is drastically reduced, about 200 MCM, because of the extensive pumping from the river, which has caused significant environmental hazards, such as the decline of the water levels in the Dead Sea and associated sinkholes (Beyth, 2006). As shown in Table 4.3, the current extraction rates exceed the safe yield of most of these resources, causing them to become overexploited and depleted. The depletion of water reservoirs and the associated water quality deterioration, which are both the result of a growing population, are to blame for the impending water crisis in the region. This has a significant hydro-political implication on increasing the regional water stress and competition over shared water.

Cultural hydrology: volatile waterways

The relationship between man and water, and between riparian disputants, not only influences the riparians' tendency to cooperate or fight, but it also constitutes a significant implication in the way people perceive, value, and consume water to meet their needs. Water is needed for essential human needs, which is for the most part comparable within the same geographic and climatic contexts, and for economic development needs, which can vary as a consequence of socioeconomic factors in each country. Conflict over water allocation, however, often arises over essential human water needs. This vexing side of water conflicts poses the troubling question of water access, rather than needs *per se*.

Mindful of these pivotal issues, this section presents essential elements of the basin's cultural hydrology, including different water consumption trends, hydro-dependency, and economic development, water conflict, and previous water agreements. By highlighting the underlying issues that influence water-sharing equity in the basin, the overall goal of this part is to provide background information essential to the understanding of these important dynamics of water-sharing equity.

Water consumption trends, hydro-dependency, and economic development

Freshwater is not distributed evenly around the world; very few nations enjoy plentiful amounts of water beyond their actual requirements, whereas many others suffer because of lack of enough water to meet their essential needs. As shown in Figure 4.2, the Middle East has one of the lowest annual per capita water availabilities worldwide of less than $500\,m^3$, which does not allow for self-sufficiency (Libiszewski, 1995; GLOWA, 2008). This extreme water stress index puts the region at the top of the list of the world's most water scarce regions.

The main water resources in the region are the Jordan River, which is the main source of surface water, and a series of aquifers, which are unequally shared by the co-riparians. In this regard, the Jordan River basin constitutes a major area of water stress and a significant challenge to the attainability of water-sharing equity in the region. The exponentially growing population has also contributed to the exacerbation of the problem. By increasing the demand on water resources, the growing population has resulted in a condition of severe water stress, which puts the region on the precipice of water conflict. Equally, for both Arabs and Israelis, water is needed not only for human consumption, but also irrigation in agriculture and compelling economic development.

Although humans are essentially alike in terms of physiological water needs and demands,[4] actual water use may vary from one region to another, depending on a number of reasons related to climatic conditions, household sizes and income, water pricing, and the age and condition of the water infrastructure. Similarly, water consumption varies greatly across the Middle Eastern region, demonstrating huge gaps between the Arabs and Israelis. Although the Palestinians, Jordanians, as well as the Israelis, all receive well more than the survival amount of potable water (two liters per day), the amount of domestic water that the Palestinians receive, for example, which is approximately 70 liters per capita per day, does not meet the minimum threshold of 100 liters per capita per day, specified by the World Health Organization (WHO, 2003).

Because Israel has access to most of the available regional water resources, the amount of water that the Arab populations receive is drastically less than that of the Israelis. As shown in Table 4.4, the current per capita water consumption in Israel is estimated at about $360\,m^3$/year, compared with about $220\,m^3$/year in Jordan, and only less than $100\,m^3$/year in Palestine (Libiszewski, 1995). The annual domestic water consumption in Palestine does not exceed $25\,m^3$ per person, compared with approximately $46\,m^3$ per capita in Jordan, and $115\,m^3$ per capita in Israel (Libiszewski, 1995; Lipchin, 2003). These water consumption trends are a direct result of the acute inequitable access and sharing of the available regional water resources.

The water consumption of the Jordan River region indicates that agriculture alone accounts for more than 64 percent of the annual water budget, although it

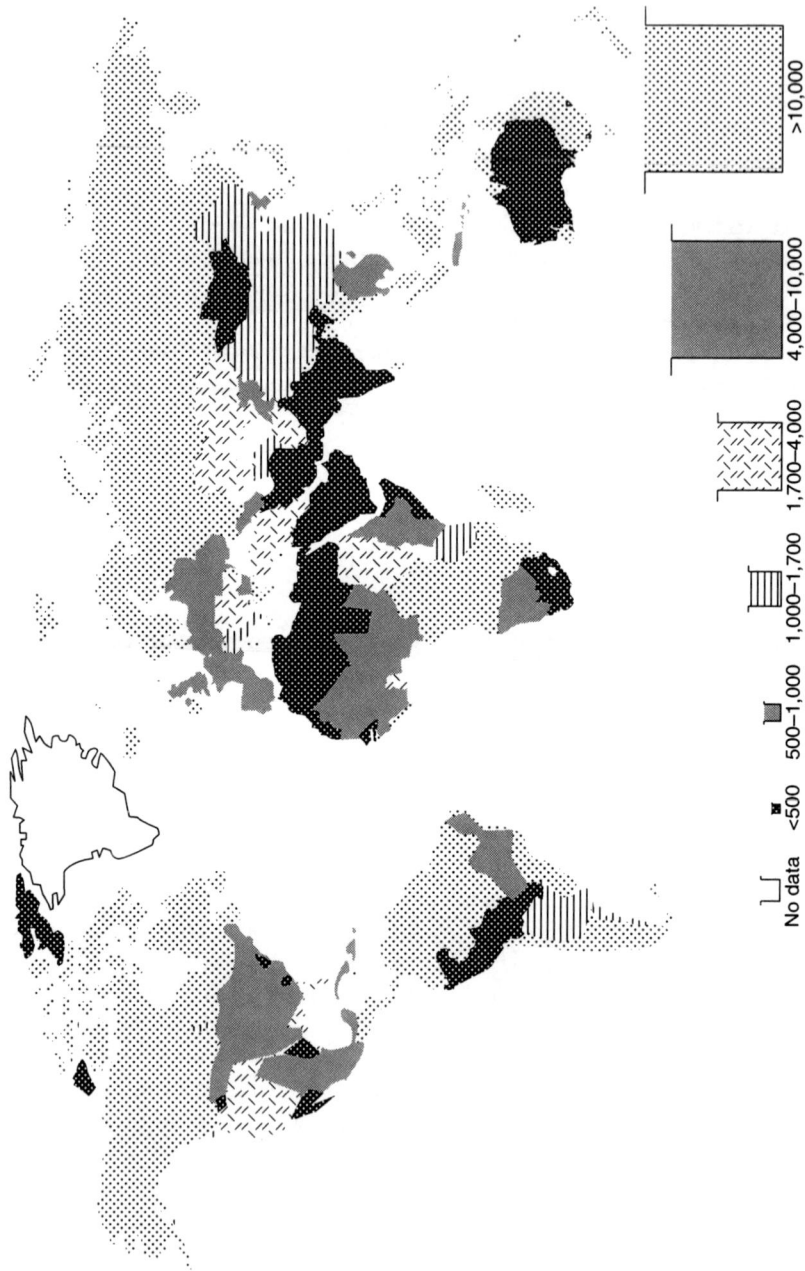

Figure 4.2 Annual renewable water (m³/person/year). Source: World Business Council for Sustainable Development (WBCSD, 2005).

Table 4.4 Annual per capita freshwater utilization in the region compared with other selected countries

Country	Per capita water use (m³/year)
Palestine	<100
Jordan	220
Israel	360
Lebanon	1,261
Syria	1,622
Iraq	3,287
Egypt	920
Saudi Arabia	118
Qatar	94
South Africa	319
India	566
US	1,677

Data sources: Libiszewski (1995); El-Fadel *et al.* (2001); Postel and Wolf (2001); and FAO-Aquastat (2002).

is not deemed one of the major sectors in the economy, followed by domestic water use (30 percent of the total annual water budget), and industrial water use (only 6 percent of the total annual water budget) (Water Care, 2004). In Israel, the total annual potential of renewable water stands at approximately 1,500–1,800 MCM (Beyth, 2006; Dinar, 2009). Agriculture consumes more than 63 percent of the country's annual water budget (Isaac, 1994; Dinar, 2009). Domestic water use consumes about 31 percent (which is expected to increase by 13–26 percent in 2020) and industrial consumption stands at only 6 percent (Libiszewski, 1995; Dinar, 2009). Jordan's overall total sustainable water supply is estimated at around 700–900 MCM/year (Libiszewski, 1995; Dinar, 2009). Agriculture consumes over 74 percent of the total water use in Jordan, followed by domestic water use, 21 percent, and industrial consumption, only 5 percent (Libiszewski, 1995; Hummel, 2006).

Population growth and water consumption are intertwined. As shown in Figure 4.3, the increase in the demand for freshwater is a consequence of natural population and economic growth (PASSIA, 2002). The estimated total renewable water supply available in the region (namely for Israel, Jordan, and Palestine) is a fixed amount of about 2,500 MCM/year (Wolf, 1995). However, water demand increases over time as the population of the region continues to grow. The current water shortage in the region exceeds 375 MCM/year, which is being pumped from the aquifers without being replenished (EXACT, 2005a). Based on population growth, the region is expected to have a water deficit of about 715 million cubic meters by 2010, 1.2 billion cubic meters by 2020, and 2.2 billion cubic meters by 2040

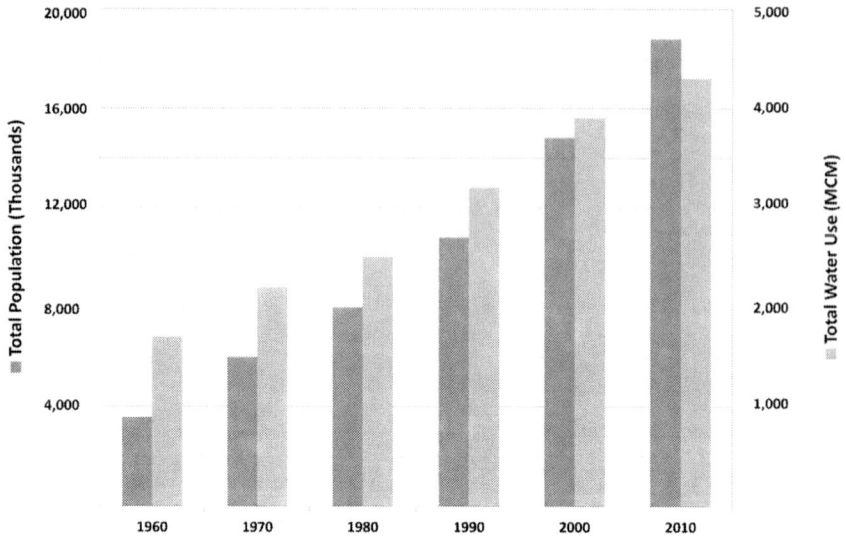

Figure 4.3 Israeli, Jordanian, and Palestinian population and water use forecasts. Data source: adapted from Executive Action Team (EXACT, 2005a).

(Water Care, 2004). The political and economical future of the region, along with the quality of the environment, depends primarily upon the ability to cope with the increasing demand for water. Population and water demand forecasts show that, as the population of the Jordan River basin continues to grow, the demand for water will continue to increase at an unsustainable rate. Some predict that the demand for drinking water will soon exceed the safe supply levels (Goeller, 1997). This increasing demand necessitates the search for alternative resources, including technologically driven and policy-driven resources, on one hand, and conservation strategies on the other.

Reflecting their highest hydro-dependency upon the Jordan River basin, Israel and Jordan have constructed the National Water Carrier and the King Abdallah Canal respectively to cater to their domestic and agricultural consumption. As shown in Table 4.5, in the early 1990s, Israel's *de facto* water use[5] was estimated at 640–690 MCM/year, while Jordan's *de facto* water use was about 120–130 MCM/ year (Libiszewski, 1995; Wolf, 1995). These regional water consumption trends provide evidence that the current situation in the Middle East represents a state of severe inequity in water sharing among the different users of the Jordan River basin. These fundamental differences in access to water and consequent consumptive water use and development are the underlying reasons for much of the tension over water supplies in the Middle East.

Economically, Israel is the most developed country in terms of its socioeconomic base and infrastructure, which allows it to extract relatively higher amounts of water than other co-riparians sharing the basin.[6]

Table 4.5 The *de facto* water utilization of the Jordan and Yarmouk waters in the early 1990s (MCM/year)

Country	Upper Jordan	The Yarmouk	Total	
Israel	ca. 550	70–100	640–690	Actual use comes from the Jordan, the Yarmouk, and Lake Tiberias
Jordan	0	120–130	120–130	These amounts have changed in the wake of the 1994 Treaty with Israel
Syria	0	150–240	150–240	Actual use comes from the Yarmouk
Lebanon	0	not riparian	0	Its use (20 MCM/year) comes from Al-Hasbani
West Bank	0	0	0	Access to Jordan River Valley is denied

Data sources: Libiszewski (1995); Wolf (1995); Elmusa (1998).

Water conflict and hydro-politics along the Jordan River

When it comes to the Middle East, water and war are two words that have been increasingly and interchangeably used in many contexts. This is because water constitutes a fundamental ingredient in the making of Middle East geopolitics (Abukhater, 2009b). Echoing the trend in many regions around the world, the issue of water scarcity has commonly become one of the most severe conundrums confronting the entire Middle East. The shortage of adequate supplies of freshwater has emerged as one of the most serious humanitarian and political predicaments in times of war and militant altercation. The impact of natural water resources on the Arab–Israeli conflict has become undeniably prominent; water is politics in the Middle East. As it permeates into every aspect of the region, water is deemed the driver of conflict as well as peace. In a region engulfed in a weighty political power struggle, water is perceived not only as a precious natural resource, but also as the sustainer of life and peace. As such, the significance of water fueled many conflicts and drove opportunities for resolution. In this semi-arid region, water is an extremely confined and precious natural resource that is believed to have exacerbated an already smouldering and protracted conflict. By the same token, water has been the reason for cooperation in the past and could possibly be the reason for a sustainable and robust peace settlement in the future.

The Arab–Israeli conflict dates back to the late nineteenth century, with the Jewish immigration to Palestine, issues surrounding land ownership, and the later establishment of the State of Israel in 1948, which triggered armed conflicts that lasted for decades[7] (Shamir and Haddadin, 2003). With the increasingly emerging idea of establishing a sovereign state in Palestine for Jewish settlers, Great Britain turned over the fate of Palestine to the United Nations, which, in 1947, decided to partition Palestine into provisional states: one Jewish and one Arab. The United Nations Partition Plan allocated 44 percent of the land to the Arab state and 56 percent of the land to the Jewish state, although the Palestinian

Arabs constituted more than two-thirds of the population and owned 93 percent of the land (Söderblom, 2003). Because of its inequitable allocation, the Plan was rejected by the Arab states and triggered the 1948 Arab–Israeli War, which resulted in the establishment of Israel and the creation of the Rhodes Armistice Demarcation Line marking Israel's borders, including historical Palestine, except for the West Bank and the Gaza Strip, which were subsumed by Jordan and Egypt respectively.

There has been a history of conflict in the region, leading in quite a few cases to military action in the Jordan River basin, and Jordan and Israel are not immune to that. In fact, conflict in this region is not an anomaly; it is, for the most part, the norm. The main conflicting issues that divide the co-riparians along the river basin are water allocation and development of the river basin system (Lowi, 1995). All five co-riparians heavily depend on the river system for their survival and for some the water of the river is linked to national security concerns. The war of June 1967, known as the Six-Day War, was a decisive event that altered the nature of the conflict as much as it altered the landscape of the region, the positioning of the co-riparians and the water allocation and utilization of the river basin. For the Arab countries, the outcome of this war was deemed to be a defeat, as they lost part of their territories and once again Israel reinforced its military superiority. The impact of this war was not confined to changing the international border; rather, it also changed the demographic characteristics and impacted on the co-riparians' access to water as well as on their relationships (Louka, 2006; Zawahri, 2009).

In the wake of this war, the Armistice Line changed, as Israel extended its boundaries and took over the Palestinian Gaza Strip and the West Bank, the Syrian Golan Heights, and the Egyptian Sinai Desert (Lowi, 1995). As the West Bank and the Gaza Strip were transferred from Jordanian and Egyptian dominion respectively to Israel, thousands of Palestinian refugees were once again displaced into the neighboring Arab states. With the exception of Sinai, which was returned to Egypt through the 1979 Israeli–Egyptian Peace Agreement, the resultant changes in the international borders created in the wake of the 1967 War largely remain intact today. This new border and territorial configuration also impacted access to water. Now the Upper Jordan River flows, almost entirely, into Israeli territory (Lowi, 1995). By controlling the Banyas River, Lake Tiberias, the Lower Jordan River, and expanding its access to the Yarmouk, Israel was able to prevent the neighboring Arab countries from accessing the Upper Jordan River, to block the River at its exit from Lake Tiberias, and exclusively use the latter for its own storage needs (Shamir and Haddadin, 2003; Zawahri, 2009). By doing so, the Palestinians and Lebanese were prevented from accessing the River basin altogether (Elmusa, 1995). With its occupation of the West Bank, Israel also gained control over the Palestinian groundwater resources, namely the mountain aquifer, which is proven significant to its water security (Louka, 2006).

Jordan was prevented by Israel from building the Maqarin dam to divert water from the Yarmouk, as stipulated by the Unified (Johnston) Plan (Elmusa, 1995). As a result, Jordan's water utilization was confined to the early 1960s water diversion from the Yarmouk via the King Abdallah Canal and the water of the side wadis (valleys). This situation created a relationship of hydrological interdependency, as well as tension, between Jordan and Israel as far as managing the river basin. As the Arab states "failed not only to back up their provocative military challenge to Israel, but even to defend themselves minimally against her response" (Kerr, 1969: 33), Israel established and re-affirmed itself once again as a hegemonic power in the region (Lowi, 1995). In addition, the construction and operation of Israel's National Water Carrier in 1964 had great implications not only for the other co-riparians sharing the river basin, but also for the region at large (Beyth, 2006). By constructing this project, which has been in the planning phase since the early 1950s, Israel was able to extract greater amounts of water and transfer it to areas outside the river basin, as far to the south as the Negev Desert.

It is believed that the current political boundaries of Israel, Jordan, Lebanon, and Syria are formed by water-related conflicts, which also precipitated the Palestinian–Israeli dispute. Conversely, water has also played a more constructive role in bringing these co-riparians together to formulate negotiated agreements, even during times of war and hostility. As such, the hydro-political arena can be characterized as a "talk-fight" situation (Gurr *et al.*, 2001). As cooperation and conflict coexist in this region, the Middle Eastern hydro-politics can best be described as "hydroschizophrenia" (Jarvis *et al.*, 2005). Because of its central and pre-eminent role in the Middle East political landscape, water alone caused a political impasse on many occasions. It was the reason for temporarily halting negotiations and diplomatic efforts on several fronts, including the Israel–Jordan Peace Treaty and the Israeli–Palestinian Interim Agreement (Oslo II) (Sosland, 2007). It also continues to be a fundamental point of contention on the Syrian–Israeli political and diplomatic front.

Previous water agreements

Because of its crucial role in survival, water can fuel particularly contentious political conflicts and if carefully managed, can also induce cooperation. To this end, Frey and Naff (1985: 67) explain, "precisely because it is essential to life and so highly charged, water can – perhaps even tends to – produce cooperation even in the absence of trust between concerned actors." This fact is clearly demonstrated throughout the history of the Middle East. As water has been a source of cooperation, there is evidence of both signed and unsigned water (mostly bilateral) agreements in the region (Wolf, 1995; Zawahri, 2009). These bilateral agreements led to a fragmented governance of a multilateral river system, and eventually caused inefficiency in managing the basin water (Phillips, 2008). The previous water negotiation and diplomatic efforts were proven unfruitful and ineffective in managing the Arab–Israeli conflict over disputed scarce water resources.

As shown in Table 4.6, bilateral agreements to govern the sharing of the Jordan's water were reached as early as 1920, involving Britain and France, and the latest was the 1994 Treaty of Peace between Israel and Jordan. Although these border disagreements between the British and the French during the demarcation process resulted in decades of bitter conflict over international borders in the

Table 4.6 Previous treaties on the Jordan River basin

Date	Treaty basin	Signatories	Treaty name
December 23, 1920	Jordan, Tigris–Euphrates, Yarmouk	France; Great Britain	Franco-British convention on certain points connected with the Mandates for Syria and the Lebanon, Palestine and Mesopotamia
March 7, 1923	Jordan, Yarmouk	France; Great Britain	Exchange of notes constituting an agreement between the British and French governments respecting the boundary line between Syria and Palestine from the Mediterranean to El Hamme
February 2, 1926	Jordan	Great Britain, on behalf of the Territories of Palestine; France, on behalf of Lebanon and Syria	Agreement of Good Neighbourly Relations concluded between the British and French governments on behalf of the Territories of Palestine, on the one part, and on behalf of Syria and Great Lebanon, on the other part
June 4, 1953	Yarmouk	Jordan; Syria	Agreement between the Republic of Syria and the Hashemite Kingdom of Jordan Concerning the Utilization of the Yarmouk waters
October 26, 1994	Jordan	Israel; Jordan; Lebanon; Syria	Johnston negotiations
October 26, 1994	Araba/Arava, groundwater, Jordan, Yarmouk	Israel; Jordan	Treaty of Peace between the State of Israel and the Hashemite Kingdom of Jordan, done at Arava/Araba crossing point
September 28, 1995	Jordan	Israel; Palestine Liberation Organization	Israeli–Palestinian Interim Agreement on the West Bank and the Gaza Strip, with Annexes I to VII

Data source: adapted from *Atlas of International Freshwater Agreements* (Wolf, 2002).[8]

Middle East, the 1994 Treaty of Peace between Israel and Jordan opened the door to prospects of future cooperation and diplomatic relations. As the groundwater issue of the Palestinian communities was not addressed in 1994, this issue remains to be resolved bilaterally in the final round of negotiations (Daoudy, 2009).

It is believed that the current Middle East peace process to settle the Arab–Israeli conflict officially started on the White House lawn on September 13, 1993, with the handshake between the Israeli prime minister, Yitzhak Rabin, and the Palestinian president, Yasser Arafat. However, considering the long process of negotiation between the disputing parties prior to that (namely Israel and Jordan), this belief might appear to largely represent a common misconception, as it seems oblivious to the extensive negotiation rounds and the several hydro-political and geopolitical events that occurred prior to this monumental feat.

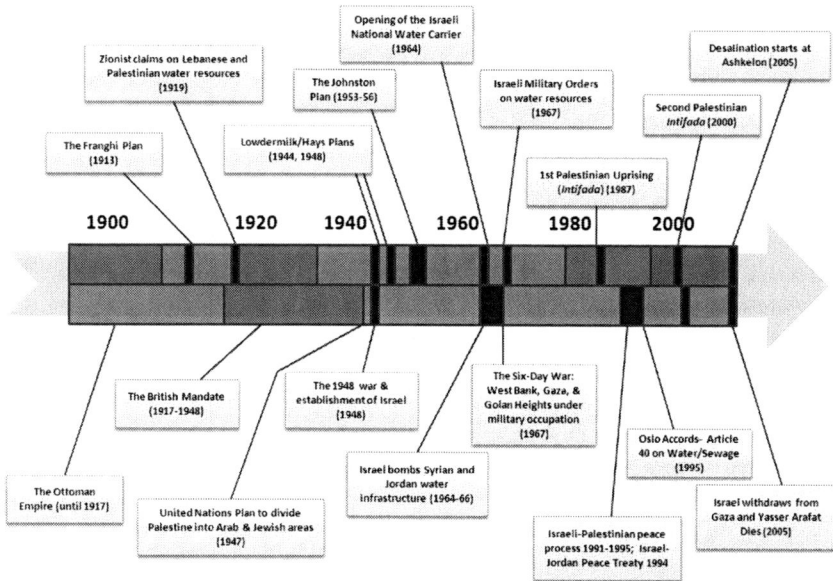

Figure 4.4 A timeline demarcating the hydro-political and geopolitical events preceding and subsequent to the 1994 Peace Treaty. Source: adapted from Phillips *et al.* (2006).

These hydro-political and geopolitical events preceding and subsequent to the 1994 Peace Treaty are shown in Figure 4.4 and date back to the early 1900s.

The Arab–Israeli negotiation process underwent many stages that date back prior to the 1950s. Initially, because of the political atmosphere that was less than conducive, Israel and Jordan did not cooperate openly over sharing water resources. Instead, both countries resorted to a series of secretive talks that date back to the 1940s, to avoid criticism and ease pressure from their respective constituencies. With intervention from the United States, both countries were able to manage the sharing of the Yarmouk water through extensive secret talks (Sosland, 2007). This situation, which continued to dominate the countries' political relationships for many years, can best be described as constructive confrontation, as the conflict was not deemed "ripe" for negotiation (Burgess and Burgess, 2007).

Throughout history, several water schemes were prepared to utilize the waters of the Jordan River, starting as early as 1899 with the Zionist Organization and later under the Mandate (Naff and Matson, 1984). In the post-1949 period, Israel and the neighboring Arab countries were unilaterally formulating and implanting national water development plans to improve their economies and address the issue of water needs for immigration and Palestinian refugees in adjacent Arab countries after the establishment of the State of Israel in 1948 (Lowi, 1995; Shamir and Haddadin, 2003). During these years of no direct communication, both the Arabs and Israelis unilaterally proposed many conflicting plans to govern and manage the utilization of the the Yarmouk water. These unilateral proposals included the McDonald Plan and the Bunger Plan (both proposed by Jordan), the All Israel Plan, and the Syrian–Lebanese Plan (Haddadin, 2002a). The key difference between these plans is the emphasis on in-basin use *vis-à-vis* inter-basin transfer proposed by the Arab states and Israel respectively[9] (Haddadin, 2002a; Shamir and Haddadin, 2003).

These plans, as they provided national solutions to a regional problem, failed to address the overall regional dimension of water and caused additional tension. As a result, clashes occurred along the borders between Israel and its neighboring Arab countries in the Armistice Demarcation Zone (ADZ) that was established by the Armistice Agreement in 1949 (Lowi, 1995; Haddadin, 2002a). As the parties resorted to diplomacy, a number of plans were proposed to manage the allocation of the shared waters, promote the development of the Jordan River Valley, and settle the issue of the Palestinian refugees. These plans include the Main Plan, the Cotton Plan, the Arab Plan, and the Unified (Johnston) Plan. Allocations of water based on these plans compared to the Unified (Johnston) Plan are presented in Table 4.7.

The Main Plan, which constituted the basis of the later shuttle diplomacy, was carried out by Chas Main and the Tennessee Valley Authority (TVA) upon the request of the UNRWA[11] in 1953 (Shamir and Haddadin, 2003; Wolf and Newton, 2008). As it excluded the Litani River from the system, the Main/TVA Plan proposed the construction of a series of small and medium size dams on the Hasbani, Dan, Banias, and Maqarin and the utilization of Lake Tiberias for storage (Haddadin, 2002a). As a reaction to the Main Plan, Israel proposed the Cotton Plan in 1954, which incorporated the Litani River to augment the size of the supply and, unlike the Main Plan, required out-of-basin transfers and identified Lake Tiberias as the main storage facility (Wolf and Newton, 2008). The Arab Plan, which was offered by the Arab League Technical Committee in 1954 as a counterproposal to the Main/TVA Plan, excluded the Litani River, promoted in-basin use, and rejected storage in Lake Tiberias, which is entirely located in, and

Table 4.7 Water allocations of different development plans in MCM/year

Plan	Israel	Jordan	Lebanon	Syria
Main	393	774	-	45
Cotton (Israel)[1]	1,290	575	450	30
Arab	182	698	35	132
Unified	400[2]	720[3]	35	132

[1] The Cotton Plan included the integration of the Litani River into the Jordan basin.
[2] The Unified Plan allocated Israel the "residue" flow, that is, what remained after the Arab states had drawn their allocations, estimated at an average of 409 MCM/year.
[3] Two different summaries were distributed after the negotiations, with a difference of 15 MCM/year, on allocations between Israel and Jordan on the the Yarmouk. This difference was never resolved and was the focus of Yarmouk negotiations in the late 1980s.

Data sources: adapted from Naff and Matson (1984); Wolf and Newton (2008).[10]

controlled by, Israel (Shamir and Haddadin, 2003). These plans were envisioned to have provided a mechanism to manage the use of shared water resources. However, these separate and conflictive plans were the source of tension in the region that eventually called for reconciliation. The disputing parties, as a result, resorted to diplomacy as a way of moving forward beyond these plans.

Attempts at reconciliation produced a diplomatic process and a series of cooperative (mostly bilateral) arrangements, which started in the 1950s and continue until today. The first multilateral cooperative venture was led by the United States in 1953–55, known as shuttle diplomacy, which was conducted by the US special envoy to the Middle East, Ambassador Eric Johnston (Lowi, 1995; Shamir and Haddadin, 2003). In the wake of the establishment of the State of Israel in 1948, the co-riparians sharing the river system started to undertake unilateral actions for the development of the river system. In particular, the two main co-riparians most dependent on the Jordan River system, Israel and Jordan, had different and contradictory visions of how the river ought to be developed and each objected to the other's plans (Shamir and Haddadin, 2003; Zawahri, 2009). As the regional stability was at stake, US diplomacy took on a more proactive role in promoting a regional approach (Sosland, 2007). The involvement of the United States was motivated by many reasons, namely protecting its interests in the region, safeguarding Israel's security, and mitigating communist influence in the region (Shamir and Haddadin, 2003). The level of importance that the mission held for the US was expressed by President Eisenhower, who described the trip to be of "primary importance to the United States" (Slany et al., 1986, cited in Sosland, 2007). Cognizant of the political ramifications and impending potential for a macro-political regional conflict of such unilateral actions, the United States led this regional reconciliation effort on October 7, 1953 (Louka, 2006; Sosland, 2007). Through his shuttle diplomacy, Ambassador Eric Johnston met separately with representatives of Israel and the Arab League Technical Committee (under Egyptian leadership) to manage the sharing and development of the Jordan River Valley, focusing on each country's irrigation and agricultural development needs, to avert potential conflict (Haddadin, 2000; Shamir and Haddadin, 2003; Louka, 2006).

This shuttle diplomacy effort was conducted in an attempt to develop a comprehensive integrated plan for the regional development of the Jordan River system and to settle the regional dispute through the multilateral allocation of the river water, while promoting cooperation and economic stability (Elmusa, 1995; Haddadin, 2000; Sosland, 2007). The objective of the mission as stated by the secretary of state, John Foster Dulles, was "to secure agreement of the states of Lebanon, Syria, Jordan, and Israel to the diversion and use of the water of the Jordan River basin" (Slany et al., 1986, cited in Sosland, 2007). Akin to the Main Plan, Johnston's mission ignored the political boundaries to avoid political complications related to the Arab states' unwillingness to recognize Israel (Lowi, 1995; Shamir and Haddadin, 2003). A number of points constituted the foundations for Johnston's negotiations. These included the utilization of the TVA/Main Plan as the basis for water distribution, the adjustment of the Armistice Line to mitigate

Israel's exclusive control over Lake Tiberias, the elimination of the demilitarized zones in the Jordan Valley, and the exclusion of the Litani River as an exclusively Lebanese river (Shamir and Haddadin, 2003). Several factors were of particular consideration to Johnston in his water allocation scheme, including:

1 the area of the arable land;
2 the assumed cropping pattern;
3 the cropping intensity; and
4 the water duty per unit of cultivated area.

(Shamir and Haddadin, 2003)

A number of issues, however, posed major challenges to this cooperative effort. These challenges were related to:

1 the quantitative division (water quotas) of such a small stream with a very limited discharge among the different co-riparians;
2 incorporating the Litani River into the Jordan system as a way to augment shared water;
3 the intended Israeli use of the Jordan River water for areas located outside the basin;
4 the utilization of Lake Tiberias as a floodwater storage area; and
5 international supervision.

(Lowi, 1995)

Attempting to reach a unified plan able to allot water quotas for the co-riparians was challenged by the insignificant amount of water available for them in the river. As such, finding ways to augment such a small pie was warranted. An idea to augment the river system water supply, proposed by Israel and rejected by the Arab states, was to include the Litani River, which originates and flows through Lebanon, into the system (Shamir and Haddadin, 2003; Zawahri, 2009). After four rounds of shuttle diplomacy, the Litani River was excluded from the negotiations on the basis that it is a Lebanese river and separate from the Jordan River system. A second idea to augment the water was to use Lake Tiberias to store the Yarmouk floodwater, accompanied by dam construction along the river. This proposal of using Lake Tiberias as the main storage reservoir was mutually rejected by both Arabs and Israelis for different reasons (Shamir and Haddadin, 2003). For Israel, this meant compromising its sovereignty of the lake, whereas, for the Arab states, this meant dependency on the Israeli controlled Lake Tiberias to obtain water for their own use, which constituted a threat to their water security (Haddadin, 2002a). By means of negotiation, both sides, however, agreed to the dam construction proposal and the use of the lake for less floodwater storage (about 700 MCM) than was originally proposed (Shamir and Haddadin, 2003).

The use of water also constituted a major point of contention. The Arab states intended to utilize the water for intra-basin water use, whereas Israel needed the water outside the basin, namely in the coastal plain and the Negev Desert (and intended it for inter-basin water transfer). Although the intended Israeli water use was viewed by the Arab states as out-of-basin water transfer, and therefore was opposed at first, they agreed to allow Israel to transfer water for its needs outside the basin. The negotiations also tackled the issue of establishing an institution to govern the water management of the river basin and monitor the compliance of Treaty implementation. After four rounds of shuttle diplomacy and a great deal of negotiations and compromises, Johnston was able to develop a formula, known as the Unified Plan,[12] for water allocation among the different co-riparians.

Although the technical committees from both sides accepted the Unified Plan, the Arab League Council failed to accept the Plan in October 1955, as accepting it would have implied their recognition of Israel (Zawahri, 2009). Israel resumed its diversion plans of the Jordan waters from the northwest corner of Lake Tiberias, via its National Water Carrier, which became fully operational in 1964 (Shamir and Haddadin, 2003; Beyth, 2006). The Johnston mission ended because of the challenges imposed by complex international and domestic geopolitical realities (Sosland, 2007). As the Plan was never ratified by any of the parties, it never became the "Johnston Agreement" (Sosland, 2007; Wolf and Newton, 2008). The failure to ratify the Johnston Plan was related, first, to the macro-political nature of the conflict, namely the Arab states' view that recognizing the right of Israel to exist may compromise the Palestinian cause, and, second, to the high cost of adopting the Plan when compared with its insignificant benefit[13] (Zawahri, 2009).

Although the Plan was never ratified, it was accepted by all Arab states and became a customary law in the region (Elmusa, 1995). Both sides have generally adhered to its technical details and water allocations as it constituted the basis for the Jordan River Valley development thereafter (Louka, 2006; Wolf and Newton, 2008). This was the case until the 1967 War, in which Israel expanded its boundaries, water consumption and utilization (Elmusa, 1995). Moreover, the Plan laid the foundation for future negotiations and is still referred to as the basis for any potential water-related negotiation along the Jordan River Valley, although its exclusion of groundwater resources would later constitute a major issue (Wolf and Newton, 2008). In particular, the Plan was the catalyst for many reconciliation attempts, namely the Picnic Table Talks between Israel and Jordan, the "water-for-peace" desalination projects during the 1960s, and the negotiations over the the Yarmouk and the Unity Dam that took place during the 1970s and 1980s (Wolf, 2000). For Israel and Jordan, the Plan had a particular importance serving as the guidepost for the development of the river basin. Because Israel and Jordan were the greatest beneficiaries of the Plan, as well as the most dependent co-riparians on the Jordan water, they made their intention to unilaterally comply with it clear to the United States through secret notes (Sosland, 2007). The commitment of Israel and Jordan to abide by the Plan was in exchange for economic assistance from the United States in the development of the National Water Carrier in Israel and the East Ghor Canal in Jordan (the King Abdallah Canal) (Shamir and Haddadin, 2003; Sosland, 2007). Aid from the United States was conditional, based on the extent to which the two parties abided by the amounts specified by the Plan. To that end, the United States indicated in a letter sent to Jordan that, "assistance will be extended provided there is an explicit undertaking from the Hashemite Kingdom of Jordan that it will not draw from the Yarmouk River more than the share allotted to it under the [Johnston] Plan" (cited in Sosland, 2007). Overall, the Plan was successful in "damping down" the hydro-political tension over the waters of the Jordan (Sosland, 2007).

The mission was motivated by a number of events. As the Arab–Israeli conflict intensified, so did the Soviet–American rivalry, which resulted in an arms race in the Middle East (Sosland, 2007). These profound geopolitical discourses had a defining impact on the regional stability and security, which in turn led to a shift in US diplomacy toward a more proactive role, known as the "piecemeal approach,"[14] in settling the Arab–Israeli dispute (Sosland, 2007).

Analysis of the Israel–Jordan Peace Treaty, 1994

On October 26, 1994, the State of Israel and the Hashemite Kingdom of Jordan signed a peace treaty that included a water agreement. This section provides an analysis of the Treaty, paying significant attention to issues related to both the process and the outcome. The water agreement is negotiated between Israel and Jordan and involved the delegations shown in Table 4.8.

Intent of the Treaty

The main focus of all diplomatic efforts in the Middle East has been, for the most part, targeted at the need to achieve viable and sustainable peace in the region, while recognizing equal partnership.[17] Echoing the trend of these diplomatic efforts, Israel and Jordan have reached a resolution over dividing and sharing crucial water resources as part of the Israel–Jordan Peace Treaty of 1994. Terminating the state of belligerency between the two states, Israel and Jordan agreed to accept this peace agreement based on the UN Security Council resolutions 242 and 338 to achieve a just, lasting, and comprehensive regional peace and a mutual understanding of each other's needs and interests (Treaty of Peace, 1994: Preamble). Both parties, realizing the importance of living in peace side by side, developing and maintaining cooperation, and ensuring national security and sovereignty, acknowledged the provisions of the Charter of the United Nations and the principles of international law to govern their future relations (Treaty of Peace, 1994: Article 2). Complying with this Treaty, therefore, means that both states must recognize and respect "each other's, as well as every state in the region's, sovereignty, territorial integrity and political independency" (Treaty of Peace, 1994: Article 2). A great deal of emphasis is placed on maintaining national security and the commitment to eliminating hostility in the region[18] (Treaty of Peace, 1994: Article 4). However, the treaty formation was not guided by the principles of the international water law, as Israel preferred a more pragmatic approach to negotiation.

In addition to peace and security reasons, the Treaty attempts to put forth an allocation scheme to resolve the shared international waters in dispute "in

Table 4.8 Stakeholders and delegations involved in the water negotiation[15]

The Israeli delegation	Mr. Noah Kinarti: chief water negotiator on behalf of Israel Dr. Uri Shamir: member of the Israeli negotiation team Mr. Daniel Rizner: a lawyer in the Israel Defense Forces and the legal adviser of the Israeli negotiation team Ambassador Elyakim Rubinstein: Government Secretary (later Attorney General) for Israel and chairman of the Israeli delegation for negotiations with Jordan Mr. Yitzhak Rabin: Israel's prime minister
The Jordanian delegation	Dr. Munther Haddadin: former president of the Jordan Valley Authority, served as the senior water negotiator on behalf of the Jordanian side in the Israel–Jordan peace talks (also the head of the water resources delegation) Dr. Fayez Tarawneh: ambassador (later Jordan's prime minister), head of the Jordanian delegation to the Regional Economic Development Committee of the Israel–Jordan peace talks Mr. Awn al-Khasawneh: the legal expert and adviser to the Jordanian delegation to the peace negotiations between Israel and Jordan Dr. Abdul Salam Majali: head of the joint Jordanian–Palestinian delegation to the Madrid peace negotiation and of the Jordanian delegation until 1993, when he became Jordan's prime minister Prince El Hassan bin Talal: the Crown Prince of Jordan King Hussein bin Talal: the King of Jordan
The joint Palestinian–Jordanian delegation	The delegation, composed of 14 members (half of whom were Palestinians residing in the Occupied Territories outside Jerusalem and the other half were Jordanians), attended the International Peace Conference in Madrid in 1991, where the Palestinians had the chance to demonstrate good faith in, and strong commitment to, peace[16]

Data sources: Haddadin (2002b); Majali *et al.* (2006).

their totality" by recognizing the need to conclude a "practical, just, and agreed solution" (Louka, 2006). Article 6 of the Treaty provides general principles and guidance for a comprehensive and lasting settlement over water allocation, regarding water as a basis for advancing cooperation between the two countries. This includes the mutual recognition of the "rightful allocation" for both countries in the quality and quantity of surface and groundwater, including the Jordan and Yarmouk river waters and the Araba/Arava groundwater. Annex II of the Treaty outlines details about the target source and location of water transfer, the amounts of water diverted for each party, the quality, the times of the year, and the financial arrangement (Shamir, 2008).

Process equity and treaty formation

Since the Johnston mission in the 1950s, issues related to water sharing and management of the Jordan River valley remained unresolved until the 1990s Madrid Peace Process, at which point disputing co-riparians entered into direct channels of negotiation and diplomatic discourse (Sosland, 2007). The Middle East era of diplomacy and normalization began in the wake of a series of decisive events and profound changes that took place in the region and around the world concluding in the 1980s (Haddadin, 2008). Most profoundly, the end of the Iran–Iraq war, the collapse of the Berlin Wall, the establishment of the Arab Cooperation Council and the Maghreb Union, the Iraqi invasion of neighboring Kuwait in 1990, and the consequent Gulf War, all presented remarkable geopolitical discourses that brought peace to the Middle East (Haddadin, 2002a). These events provided opportunities for the United States to reinitiate diplomatic interest in resolving the Arab–Israeli conflict (Sosland, 2007). During the eight months following the Gulf War, Secretary of State James Baker conducted eight trips[19] to the Middle East to set the stage for resolving the Arab–Israeli conflict (Haddadin, 2008). It became clear that the Arabs and Israel, with the help of the United States and the Soviet Union, were about to enter into serious negotiation in order to settle their disputes and normalize relationships (Haddadin, 2002a).

However, the question of bilateral vis-à-vis multilateral negotiation became one of the first issues to challenge the negotiation. As they are relatively easier to negotiate and faster to ratify, bilateral water agreements are historically more common than multilateral agreements, even when governing multilateral basins, both in the context of the Middle East as well as those of global stature (Wolf, 1998). This common reliance on bilateral agreements hinders the parties' ability to attain integrated river basin-wide management (Elmusa, 1995; Wolf, 1998). Although the Jordan River basin involves multiple parties that would normally be involved in a multilateral negotiation, Israel insisted on negotiating with each riparian country bilaterally. This Israeli proposed negotiation tactic of bilateral negotiation provided Israel with the advantage of isolating each individual co-riparian and therefore addressing water-related issues in separation from the other co-riparians, which gave it more control over the different negotiation trajectories (Phillips, 2008). As a result of this, Israel remains the key beneficiary in these segmented bilateral accords, as they place Israel at the hub of the river management system, rather than as a party in a larger dispute (Elmusa, 1995). On the other hand, this bilateral negotiation strategy helps to avoid coalition formation among several parties that could be detrimental to the overall negotiation processes and outcomes.

The Peace Process officially started with the Madrid Conference, which brought together all involved stakeholders, including Israel, Lebanon, Syria, Jordan, and Palestine.[20] Under the auspices of the Madrid Peace Process, the negotiation between Israel and its neighbors followed two courses: (1) a multinational conference involving international participation that focused on issues of regional importance; and (2) separate, yet simultaneous, bilateral negotiation tracks between Israel and each of its neighboring countries to address issues of concern[21] (Haddadin, 2002a; Fischhendler, 2008b). Constituting one of the key pending negotiation issues, water was negotiated both in the bilateral groups and in one of the multilateral groups[22] (Elmusa, 1995). It is believed that, until the formal negotiation process, which began in 1991, political and resource issues, reflecting the realms of "high" and "low" politics respectively, were handled in separation from one another (Wolf and Newton, 2008). Arguably, this separation is considered by many to be the reason for the failure of these previous diplomatic initiatives (Phillips, 2008; Daoudy, 2009). For instance, the diplomatic attempts that tended to strictly address water resources allocation, such as the Johnston negotiations of the 1950s, the "water-for-peace" initiative in the late 1960s, negotiations over the the Yarmouk in the 1970s and 1980s, and the Global Water Summit Initiative of 1991, failed in settling the conflict as they were conducted almost entirely in separation from the larger political discourse (Wolf and Newton, 2008).

Within the course of negotiation, however, these high and low politics issues were combined when they appeared tangible. During the negotiation process, the two parties were heavily involved in parallel second-track diplomacy efforts that focused not only on high politics, such as security, mutual recognition, land, and borders, but also on low politics, such as issues related to water, natural resources, refugees, and environmental and regional development. Solving issues of low politics,[23] including water, is believed to have helped in facilitating negotiation over issues of high politics (Priscoli and Wolf, 2009). As shown in Figure 4.5, these include five multilateral groups (representing issues of regional importance), which emerged from the Moscow Conference, including water, environment, refugees, regional security and arms control, and regional economic development (Haddadin, 2002a; Fuwa, 2003; Shamir and Haddadin, 2003; Fischhendler, 2008b).

Composed of seven rounds of negotiations (Fuwa, 2003), the main purpose of these multilateral talks was to facilitate smooth and constructive dialogs among participants. This helped in providing a form of conciliation and precisely

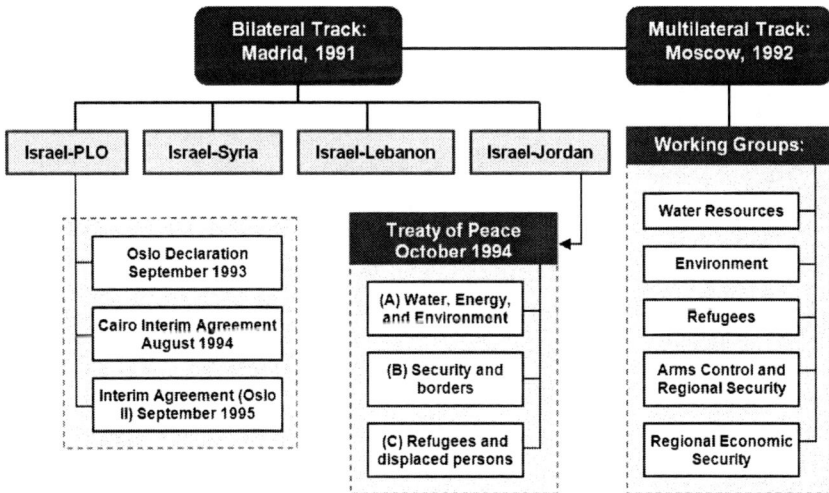

Figure 4.5 The structure of the 1990s Middle East Peace Process. Source: adapted from "Contention to Cooperation: A Case Study of the Middle East Multilateral Working Group on Water Resources" (cited in Sosland, 2007).

identifying major issues, interests, needs, and concerns of both parties, which were consensually outlined and made clear to everyone. These pre-negotiation efforts helped in meeting the "other" party, understanding how history is politically manipulated in order to examine current situations and explore possibilities and potential for cooperation. This also helped the parties overcome fears and eradicate ungrounded rationales for the annihilation of the "other," while utilizing and embracing the diversity of the parties involved. During these parallel negotiations, both Israel and Jordan developed more acceptable common ground and realized the importance of mutual understanding and advancing a "resolutionary" process of peace negotiation and normalization.

These negotiations involved multiple parties, including US Secretary of State James Baker, donors from the United States, the European Union (E.U.), Japan, and Northern Europe, and other participating countries from the Middle East and North Africa (Fuwa, 2003). With representatives from all key parties (except Syria and Lebanon), along with others who acted as sponsors, facilitators, and donors, these multilateral groups continued to conduct several rounds of negotiation in different locations: Moscow, January 1992; Vienna, May 1992; Washington, September 1992; Geneva, April 1993; Beijing, October 1993; Muscat, April 1994; Athens, November 1994; and Amman, June 1995, until their termination in 1996 (Haddadin, 2002a; Shamir and Haddadin, 2003). The refusal of Syria and Lebanon to participate was because they felt that it was premature to normalize relations with Israel before resolving core issues and reaching peace first (Dauody, 2009; Haddadin, 2002a). Jordan obviously had a different negotiation strategy and stance on this issue.[24] When asked about the Jordanian expectation of normal cooperation with Israel, Dr. Haddadin[25] responded, "the sequence is restoration, mitigation, and cooperation," meaning "restoration of rights, mitigation of adverse impacts resulting from the creation of Israel, and then cooperation would start" (Haddadin, 2002a).

As shown in Figure 4.6, the agenda of the Multilateral Working Group on Water Resources incorporated four elements of regional importance:

1   the enhancement of data availability;
2   water management and conservation;
3   enhancement of water supply; and
4   concepts of regional cooperation and management.

(Israel Ministry of Foreign Affairs, 2000)

Because making informed decisions regarding future water planning and management requires accurate and reliable data regarding water availability and consumption trends, sharing this data was crucial not only in enhancing the accuracy and success of water-related decisions, but also in building trust among the parties (Sosland, 2007).

Unlike the bilateral negotiations, the work and "painstaking negotiations" of the Multilateral Working Group on Water Resources provided the teams with the opportunity to identify major problems, substantively search for solutions and

```
                    ┌─────────────────────────────┐
                    │  Multilateral Working Group  │
                    │      (Water Resources)       │
                    └─────────────────────────────┘
```

Enhancement of data availability	Water management & conservation	Enhancement of water supply	Regional cooperation & management
Regional water data banks	Water management practices and conservation	Regional water supply & demand study	Water sector training program
Availability/ exchange of water data	Public awareness and water conservation	Middle East Desalination Research Center	Declaration on principles for cooperation on water-related matters & new & additional water resources
Support decision making (local and regional)	Comparative study of water laws and water institutions in the region		Water atlas
			Waternet

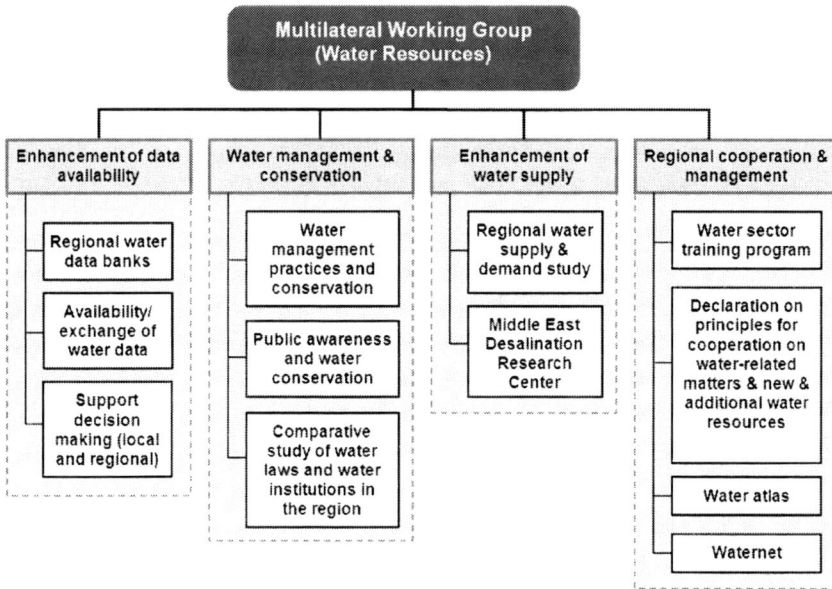

Figure 4.6 The project portfolio of the Multilateral Working Group on Water Resources. Source: adapted from Israel Ministry of Foreign Affairs (2000) (cited in Sosland, 2007).

address future water issues[26] (Sosland, 2007). The goal of these five multilateral groups, which were conducted in tandem with the bilateral negotiations, was to set the foundation for the latter and reinforce peace.[27] On the other hand, the goal of the parallel bilateral arena was to establish direct channels of communication, where parties can address and negotiate their specific needs and interests head on. According to Shamir and Haddadin (2003), the two tracks of negotiation, serving different yet complementary purposes, worked for the most part in harmony, except for a few instances of clashing, which hindered the efficiency of the negotiation.[28] These parallel negotiation tracks were proven to be prudent to the advancement of cooperation and creative negotiation outcomes (Susskind, 1994).

Although the two bilateral and multilateral tracks of negotiations were meant to complement one another, clashes occurred and more coordination between the two arenas was needed on certain issues (Shamir and Haddadin, 2003). For example, the two tracks of negotiations were challenged by the issue of the Palestinian representation, as the Palestine Liberation Organization (PLO) was not recognized by either Israel or the United States. With the Israeli refusal to conduct direct negotiations with the Palestinian delegation as a formal partner, and lack of coordination between the two arenas, the two parties convened in separation from and in tandem with the Israeli–Jordanian meetings. King Hussein's decision to disengage from the Palestinian West Bank made it clear that the Palestinians were to be represented in separation from the Jordanian delegation. This created a dilemma that needed an acceptable solution to Israel as well as the Palestinians.

US Secretary of State James Baker conducted shuttle diplomacy that lasted over eight months to resolve the issue of the Palestinian representation. He was successful in brokering an agreement to allow the Palestinians to participate in the 1991 Madrid Conference under the Jordanian state umbrella, which resulted in the creation of a joint Palestinian–Jordanian delegation[29] (Shamir and Haddadin, 2003). However, the joint Palestinian–Jordanian delegation separated at the beginning of the second round in Washington and two tracks of negotiations emerged as a result; the Jordanian negotiation track and the Palestinian negotiation track[30] (which will be discussed in more detail later in this chapter) (Haddadin, 2002a).

These preceding events set the stage for a bilateral diplomatic front and a series of meetings between Jordan and Israel, which successfully continued until the Israeli–Jordanian Peace Treaty of 1994 was accomplished. The process of negotiation and the preceding diplomatic initiatives leading to it were generally bilateral in nature and took place along two levels – the technical and political levels – which, for the most part, were separated from each other. As illustrated in Figure 4.7, the analysis of the negotiation process is presented in three different stages, which include the pre-negotiation, negotiation, and post-negotiation processes. This partitioning of the stages of negotiation has been found to be useful in the analysis of these cases.[31]

The pre-negotiation process

During the pre-negotiation process, all affected parties must be identified and included in the process to avoid the adverse impacts that might arise because of the *ex parte* syndrome by excluding important stakeholders. Based on this, it is obvious that the Treaty ignored important stakeholders, as Israel and Jordan are not the only two parties involved in the dispute over the water resources of the Jordan and Yarmouk rivers.[32] Israel's strategy was to refuse to collectively negotiate with all Arab riparians and instead make separate bilateral treaties with each individual country to isolate each party and maximize its own gain (Elmusa, 1995).

There have been a number of negotiation rounds prior to the official negotiation process that ensued in the 1990s, which resulted in the 1994 Peace Treaty. Before the beginning of the official bilateral negotiation, various attempts were made to reach a negotiated settlement, none of which yielded a discrete resolution.

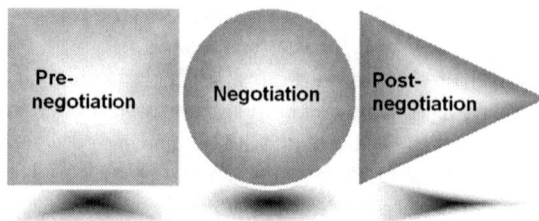

Figure 4.7 Negotiation process stages.

This situation had changed by the 1980s, with significant geopolitical events unfolding in the Middle East, and an agreement seemed plausible. As per the Truce Agreement of 1949 between Israel and Jordan, and under the auspices of the United Nations Truce Supervision Organization, joint meetings between the Jordanian and Israeli representatives were conducted to resolve the sharing and operation of the the Yarmouk (Shamir and Haddadin, 2003). The pre-negotiation process over water was purely technical in nature and started years before it was actually announced to the public. It is widely documented that the negotiation over water between Israel and Jordan began in the form of "Picnic Table Talks" to determine and manage the day-to-day hydrologic operations of the Yarmouk (Wolf, 1995), where the two parties, mainly technocrats, scientists and engineers, met on the technical level, forming joint fact-finding teams (JFF).[33] These technical teams met at Adassiya/point 121, the main diversion point, to discuss solutions to raise the water level and allow for more diversion flow into the King Abdallah Canal (KAC), formerly known as the East Ghor Canal (Shamir and Haddadin, 2003). This was accomplished by placing a sand bar and later sand bags across the riverbed (Haddadin, 2002a). These meetings to adjust the water level, which were driven by the water shortage problem in Jordan, produced *ad hoc* agreements to manage the timing and volumes of water.[34]

Although Dr. Haddadin sees no connection between the "Picnic Table Talks" of the 1980s and the 1990s Middle East Peace Process, what happened in the technically driven Picnic Table Talks during the years leading up to the 1994 Peace Treaty directly and fundamentally impacted the water-related component of the Treaty. Ideally, one would assume that these joint technical committee talks that lasted for many years helped in solidifying a factual basis for subsequent political negotiations and building trust and confidence between the two sides. First, these pre-negotiation stages were an important precursor to the broader political process leading to the formation of the Peace Treaty. Many experts argue that they were necessary to promote trust and confidence-building measures (CBMs) and create an atmosphere of mutual respect conducive to working relations (Shamir and Haddadin, 2003). However, the nature of the current relationship between Israel and Jordan can hardly be characterized as one of mutual trust.[35] Although the Treaty normalized the relations between them and prevented a larger armed conflict from occurring, it did little with regards to building real trust and confidence. This issue became more apparent in the lack of long-term negotiations, even when parties experienced disagreement and institutional memory was lost. This is evident by the 1999 drought event that led to a great political drama with finger pointing and accusations on both sides, which was a detriment to the mutual trust between the two co-riparians[36] (BBC News, 1999). Second, offering economic incentives and addressing economic issues through reconstruction and development are key components of the pre-negotiation process.[37] Stabilizing the economy not only tends to mitigate poverty – a major reason to clash – but it also boosts political, economic, and social reforms. This helps in creating opportunities for multilateral dialog and discussion. These economic incentives were overlooked in the pre-negotiation process but were adequately addressed during the negotiation, as the two parties participated in frequent meetings of the Trilateral Economic Committee in Washington, DC, and later in the region itself. Third, these technical talks resulted in a factual basis being laid regarding current water use, future water needs, and deficit.[38] There was some agreement on this factual basis, as both sides were cognizant of their water use and needs owing to the great importance of water in Israel and Jordan. Unlike the Palestinians, who rely far more heavily on water-related data provided by the Israeli side, neither the Israelis nor the Jordanians questioned the validity of the water-related data at this stage.[39]

It is crucial, however, to differentiate between negotiations that are concerned with the technical level, which started as early as 1953 with the shuttle diplomacy, and those that are concerned with the political level, which started in the wake of the International Peace Conference in Madrid in 1991. The timing of the political side of the negotiation was a matter of great concern to the Jordanians. Negotiating and signing an agreement before resolving the Israeli–Palestinian conflict was very difficult and would be frowned upon by other Arab countries. As a result, the Jordanians were very skeptical about publicly finalizing a peace agreement with Israel until the Palestinian issue was resolved, or at least addressed in any meaningful manner. It was not until 1991, namely the Madrid Conference, when the Jordanians felt that an opening was available to broker a peace agreement with Israel.[40] It provided the Jordanians and the Israelis with a legitimate platform for timely and acceptable peace talks that otherwise would have been seen as a betrayal of the Palestinian cause. As soon as the Palestinian–Israeli peace talks were in place, the bilateral talks between Israel and Jordan were well under way. Jordan's negotiation strategy was reflected in Dr. Majali's own words when he explained that "there should be no entertaining of a bilateral peace treaty with Israel outside of a comprehensive peace. We shall therefore urge the Israelis, despite all the current and past difficulties, to make progress in their negotiations with the Palestinians in particular, and on the other two tracks as well" (cited in Haddadin, 2002a).

According to the Israel Ministry of Foreign Affairs (1995), these bilateral talks continued for almost two years in Washington and resulted in the signing of the Israeli–Jordanian Common Agenda on September 14, 1993, only one day after the signing of the Declaration of Principles (DOP) between Israel and the PLO, which constituted the framework for future relations between the Israelis and Palestinians.[41] The handshake between the Israeli prime minister, Yitzhak Rabin, and the Palestinian president, Yasser Arafat, signaled the go-ahead for the signing of the Common Agenda between Israel and Jordan, which was based on the Madrid Letter of Invitation. This agenda, which constituted the blueprint for the Peace Treaty, addressed items related to security, water, refugees and displaced persons, borders and territorial matters (Lillian Goldman Law Library, 2008). The signing of the Israeli–Jordanian Common Agenda marked a giant leap forward for the diplomatic efforts led by the two countries as they prepared to enter into intensive rounds of negotiations and, by extension, for the Middle East Peace Process as a whole. In addition, the Oslo Accords motivated Israel and Jordan to conduct public meetings between King Hussein and Prime Minister Rabin to finalize a durable peace agreement. The first of these public meetings was initiated by the US president, Bill Clinton, in Washington, DC, on July 25, 1994, and led to the Washington Declaration, which was signed by both parties. This had a significant impact on opening direct channels of communication between Israel and Jordan and led to continued diplomatic efforts.[42] These diplomatic efforts continued and culminated in the ceremonial signing of the Israel–Jordan Peace Treaty on October 26, 1994, in the Araba/Arava Valley just north of the cities of Eilat and Aqaba near the Jordanian border[43] (Shamir and Haddadin, 2003).

The negotiation process

Jordan and Israel had been negotiating for about three years before reaching this agreement, which led to conflict prevention and collaborative water management (Fuwa, 2003). The Treaty negotiations, which were political in nature, scope, and focus, were direct (informal and formal) bilateral negotiations with generally no meditation or third-party involvement (Shamir and Haddadin, 2003). However, the early stages of the negotiation process can be said to have had some sort of facilitation from the United States and the Soviet Union, who initially acted to instigate the process and offer the parties informal advice. The diversity of the negotiation groups allowed the parties to consider a broad range of issues that permeated across different topics and, in some cases, combined certain groups that were related to, and overlapped with, one another, such as land/border, water, energy, and the environment[44] (Shamir and Haddadin, 2003). The combination of these overlapping groups was done as a way to rejuvenate the process and bring a more dynamic atmosphere to the negotiation, whenever it seemed to stall.[45]

The Treaty was negotiated over multiple stages of bilateral negotiations that took place in different locations, starting in Madrid, Washington, DC, and later in the Middle East. These stages included 12 rounds of bilateral negotiations, the first of which started at the Madrid Conference while the other 11 rounds took place over a period of two years in Washington, DC.

First: the Madrid Conference – the first round of bilateral negotiations   The first round of bilateral talks was held the day following the opening of the International Madrid Conference of 1991, which was the starting point of open bilateral talks between Israel and Jordan (Sosland, 2007). During these negotiations, three negotiation teams representing the three parties were formed (Israel, Jordan, and the Palestinians), each of which was composed of five members that entered into direct negotiations with little intervention from any third party. According to Haddadin (2008), the impact of power imbalance was neutralized to lead a fair and constructive negotiation. "If this is a military confrontation, I am the wrong guy to talk to; this is not my specialty. This is a battle for peace," he said as he approached negotiation with his Israeli counterparts (Haddadin, 2008). This framing negotiation strategy helped in moving the conflict from intractability to cooperative discourse.

The United States and the Soviet Union acted as the sponsors of the peace process and were involved at the early stage of negotiation as mediating parties to facilitate face-to-face negotiation and form common ground to resolve the conflict (Shamir and Haddadin, 2003). However, as negotiation continued, the role of third-party mediation was diminished and the two parties led purely direct bilateral talks without any sort of mediation or facilitation (Haddadin, 2008). Although the two parties sought some informal help from the sponsors, it was evident that both Israel and Jordan had a mutual understanding and willingness to enter into a direct bilateral negotiation with little intervention from a third party.[46]

The negotiation process seemed sluggish at the very beginning and challenges started to rise as soon as the teams finished delivering their speeches (Haddadin, 2008). The expected outcome and setting of the negotiation was a point of contention between the Arab and Israeli delegations that needed to be addressed and resolved to be able to continue with the other negotiation matters. Although both parties were aware of the importance of normalization, each sought a different negotiation outcome. The Israeli negotiator was very keen to normalize relations, while the Arab negotiator was hesitant to normalize before achieving real success over important matters, such as borders, Palestinian refugees, and water (Daoudy, 2009). Although earlier attempts to agree on a common outcome for the negotiation had so far floundered, the option of mediation appeared to be gaining some support. As the parties failed to agree on a common goal for their negotiation during these bilateral talks, the negotiating parties resorted to the United States and the Soviet Union to find an agreeable solution, who in turn suggested that these direct bilateral negotiations were to be continued in Washington, DC, on December 4th, 1991 (Haddadin, 2008). Both sides agreed to the proposal.

Second: Washington, DC, bilateral talks – the second through twelfth rounds   As the delegations started their preparation for the *second* round of bilateral talks, the Jordanian delegation was expanded to include 14 delegates and nine supporting staff members. As the joint Arab delegation prepared to leave Amman on Monday, December 2nd, 1991, not much collaboration occurred regarding the position or the nature of their relationships during the negotiation. This vaguely defined relationship created a window for conflict and disagreement. The first rift between the joint Arab delegation surfaced as soon as they arrived at Washington, DC, as the Jordanians and Palestinians stayed in two different hotels.[47] As such, the process lacked intra-group consensus building strategy. As the Palestinians were eager to have their own independent representation in negotiating with Israel in separation from the Jordanian track, their participation under the Jordanian umbrella was viewed as undermining their role. Similarly, the Jordanians were also eager to resolve their issues with Israel independently to keep the negotiation simple and focused on Jordanian affairs.

The two groups arrived on time and took their seats in the negotiation wing on the ground floor of the State Department, only to find out that the Israeli delegation had not made it yet.[48] To their surprise, the Israeli group arrived a week later, on Sunday, December 8th. This incident had several significant implications. Not only did it send a message about Israel's disinterest and lack of enthusiasm to enter into talks that might lead to a negotiated settlement with the Arabs, but it also had a negative psychological impact on the Arab delegation and the negotiation as a whole. The impact of this incident on decisions pertinent to the timing and the format of negotiation became apparent as soon as the Israeli delegation arrived and even before the formal negotiations started. Another time was scheduled for the meeting and, this time, all negotiating teams were able to make it on time. Disagreement, however, surfaced again regarding the meeting time.[49] Nonetheless, the groups began the second round of the bilateral talks on Thursday, December 10th, 1991, which continued for a week.

In the absence of neutral, equidistant third-party mediation, the negotiation was challenged by the issue of representation and the format and setting of the negotiation. Israel proposed to divide the negotiation based on the topic of discussion. The Israeli proposed format for negotiation would result in two tracks, one Israeli–Palestinian and another Israeli–Jordanian, within which both Jordanian and Palestinian groups would be involved.[50] The approach suggested by the joint Arab delegation was to have two separate tracks; one Palestinian and another Jordanian. Their proposal also recommended that one or two Palestinian members attend the Israeli–Jordanian negotiation and one or two Jordanian members attend the Israeli–Palestinian negotiations, without any Jordanian intervention in the Palestinian affairs.[51] This divergence of the negotiation process hindered the negotiation teams' ability to reach an agreement over the format of the negotiation even before entering the negotiation room. Despite their efforts, albeit futile, the parties failed to reach an agreement on this matter. This indicates that the parties missed a chance to build trust and confidence, which seemed problematic at the beginning. In a subsequent meeting with Ambassador Richard Armitage, the Jordanian water delegate discussed different topics including ways to narrow the gap between the two parties, the Syrian dam building activities that were in violation of the 1987 Jordanian–Syrian Agreement, and the role that water could play in the upcoming multilateral conference in Moscow[52] (Haddadin, 2002a).

The *third* round of bilateral negotiations started on January 13th, 1992, although scheduled for January 7th.[53] During the four-day period of this round, the parties were able to overcome the aforementioned major obstacles incurred during the second round of bilateral negotiations.[54] Meanwhile, it was agreed that a general meeting involving all parties (13 Israeli representatives, 11 Jordanian representatives, and eight Palestinian representatives) would be conducted to discuss issues of mutual interest (Haddadin, 2002a).

However, according to Haddadin (2008), no water negotiations were allowed in the bilateral Israeli–Palestinian track of negotiation, as water was deferred to the final status negotiations[55] (Shamir and Haddadin, 2003). As indicated later in the first paragraph of Article 40 "Water and Sewage" of the 1995 Oslo II Agreement, "the Palestinian water rights in the West Bank . . . will be negotiated in the permanent status negotiations and settled in the Permanent Status Agreement relating to the various water resources" (The Israeli–Palestinian Interim Agreement on the West Bank and Gaza Strip, 1995). For that reason, the Palestinians were eager to discuss water-related matters (a bilateral negotiation item) during the multilateral negotiations, which created a clash between the Israeli and Palestinian delegations in many multilateral tracks (Haddadin, 2002a).

On the Israeli–Jordanian track, both parties presented a different "vision of peace" in their attempt to jointly adopt an agreed upon formula for the negotiation agenda. The process failed in adopting an effective negotiation strategy of setting a common meta-goal by reframing the issue as a shared problem and opportunity for peace. A wide gap persisted between the two proposals and the negotiation seemed to be approaching a halt. A Jordanian proposal, which was discussed and agreed upon by both parties, suggested combined informal talks on water, energy, and the environment.[56] This strategy not only helped in easing the tension, integrating different topics, and discussing issues of mutual interest, but it also enabled the teams to be aware of the negotiation progress of the other groups.[57] They reached an agreement on preserving and enhancing the quality of the shared water, namely bringing the Yarmouk and Jordan rivers to acceptable standards that "allow their unrestricted agricultural use," by prohibiting the disposal of untreated municipal and industrial wastewater into the courses of these rivers, as was later indicated in Article 3 of the Peace Treaty (Shamir and Haddadin, 2003).

The *fourth* round of bilateral negotiations, which started on February 24th and continued until March 4th, 1992, picked up from where the third round had left off by focusing on drafting a mutually agreed upon agenda for negotiation. The parties decided to identify agreed upon items and separate them from the items of contention. A "Single-Text Negotiation," which was developed almost entirely by Dr. Haddadin, was used for this purpose, where the parties used one text to reconcile their disagreements[58] (Shamir and Haddadin, 2003). In this Single-Text Negotiation (STN), no specific water shares were suggested and instead these were subject to negotiation (Haddadin, 2008). The parties utilized a simple technique to keep the agreed upon text untouched, by providing their own suggested text side-by-side and using square brackets to highlight points of contention, which were also highlighted in bold text (Shamir, 2008). This provided the parties with an efficient way to reconcile and validate their different positions.[59] Another efficient technique that was utilized during the different stages of negotiations was the use of "non-papers" as a way for each party to propose their ideas and needs to the other party for consideration (Shamir and Haddadin, 2003). Once these items were agreed upon by the other party, they were added to the Single-Text Negotiation (Shamir, 2008). The combined expert group on water, energy, and the environment mutually agreed on the following text for the water item of the Common Agenda:[60]

3.   Water
Securing the rightful water shares of the two sides in the Yarmouk River and the Jordan River.
Searching for ways to alleviate water shortage.

(Haddadin, 2002a: 293)

Unlike the Jordanians' proposed text, which focused on the rightful share of the co-riparians, the mutually agreed upon text only considered the rightful share of the two sides in reference to Israel and Jordan, disregarding other co-riparians who also share the river basin. For that reason and the pronounced exclusion of other parties' mutual interest, the text was kept from the public.[61]

The *fifth* round of bilateral negotiations started on April 20th, 1992, at the Willard Hotel, during which Jordan drafted a more detailed proposal of the Common Agenda, which was handed to the Israeli side for negotiation and approval.[62] Negotiation over water constituted a significant part of this round, as the delegations laid out their intentions for the water negotiations in their opening statements. The text of the Common Agenda on water introduced the term "rightful water shares" without much elaboration of what this term would really mean in practice. When asked about the meaning of the term "rightful water shares" for Jordan, Dr. Haddadin explained that "a compromise solution was reached in 1955 through the American envoy, Ambassador Eric Johnston. Jordan committed herself to her share under that compromise, and Israel did likewise as a condition to have her water projects in the basin financed. We do not intend to turn our backs on commitments we made unless Israel turned its back on her commitments first, in which case we will have to start from the very beginning" (Haddadin, 2002a: 293–294). He further argues that Jordan's negotiation strategy over water was to maintain the same water share in 1955, including both the East Bank and the West Bank[63] (Haddadin, 2008). The Jordanian negotiator thought that the Johnston Plan was a good start for the water negotiation, whereas the Israeli negotiator was less than enthusiastic to accept the offer (Elmusa, 1995). Nonetheless, the informal negotiation ended with some progress on the drafting of the items of the Common Agenda.[64]

The *sixth* round of bilateral negotiations took place from August 24th to September 23rd, 1992, and witnessed substantive progress (Haddadin, 2002a). This is because certain political dynamics of the Middle East had changed and a more conducive atmosphere for peace was in sight, as the Labor Party (led by Mr. Yitzhak Rabin) won the Israeli election.[65] The negotiators were able to capitalize on the good relations between Mr. Yitzhak Rabin and King Hussein of Jordan to achieve progress in their negotiations. Although progress was made on the water item of the Common Agenda and the issue of the Dead Sea, which were proposed by the Jordanians and updated and approved by the Israelis, other items (namely security, refugees, and the occupied Jordanian territories) remained unsettled (Haddadin, 2002a). The Syrian withdrawal from the Yarmouk was of concern to the Israeli negotiators, who expressed their discontent at the bilateral nature and outcome of the 1987 Jordanian–Syrian agreement. Dr. Haddadin was quick to point out that the bilateral nature of the negotiation was an Israeli preference and that the 1987 Jordanian–Syrian agreement, which was predominantly concerned with the Wehda Dam, had not been implemented at the time (Haddadin, 2002a).

The issue of Palestinian water rights was briefly discussed, followed by a detailed discussion of the quantity and quality of water from the Yarmouk allocated to each party in accordance with the Johnston Plan. In this discussion, the Israeli side downplayed the relevance of the Johnston Plan and was less than willing to accept it, whereas the Jordanians were eager to utilize it as the basis for their negotiations[66] (Elmusa, 1995). However, evidence suggests that the two parties reached an implicit consensus over the relevance and use of the Plan as the background for negotiations (Elmusa, 1995). As the Israeli delegation was eager to discuss matters related to the environment and energy, the Jordanian delegation was interested in discussing water-related matters.[67] This presented an obstacle but also an opportunity to employ a key negotiation strategy of trading across issues that were deferentially valued. After extensive rounds of negotiations, the Israeli side submitted two documents ("non-papers") to the Jordanian delegation. The first document confirmed the agreed upon amount of flow from the Adasiyya diversion point into Israel (which is 30 percent of the summer flow), whereas the second proposed the construction of a temporary diversion weir at Adasiyya to improve the diversion efficiency (Haddadin, 2002a).

The *seventh* round of bilateral negotiations started on October 21st and lasted until November 19th, 1992.[68] As this round coincided with the first anniversary of the Madrid Conference, the parties were eager to achieve significant progress. Although the Syrian and Lebanese tracks experienced deterioration, the Israeli and Jordanian delegations focused on drafting the Common Agenda items, which included two sections. Section A outlined the intent and purpose of the negotiations and Section B outlined the nine "Components of Jordan–Israel Peace Negotiations."[69] After extended discussion, the parties approved the Common Agenda and agreed to keep it away from the press.[70] This round saw another important development, which was the replacement of Dr. Haddadin on the bilateral delegation by Omar Dokhgan and Mohammad Keilani.[71] Although the real reason why this change occurred is largely unknown, some tie it to the Jordanian leaders' dissatisfaction with the way the negotiation process was conducted and the results it produced for Jordan. Nonetheless, this change was a strategic move on the part of Jordan to inject the stagnant negotiation process with a more fluid dialog through the appointment of new negotiators.

The *eighth* round of bilateral negotiations was short and the negotiation was impacted by Israel's deportation of 300 Palestinians from the West Bank and Gaza to Marj al-Zuhour in South Lebanon. The issue of Jerusalem remained a subject of subsequent discussion and refinement on the Common Agenda, which, in spite of its approval, was not officiated until progress on the Palestinian track was to be achieved. As the delegations prepared for the upcoming negotiation rounds, Dr. Haddadin was reinstated on the bilateral delegation. The delegation conducted four meetings to formulate Jordan water negotiation strategy, which was reached on April 4th, 1993[72] (Haddadin, 2002a).

Further informal bilateral negotiations – namely water, energy, and the environment group – continued during the *ninth* round of bilateral negotiations, held on May 3rd, 1993, in which the Jordanian delegation laid out their negotiation proposal and priorities. These included surface water basins, shared groundwater aquifers, alleviation of water shortage with efficient management, potentials for regional cooperation within a regional context, and the Jordan Rift Valley development[73] (Haddadin, 2002a). Both Jordan and Israel presented and discussed quite conflicting negotiation priorities.[74] The negotiation teams agreed to establish three working groups[75] (without giving priority to one over the other). The Israeli side accepted (with no substantive amendments) most of the items in the Jordanian proposal for negotiation, but objected to certain items, namely the rightful water shares for both parties on the Jordan and the Yarmouk (Haddadin, 2002a).

In the subsequent *tenth* round of bilateral negotiations, held on August 31st, 1993, the Israelis agreed on the proposed items and further proposed to change the text from "the rightful shares for the two parties" to "the rightful shares for all parties," which reflected the original Jordanian proposal. This new language was negotiated and sought by the parties as a way to safeguard the negotiation from any adverse impacts that could ensue as a result of using the alternative, commonly used term "water rights," which signifies an argument of entitlement and could potentially sabotage the negotiation.[76]

The *eleventh* round of bilateral negotiations took place in September 1993, and focused on the Common Agenda, which was ready for months awaiting a significant breakthrough in the Palestinian–Israeli negotiation.[77] The Israel–Jordan Common Agenda was signed and initiated by the US State Department on September 14th, 1993, one day following the signing of the Oslo Accord between Israel and the PLO. As it seemed easier for Jordan to proceed on its own in separation from the Palestinian delegation, Jordan decided to disengage from the Palestinian negotiation team. This decision was based on King Hussein's request to disengage, excluding the Palestinians from involvement in such a historic deal.[78] According to Haddadin (2008), this constant involvement of the king throughout the course of negotiation provided the kind of leadership that set the political direction for the negotiation. He further adds, "we did not want to interfere in other parties' affairs, namely the Palestinians . . . so I negotiated on behalf of Jordan not Palestine." As such, Israel negotiated two quite separate tracks: Jordanian–Israeli and Palestinian–Israeli, which remained to be conducted in separation from one another from this point on[79] (Shamir, 2008). During the *twelfth* round of bilateral negotiations, group A met on February 14th, 1994, in Washington, DC, under the umbrella of the Trilateral Economic Committee. Progress was made and agreement was achieved on the sub-agenda for negotiation on water, energy, and environment.[80]

Third: the Middle East bilateral arena The negotiations took a more substantive and serious turn as they moved to the Middle East. The *first* round of bilateral negotiation took place on July 18th, 1994, in Wadi Araba, a neutral location between the two countries.[81] Although there was some tension, as this was the first meeting in the region, both delegations were very optimistic. This was because a promising and much anticipated set of historic events were about to take place, namely the upcoming Trilateral Economic Committee and the White House meeting between King Hussein and Prime Minister Yitzhak Rabin under the sponsorship of President Bill Clinton. These events provided an atmosphere that was more suitable for negotiation and a great momentum for the peace process. As the delegations and various expert groups started to discuss items on the

Common Sub-agenda, the difference in negotiation strategies between the two parties surfaced again in the meeting of group A on water, energy, and the environment. Jordan wanted to negotiate water-sharing issues, whereas Israel wanted to focus on a water transfer project across the shared border. The next day, the Trilateral Economic Committee convened at the Dead Sea Spa Hotel in the Rift Valley.[82]

The Washington Declaration was a major cornerstone that set the stage for achieving peace between the two countries.[83] A rich round of negotiations was held in August at the Moria Plaza Hotel on the Dead Sea, where progress over water-related matters was attained (Shamir and Haddadin, 2003). In this round, Jordan emphasized the independence of the Palestinian negotiation, affirming that any agreement between Israel and Jordan should not impact the water rights of other co-riparians.[84] The parties were able to tackle water-related issues in much detail. However, they disagreed regarding quantities of water sharing.[85] Without the work of the combined expert groups, it was difficult for the parties to reach a consensus over figures of current use and future requirements with such divergence in their contradictory positions. Because water can be closely related to other issues, such as the environment, land, and borders, this strategy of combined expert groups was successful in easing the tension and focusing on interests rather than positions[86] (Shamir and Haddadin, 2003).

First, the negotiations benefitted from the water–environment nexus by mutually considering issues of water quality and energy that impacted both sides. As the two sides continued their meetings at the Dead Sea Spa Hotel, they were able to discuss water development projects, such as the Red Sea–Dead Sea Canal and the Mediterranean–Dead Sea Canal.[87] Second, the negotiation benefitted from the water–land (border) nexus with regard to the issue of the shared border between the two countries.[88] Mr. Kinarti, alluding to Israel's intention to trade water for land, explains, "the solution for water in the north and south will be related to issues of borders" (cited in Haddadin, 2002a). To that end, Jordan claimed sovereignty over a small farmland called the Baqura, later called the "Peace Island," east of the Jordan River, which previously was under Israeli control and was used by Israeli farmers (Shamir and Haddadin, 2003). Similarly, Jordan claimed sovereignty over a stretch of land in the Araba/Arava Valley, where Israel drilled 14 wells for irrigation purposes.[89] As land was returned to Jordan in the Baqura and the border was moved westwards into the Araba/Arava Valley, Israel, according to the agreement, maintained its use of both locations. Farmers were allowed to cultivate the Baqura for 12 years, and continued to utilize the wells in the Araba/Arava Valley. Israel, according to Article 4(1) of the Agreement, would maintain the use of its existing wells on the Jordanian side of the border and would further increase its uses by up to 10 MCM over five years, according to Article 4(3) of the Agreement (Fischhendler, 2008a,b). Therefore, both parties' interests were satisfied: Jordan was able to claim its sovereignty over the land, whereas Israel was able to use the land in the north and the water in the south (Shamir and Haddadin, 2003).

On the political level, Israel and Jordan drafted a proposal and counter-proposal respectively of a draft peace treaty on September 26th, 1994 (Haddadin, 2002a). Although surprised to know that a peace treaty was in the making, Dr. Haddadin proceeded to prepare a draft of the Water Article to be incorporated into the Treaty, which was reviewed and accepted by the Israeli side during their negotiation in Aqaba on September 29th, 1994 (Haddadin, 2002a). Aiming to exclude language of controversy and referring to a more detailed Annex, this part of the Treaty was written in general terms, leaving specific details regarding water allocation to be discussed and mutually agreed upon in upcoming rounds of negotiation. Little progress, however, was accomplished with regard to controversial political issues, such as the issue of refugees and displaced persons and shared borders, which remained unresolved until the leaders of the two countries, King Hussein and Prime Minister Rabin, arrived in Aqaba and joined the negotiation. Given the good relations between the two leaders, their involvement was crucial in negotiating and reconciling these political issues and opening the door to a comprehensive peace treaty that included water-related components.

As the political atmosphere seemed conducive, the parties were able to capitalize on the involvement of the two leaders and readily move ahead with a treaty. However, the details and whereabouts of the water allocation component of the Treaty were not complete yet. At that point, realizing the importance of taking advantage of this political breakthrough, the Jordanian legal expert and negotiator, Awn al-Khasawneh, asked Dr. Haddadin to immediately write the details of the Water Annex of the Treaty, with little time given to him to do so. "You have to prepare the Water Annex to the Peace Treaty within 24 hours," demanded Mr. al-Khasawneh. Dr. Haddadin, who was taken by surprise, asked, "Who said so?" "The Crown Prince," answered Mr. al-Khasawneh. Dr. Haddadin explained, "But I am not carrying any documents related to the subject, no references." Mr. al-Khasawneh responded, "You have composed all the documents, you do not need references" (Haddadin, 2002a: 375). A day later, on October 3rd, 1994, the draft Annex II – "Water Related Matters" – was completed. The development achieved regarding water issues was not as quick as the sudden political progress that took place in the wake of the leaders' meetings.[90]

Starting on October 10th, 1994, the last stretch of "hard" negotiations was held in Eilat, Aqaba, and Amman with the supervision of Prince El Hassan (Shamir and Haddadin, 2003). The goal was clear: to reach a comprehensive peace treaty and end hostility and bloodshed. As the two delegations tackled topics related to bilateral negotiations, item six on the Common Agenda, more serious issues related to water allocation, storage, groundwater, and water quality remained unresolved.[91] After a sudden meeting at Hashimiyya Palace in Amman between the two leaders and other high-level officials, which was a great shock to the uninformed and uninvolved Jordanian water negotiators, the Israeli delegation was able to offer the Jordanians a satisfactory proposal on water. This included a three-component water package (50 MCM each) and a minor adjustment to the border, which, according to Haddadin (2002a), was well received by the Jordanians. This offer was acceptable as it was combined with the border adjustment, which constituted a strategic gain for Jordan. Although the sources of the two water components were known (the first from the current Israeli use; the second from the joint dam construction), the third was undetermined. Nonetheless, the deal, which could include an additional 50 MCM and border adjustment, received qualified backing from the Jordanian negotiator and support from the Jordanian high-level officials.[92]

During these last meetings, the two sides were able to resolve their differences in separation from the larger political issues and other regional parties, namely Syria and the Palestinians. The negotiating teams were cognizant of the difficulties in excluding these crucial regional players; former US Secretary of State, Henry Kissinger, astutely noted that in the Middle East "no war is possible without Egypt, and no peace is possible without Syria" (cited in Daoudy, 2008). Although critical to the regional water allocation and the larger peace process, Syria was not involved in the negotiation of the Treaty between Israel and Jordan. This was a particularly tactical negotiation strategy, not to undermine the role of Syria but rather to avoid complications and further stalemate due to the current Syrian water utilization of the Yarmouk and the issue of the Golan Heights.[93] However, the uncertainty surrounding the yield and hydrologic characteristics of the Yarmouk imposed by the uncontrolled Syrian utilization caused difficulties during the negotiation apropos of predicting and managing the river flow (Shamir, 1998; Haddadin, 2002a). Similarly, the Treaty deliberately made no mention of the segment of the Jordan River bordering the eastern side of the Palestinian West Bank, since there was no agreement over the issue of the Palestinian water rights (Shamir and Haddadin, 2003). According to the Jordanian negotiator, the interests of the Palestinians were in fact considered in the bilateral Israeli–Jordanian negotiation but were deliberately postponed to later rounds of negotiation (Haddadin, 2002a). It is, however, clear that dividing the Jordan water between only Israel and Jordan would inevitably determine the shares of both Syria and Palestine, even though no agreement had been reached regarding their specific shares in the bilateral Israeli–Jordanian negotiation. In addition, the exclusion of both Syria and Palestine created additional challenges for the acceptability of the Treaty to the general public.

The issue of arbitration was raised by the Jordanian team as an option to be considered in the Treaty. Although it had used arbitration to settle its dispute with Egypt on Taba, Israel did not see arbitration as a practical or relevant tool in addressing their conflicts and rejected the proposal altogether[94] (Haddadin, 2002a; Shamir, 2008). However, post-settlement contingency is identified by the Treaty in Article 29, entitled "Settlement of Disputes," which stipulates that any disagreement regarding the application or interpretation of the Treaty will be resolved amicably via negotiations (Haddadin, 2008) and "any such disputes which cannot be settled by negotiations shall be resolved by conciliation or submitted to arbitration" (Treaty of Peace, 1994: Article 29).

The two sides, however, managed to rectify the issue of refugees, which was one of the most difficult negotiation matters.[95] The friendly relations between the two leaders again smoothed the way of negotiation whenever conflict reared its head, namely with regards to the border modification, the additional 50 MCM/year, and ownership *vis-à-vis* the operation of wells in Wadi Araba. Of particular importance was the statement made by Mr. Rabin to King Hussein that "Israel is not interested in one drop of Jordan's water" (Haddadin, 2002a), which gave the Jordanian team enough motivation to pursue an acceptable settlement, and the Israeli team a frame of reference in their negotiation. As pressure mounted to conclude the Peace Treaty, other issues remained unresolved, such as the exact wording regarding the border between Jordan and the West Bank, which, if accepted as an international border, would then legitimize the occupation of the West Bank. Regardless, the water agreement was finalized and initialed by Munther Haddadin (the Jordanian negotiator) and Noah Kinarti (the Israeli negotiator) in the Hashimiyya Palace in Amman on October 17th, 1994 (Haddadin, 2002a). On the same day, Mr. Rabin, prime minister of Israel, and Dr. Abdul Salam Majali, prime minister of Jordan, in the presence of King Hussein and the press corps, initialed the Treaty[96] (Haddadin, 2002a).

Challenges to negotiation   The makers of the Treaty were puzzled by numerous challenges during the negotiation, as was evident by the Treaty outcome and current hydro-political circumstances which have persisted and continued in the post-negotiation stage. First, the issue of the Palestinian water rights, which was deferred to the final round of negotiations, remained an issue of concern to the Jordanian negotiators.[97] Ratifying this Treaty between Israel and Jordan has meant that any future deal between Israel and the Palestinians will most likely exclude Jordan from the negotiation, which will be of a bilateral nature. Second, the fact that the Jordanians conducted their negotiations with Israel on the basis of the Unified (Johnston) Plan and the principles of international law appeared to be problematic for them. Reducing the current Israeli use of the Yarmouk from 75 MCM to 25 MCM in accordance with the Unified (Johnston) Plan and convincing Israel to accept the terms of international law were nearly insurmountable tasks. In addition, the Plan was developed for a different situation, predicated on a set of different assumptions, and was meant to offer solutions to serve different purposes.[98] This created a stumbling block for Jordan as it tried to make the argument for acquiring water for municipal uses in Amman and augmenting its share for irrigation. Although the Plan allocated water for the West Bank and East Bank together as one entity in the original Jordanian water share, Jordan had to forgo the share of the West Bank, since it is currently a separate entity, to be able to receive its own share, according to the Johnston Plan water-sharing formula (Haddadin, 2008). To that end, Jordan separated the shares of the East Bank (479 MCM) and West Bank (275 MCM) (Haddadin, 2002a; 2002b).

Israel on many occasions tried to deny the applicability of international law.[99] Shamir (2008), exemplifying and explaining the Israeli position of evading international law, contends that it "does not provide a template for resolving issues." Advocating a more pragmatic approach to conflict resolution that is based on needs only, he further explains that "the first thing" to be incorporated into binational water negotiations to move from a non-cooperative to a cooperative state and enhance the fairness of the outcome and balance the agreement "is to agree from the very outset that the agreement must be pragmatic and practical, rather than theoretical or conceptual (by relying on international law)." However, Zeitoun (2008b), akin to many others, emphasizes the importance of international water law as the basis for negotiation, as he believes that, since needs change with time, they are not deemed to be an adequate basis for water allocation. Instead, he argues that water negotiation can and should be governed by principles of international water law as they define what is equitable and reasonable as far as water allocation is concerned and indirectly address the issue of water needs as well. In addition, in the absence of international law, which can govern and manage the parties' behavior and neutralize power imbalance, the most powerful player will always win the game (Haddadin, 2002a). This divergence in the positions of both countries regarding the issue of international law applicability created a rift and caused the two parties to, throughout the course of their negotiation, frequently revert back to the same talk-fight dynamics that had led them to clash in the first place.

Third, the unregulated Syrian water use of the Yarmouk made it hard for Jordan to anticipate and plan for the amount of water available for Jordanian extraction and storage.[100] Therefore, rendering a useable share for Jordan meant negotiating an alternative storage location with Israel that could be used as a temporary storage facility (Lake Tiberias) until the issue of the occupied Golan Heights has been resolved (Haddadin, 2002b). However, using Lake Tiberias as a storage facility was not brought up in the negotiation, as Israel refused this arrangement prior to the 1967 War. Fourth, the issue of groundwater in Wadi Araba constituted a major challenge (Fischhendler, 2008a). In these locations, Israel drilled wells that were being extracted for agricultural purposes, even though this land was to be handed over to Jordan. As such, the challenge for the two parties presented itself in Israel's need to continue its utilization of these wells and farms, notwithstanding the Jordanian unfettered control of the land. Fifth, because of Israel's objection since the 1960s, Jordan was unable to construct the diversion weir on the Yarmouk at Adasiya. Without this weir, Jordan will be unable to capture its "rightful share" of the Yarmouk, which will be lost to Israel as it flows into the Jordan River. Another challenge is related to the water quality of the Lower Jordan River. The fact that Israel discharges municipal waste water and diverts the saline springs into the Lower Jordan River caused the degradation of the water quality of the river (Elmusa, 1995). Because of the lack of adequate river flow, accompanied by the non-point source pollution from the agricultural return flow, restoring and preserving water quality of the system becomes even more challenging.

These challenges still remain unresolved, as both sides failed to collaboratively address the issue of the Syrian water allocation in a strategic fashion, disregarded the pending Palestinian water crisis, and solidified and legitimized the *status quo ante* with regard to Israel's water use. The strong persistence of the Israeli negotiator to maintain Israel's current water use, if not acquire more water, acted as a disincentive to the Jordanian negotiator, who in turn insisted on securing only Jordan's "rightful share" of water.

Post-negotiation and Treaty implementation

The post-negotiation stage includes assessing, facilitating, and overseeing implementation and enticing all parties to maintain continuous relations with each
other. Once an agreement is reached and formalized, the parties must develop
mechanisms for facilitating implementation and strategies to help put them to use
thereafter. This calls for an ongoing assessment process. Developing a monitoring
strategy and performance measures to ensure successful implementation of the
agreed upon plans, revisiting the issues, and continuing cooperation must all be
part of an ongoing assessment process. Except for the establishment of the Joint
River Committee, there seems to be no indication of any arrangements of this sort
apropos of the post-negotiation process in the Treaty.

The establishment of the Joint Water Committee (JWC)   The importance of this
Treaty stems primarily from the fact that it formally provided an end to the water
conflict between Jordan and Israel through bilateral negotiations (Haddadin,
2002b). The Treaty offered a viable shift from the secret informal water institution, which included elements of conflict resolution and management, to a formal
water institution (Zawahri, 2004). As specified by Article 6 of the Treaty, the two
parties agreed to establish a Joint Water Committee to ensure future cooperation
on data exchange, monitoring water quality, mitigating appreciable harm, and
advancing treaty implementation[101] (Haddadin, 2002a). The JWC is composed
of three members from each country and seems to be less formal as it formulates
its own agenda, procedures, and ground rules (Shamir and Haddadin, 2003). The
committee convenes as necessary and alternates its location between the two
states' major cities[102] (Zawahri, 2004).

Water became of particular significance to both countries as the region experienced periods of summer water shortages in 1994–1996. The impact of the water
shortage crisis, which was the result of climate change, the increase in upstream
water consumption, and the reduction in water availability in the Yarmouk by
an estimated 50 percent of the total average flow was exacerbated by the severe
water shortages in Amman, where summer water supply was intermittent and
confined to twice a week (Biswas and Bino, 2001). Because of the issue of destructive ambiguity[103] that the Treaty seems to suffer from, as will be discussed further
in the content analysis section of this chapter, the implementation of the Treaty
was likely to be challenged by the issue of uncertainty and contradictory interpretations of its vague terms. For example, the acquisition of the third component
of the Jordanian water package remained unclear even after the ratification of
the Treaty, although it was implicitly known to the negotiators that Israel would
provide this additional water (Elmusa, 1995). This posed a tremendous challenge
for the JWC during the very first year of implementing the Treaty, as details of
where and how this additional amount could possibly be acquired and transferred
to Jordan remained unclear, especially when there were not any other unexploited
shared water resources in the region. This challenge presented itself very profoundly as the acute water shortage in Amman worsened and more water was

needed. This demonstrates the gap between agreements in principle, which might seemingly appear to be successful, and agreements in practice, which can be profoundly problematic as a result of the ambiguity of such practical matters.

The source and financial configuration of this water were of particular importance. For the source of this water, Jordan sought to divert water from the Lower Jordan River, whereas Israel insisted on the desalination of the Mediterranean water, the cost of which was to be covered by Jordan.[104] The proposal, which suggested the desalination and transfer of this water to Jordan, was opposed and rejected by the Israeli end users (summary of the discussion from the meeting of the Steering Committee for Peace Talks, July 4, 1996). The work of the JWC, which was formed to address these implementation issues, seemed unproductive and was further handicapped by the impasse on the additional 50 MCM in the north. This also impacted the implementation of other water transfers elsewhere, namely in Wadi Araba/Arava in the south (Haddadin, 2002a; Shamir and Haddadin, 2003). Throughout the intermittent meetings of the JWC, the two sides continued to offer solutions that were not acceptable to the other side.[105] As they were unable to resolve the issue at hand, this additional 50 MCM that was stipulated by the agreement was not delivered to Jordan, which caused further tension (Balkind and Ben Meir, 1996).

With the futile attempts to resolve the additional water issue, coupled with growing water stress and the political turmoil and deterioration in the two countries' diplomatic relations, the two sides realized the importance of working together toward the further implementation of a number of difficult items in the Treaty (Haddadin, 2002a; Fischhendler, 2008a). Several joint ministerial meetings between the two countries were held to address this "mini-crisis," as it was described by the Israeli prime minister (Fischhendler, 2008a). However, these meetings ended with no resolution in sight and caused further deterioration of the two countries' international relations (Haddadin, 2002a). A temporary resolution was reached and an interim agreement was signed, through the mediation of the US Ambassador, Dennis Ross, to provide Jordan with 25 MCM from Lake Tiberias for three years until a desalination plant could be constructed[106] (Makovsky, 1997; Fischhendler, 2008a). However, the construction of this desalination plant never materialized, due to the disagreement on the proportions of the cost sharing of its construction and operation. In this regard, Jordan agreed to the idea of a Jordanian contribution proportionate to its GDP (Haddadin, 2008), whereas Israel promoted an equal cost-sharing proposal (Fischhendler, 2008a). As a result of this disagreement, Israel continues to bear the responsibility of transferring the 25 MCM amount to Jordan until today (Fischhendler, 2008a). However, as these issues remained unsettled, the 1998/1999 drought had a major impact on the political and diplomatic relations between the two countries, as will be discussed in the implementation assessment section below.

Treaty implementation   The Treaty so far has sustained, for the most part, robust implementation on both sides. A major problem with the Treaty is that its volumetric water allocation figures were not established based on the actual current and future needs of either nation. These figures were solely supply-based and did not consider the demand side. They were developed based on a steady state (the average carrying capacity of the river), disregarding times of drought and the variability of the river flow from one year to another. This is likely to create implementation roadblocks, as these droughts are bound to happen in the future. The Israeli negotiator argued that it would be controversial to allocate water based on the demand side because of the difficulty in projecting the future population of Israel, which receives fluctuating numbers of immigrants from other countries each year. However, many experts believe that the demand-side management is an important negotiation issue that needs to be addressed. Relying merely on a supply-side management strategy on the basis of convenience (and the belief that the matter is too complicated to address) does not constitute adequate grounds for disregarding the demand-side management during negotiation. "That's what the negotiation is for; to address these complicated issues," Zeitoun (2008b) explains in response to the Israeli negotiator's argument. Both aspects need to be addressed in light of the fact that water resources are finite in the region. This remains a question of interest and political will.

In spite of the fact that the two parties still abide by the terms of the Treaty, the implementation of the Treaty seems to be problematic and challenged along various aspects. Jordan in 1999, almost five years after signing the Treaty, received an estimated additional water of 50–80 MCM per year, which was less than the equivalent of 10 percent of its annual water demand and about a third of the water that Jordan expected in the wake of the Treaty (Innes and Booher, 1999). Because of the drought of 1998/1999, the implementation of the Treaty took a dire turn when Israel was unable to transfer the agreed upon amount of water from Lake Tiberias to Jordan. In the wake of this severe drought, marking the expiration date of the three-year interim agreement, Israel's water commissioner, Meir Ben Meir, announced his country's inability to transfer this 25 MCM amount to Jordan (Fischhendler, 2008a; Shamir, 2008) and further demanded to renegotiate the water amount that it was required to deliver to Jordan. This was an expected result of the lack of shortage sharing mechanisms, as the Treaty did not provide drought management clauses. The fact that the Jordanian water system, unlike that of Israel, does not include storage facilities to accommodate the balance of the flow makes it hard for Jordan to extract the amounts of water allocated by the Treaty and further makes Jordan more vulnerable to droughts. As a result of this, Jordan became responsible for making decisions regarding water storage and conveyance, which could require the costly construction of additional facilities to capture the rest of the flow that could dramatically vary from one year to another. There were also major quality-related problems with the water that Israel transferred to Jordan, which continued to deteriorate over time, impacting the local end users, namely in Amman (Phillips, 2008). This political miscalculation caused unsettling political drama and major tension between the two countries.[107] The mounting political pressure and the Israeli–Jordanian cooperation eventually forced Israel to abide by its water transfer obligation identified by the Treaty. In a summit meeting in Aqaba in 1997, the parties agreed that Israel would provide

Jordan with 25–30 MCM of the water stored in Lake Tiberias until a desalinated plant became operational (Shamir and Haddadin, 2003).

Implementation was also challenged by the fact that the source and the financial arrangement of the additional 50 MCM to be delivered to Jordan was unknown and undetermined by the Treaty. Details of the sources and cost sharing of the other two 50 MCM/year components were clearly defined by the Treaty: the first component was to come from existing resources, whereas the second component was to come from sources to be developed by the parties (Shamir and Haddadin, 2003). According to the Treaty, Israel and Jordan should cooperate to secure this additional 50 MCM/year of water for Jordan. It further stated in Annex II, Article 1(3) that, "to this end, the Joint Water Committee will develop, within one year from the entry into force of the Treaty, a plan for the supply to Jordan of the above mentioned additional water" (Treaty of Peace, 1994).

The gleeful celebration of the signing of the Treaty quickly turned to frowns when Israel failed to deliver on its obligation in the very first year. With the failure to implement this part of the Treaty, Jordan asked that Israel should deliver this amount from Lake Tiberias, but Israel insisted that this amount should come from both countries and that Jordan should bear the full cost of the operation (Haddadin, 2002a). With the divergence in the parties' conflicting positions, the issue remained unresolved until 1998 after the assassination attempt of Khalid Mishal in Jordan. In an attempt to contain and resolve the issue with Jordan, Prime Minister Ariel Sharon agreed to use Lake Tiberias to store 60 MCM to provide for Jordan's share of the Yarmouk floods (Shamir, 2008). Haddadin (2002a) explains that this was done in compliance with Annex II of the Peace Treaty and therefore did not change the water allocation of the Yarmouk made by the Treaty. Although many skeptics and experts continue to join the critics who accuse the Treaty of being unfair to Jordan, the Treaty implementation is still in place. Except for the incident of the 1999 drought, the implementation of the Treaty has continued to operate unshaken since then as agreed, despite the equally severe drought the region recently faced.

Content analysis

This Treaty offered a detailed and comprehensive agreement intended to settle water-related disputes and other political issues important to both parties to end a long period of animosity and war-like relations. The Treaty provided water-related stipulations included in the Preamble and in the water and environmental annexes, namely Annexes II and IV respectively. Incorporating a wide range of topics, the Treaty introduced provisions to manage the allocation of shared water, while fostering cooperation regarding issues of data acquisition and dissemination, the construction of storage facilities, the protection of water quality, and the formation of a joint water committee to facilitate implementation and management of the shared water system (Elmusa, 1995). It also provided a framework for future water policies regarding the sharing of the Jordan River, the Yarmouk, and the Araba/Arava groundwater. Article 6, the most significant part of the agreement, held that: "The Parties agree mutually to recognize the rightful allocations of both of them in Jordan River and Yarmouk River waters and Arava/Araba groundwater in accordance with the agreed acceptable principles, quantities and quality as set out in Annex II, which shall be fully respected and complied with" (Treaty of Peace, 1994: Article 6(1)).

Although the Treaty did not explicitly make reference to the principles of international water law, it utilized notions such as "rightful," "just," "equitable," and "practical" (Elmusa, 1995; Louka, 2006). Introducing this new term, "rightful allocations," was not merely for linguistic purposes, but rather it served as a physiological reference to "water rights," although the latter was not used as the allocation criteria for water in the Treaty (Shamir and Haddadin, 2003). The intent of the makers of the Treaty was to remedy the inherent adverse implication of the term "water rights," which usually alludes to the ties between water and occupied land and could easily be misunderstood and taken out of context. Hence, the purpose of using such an ill-defined and ambiguous term was to disassociate the water-related conflict from the overall conflict related to issues of high politics (Fischhendler, 2008a). However, to preserve the rights of both parties, they agreed to introduce this new and creative language as a way to combine and operationalize two important concepts: "rights" and "allocation." Except for individual interpretation provided by different water negotiators of what it means in reality, this term remained undefined and the agreement appears to be vague in this regard as well.[108]

The makers of the Treaty seemed to advocate practical, just, and acceptable solutions to the water problems that the parties faced (Elmusa, 1995). They emphasized the value of pragmatic and practical solutions in resolving water-related conflicts, without the hassle of relying on the vague and undefined terms of international law. This meant, as the Israeli negotiator explained, the ability of the two parties to work together and employ tools that could bridge the communication gap (Shamir, 2008). This included defining the quantities and the quality of the water and the exact locations and times of transfer and introducing innovative terms to smooth the way to a satisfactory settlement.[109] To that end, the Treaty, emphasizing the need to maintain the parties' interest in searching for practical and satisfactory solutions, contended:

> The Parties, recognizing the necessity to find a practical, just and agreed solution to their water problems and with the view that the subject of water can form the basis for the advancement of cooperation between them, jointly undertake to ensure that the management and development of their water resources do not, in any way, harm the water resources of the other party.
>
> (Treaty of Peace, 1994: Article 6(2))

It is worthwhile noting, however, that Israel's water resources were already developed when compared with Jordan's, which are still undergoing substantial development (Beaumont, 1997). This means that Jordan has to pay particular attention in its future water projects not to cause any harm to the already fully developed Israeli water resources. Jordan will be restricted by this clause, which will undoubtedly hinder Jordan's ability to conduct any major development of its water infrastructure along the river. Considering that Israel (with the construction of its National Water Carrier in the 1950s) has already caused significant harm to Jordanian water interests, this part of the Treaty seems to be one-sided. Israel's National Water Carrier has resulted in significant environmental damage as a result of extensive water use, namely the deteriorating Jordan River water quality and the declining rate of the Dead Sea level (Rouyer, 2000). The agreement had not considered these adverse consequences of Israel's previous, and current, water development on Jordan, nor did it attempt to provide any suggestions to make its operation more environmentally friendly or offer conditions on how to mitigate the continuous environmental harm that it imposed. Instead, it favored Israel's future water interests and restrained Jordan's most needed future water development.

Another clause that would seem to favor Israel holds that "the Parties recognize that their water resources are not sufficient to meet their needs. More water should be supplied for their use through various methods, including projects of regional and international cooperation" (Treaty of Peace, 1994: Article 6(3)). Although Jordan was able to enhance the efficiency of the operation of the King Abdallah Canal by increasing its diversion of water (Elmusa, 1995), this clause provides more security to Israel's future water acquisition. Although this clause seemingly advocates increasing the available resources to overcome the water crisis, it is on a par with Israel's long-term vision of augmenting the supply rather than merely considering the resources in dispute. As such, this clause gives Israel the ability to not only utilize the water resources of the Jordan River, but also potentially broaden its water acquisition plans to include water resources of other states in the region, namely Lebanon and Egypt. By focusing on other nations' water resources, Israel was able to divert the attention away from its excessive consumption of the Jordan River (Beaumont, 1997). This savvy negotiation strategy on behalf of Israel tends to safeguard the current Israeli utilization of the Jordan River water and meant that Israel would not be pressured to give up any of its current use of the system. By so doing, Israel successfully avoided relinquishing, or even negotiating, some of its current use of the Jordan River, which significantly exceeds the amounts specified by the Unified (Johnston) Plan.

Another major implication of such a clause is that any Jordanian attempt at future water acquisition from other countries (such as Syria, for instance) would not be practically possible without the involvement of Israel. Typically, Israel has no chance of obtaining water from other Arab countries in the region, but with this stipulation, Israel (through Jordan) can pursue transboundary water transfers. Jordan, on the other hand, has been successful in obtaining water from Lebanon and Syria. According to this Treaty, these sources of water supply for Jordan are not possible any more without the involvement of Israel as a beneficiary. This becomes more evident, considering that Israel is interested in obtaining additional water from Lebanon, Egypt, and Turkey.[110] This undoubtedly secured a window of opportunity for Israel to seek access to further water resources from other neighboring countries. The very next item in Article 6 also reinforces this conclusion.[111]

The exchange of Israel's technology and expertise in water resources management, namely its cutting-edge water use efficiency and drip irrigation technologies, is the only benefit that Jordan would have been able to glean. Unfortunately, this clause was not explicit about how this knowledge exchange could possibly be accomplished. The agreement, instead, seemed vague about how this future cooperation between the two parties should be sought and managed. It also provided no specific guidelines pertaining to preventing the contamination of water resources beyond stating so in the text of the aforementioned clause.

Annex II shows more details about the quantities of water allotted to both parties. The Annex is composed of four topics, including water derived from the Yarmouk, water derived from the Jordan River, additional water sources, and operation and maintenance. The volumetric water quantities allocated for each party will be examined in more detail in a later section of this chapter. Overall, the agreement offered Jordan a three-component package of 50 MCM each. The first portion of this package was to come from Israel's own use of the Yarmouk, which was reduced to 25 MCM per year[112] (Fischhendler, 2008b). This allocation scheme had profound practical implications on the water consumption trends of both countries. It meant that Israel is now guaranteed to receive fixed annual volumetric amounts, whereas Jordan remains uncertain of how much water it will receive per year, which is solely dependent upon the annual yield of the changing ecosystem. Israel, pumping predetermined, fixed amounts of water at very specific periods of time, has guaranteed that its yearly water supply is secured regardless of the fluctuation in water availability from one year to the next. On the contrary, by agreeing to take the rest of the available water, Jordan is burdened with a great deal of risk and the responsibility to plan for frequent droughts. This allocation scheme is not favorable to Jordan because it clearly failed to evenly distribute the risk of water shortages between the two parties.[113] Instead, according to this formula, Jordan is the risk taker, given its position of receiving residual water only after Israel claims its fixed shares.

The agreement further states that "in order that waste of water will be minimized, Israel and Jordan may use, downstream of point 121/Adassiya Diversion, excess flood water that is not usable and will evidently go to waste unused" (Treaty of Peace, 1994: Annex II). According to this final clause on the allocation of the Yarmouk water, the amounts of this excess flood water remain undetermined. According to this portion of the package, Israel would also increase its use from the Yarmouk by 20 MCM during the winter, in return for conceding to transfer to Jordan an additional 20 MCM during the summer period (Treaty of Peace, 1994: Annex II). Although this amount was to be delivered from Lake Tiberias, the name of the lake was intentionally excluded from the text of the Treaty to refute any Jordanian claim of being a riparian in this location. Instead, the Treaty identified the Jordan River as the source of this amount.[114] Moreover, the cost of the operation and maintenance of this additional transfer and any new transmission system was to be covered by Jordan (Beaumont, 1997). This meant that Jordan alone would be financially responsible for the acquisition of this amount, which violated the condition of equally sharing the benefits and costs.

In addition, this portion of the deal included an additional annual amount of 10 MCM of desalinated water to be transferred to Jordan from the saline springs, which are now part of the Jordan River, during dates that Jordan selects outside the summer period (Beaumont, 1997; Fischhendler, 2008a,b). However, this amount is conditional as well and subject to the maximum capacity of transmission, which again makes Jordan a second-hand user. Furthermore, the stipulation that this amount cannot be delivered during the summer, when it is needed the most, gives Israel higher priority during times of summer water shortages and droughts. Israel was also able to avoid any financial responsibility of funding this transfer by stating in the agreement that its role was limited to exploring the funding of any such transfer. Elmusa (1995), akin to many others, suggests that Israel should bear the cost of desalination to compensate for the damage it has caused to the Lower Jordan River by the rechanneling of these saline springs in 1964 to mitigate the salinity of Lake Tiberias.

The second component of the Jordanian package allowed Jordan to store a minimum average amount of 20 MCM of flood water in the Jordan River during the winter, and although Israel maintained its full current uses of the Jordan River, Jordan could only receive an annual amount equivalent to that of Israel as long as Jordan's use did not impose any kind of harm to the quantity or quality of the current Israeli uses.[115] A significant implication of this clause is that Israel was entitled to maintain its current use of the Jordan River (solidifying and legitimizing the *status quo ante* of Israel's current water use), which preceded that of Jordan[116] (Fischhendlar, 2007). This not only means that Jordan's use of the Jordan River is made subservient to that of Israel, but it also means that Israel has, in effect, a veto over Jordan's use of this water (Beaumont, 1997).

In addition, deciding whether Jordan's use is harmful to Israel's uses is subjective and requires further studies and assessment that could be costly. Therefore, by failing to delineate a clear definition of the term "no harm," the Treaty provides security for Israel in terms of the quantity and quality of water, paying little or no attention to the Jordanian water security (Fischhendlar, 2007). This is because the Treaty emphasizes the unconditional Israeli access to the Lower Jordan River, while providing conditional and limited water use for Jordan. Moreover, because Israel's uses of the Jordan River are unmonitored, accompanied by the fact that the Treaty did not offer mechanisms or legal grounds to monitor such uses, estimating the actual accurate amounts of Israel's water use seems insurmountable and subject to Israel's own estimates and interpretations.

The third component included an additional amount of 50 MCM, which was to be delivered to Jordan, contingent upon further discussion and decision. Due to the ambiguity of the Treaty regarding this additional water supply, Jordan is not guaranteed to receive this amount determined by the Annex. This is because the Treaty does not identify a specific source for this amount, a responsible party for this transfer, or a timeframe for when this amount should be delivered. Needless to say, acquiring this amount in such a semi-arid region is a challenge that cannot be trivialized and cannot be left to uncoordinated efforts and unspecified future negotiations. Further, since no one takes responsibility for delivering this amount, no party feels any obligation toward conceding this unclear and undefined condition. Instead, it mentions that both parties have to work together to find additional water sources to meet this additional Jordanian demand.

The agreement also stipulated that Israel will be able to retain its current uses of the wells located in Wadi Araba/Arava on the Jordanian side of the border (and to increase its water extraction by up to 10 MCM for five years), as long as this use does not appreciatively harm the quality and quantity of groundwater (Elmusa, 1995). Akin to other ambiguous terms used in the Treaty, the term "appreciatively" was not well defined (Fischhendler, 2008a).

On the whole, the Treaty is vague with regard to parties' financial responsibilities, cost-sharing mechanisms, and the technical details of different proposed projects, including location, implementation timeline, and funding sources. Instead, it tends to allude, in passing, to these issues in generic terms without providing ample details regarding the whereabouts of these future projects, which increases the ambiguity of its terms and makes it difficult to implement as a result. These issues were left unaddressed even after ratifying the Treaty and seem to pose "destructive ambiguity," which threatens the Treaty implementation and durability. This is because these controversial issues that were left ambiguous during negotiation will inevitably resurface in the future. Sooner or later, the makers of the Treaty will be confronted by these issues as they become more difficult to resolve in light of new circumstances and growing water demand. Some degree of ambiguity in treaty negotiation, formation, and content can be healthy. Deliberate and measured ambiguity can help defuse conflict, bring treaties to closure, reduce the high transaction costs of further negotiation, and promote acceptability of the treaty by both parties. This is referred to as "constructive ambiguity" by Fischhendler (2008a). "Destructive ambiguity," however, can lead to detrimental outcomes and potential conflicts, especially during implementation. These detrimental consequences in the implementation phase can occur as a result of lack of flexibility and the different and often conflicting interpretations of the treaty's ambiguous text.

Perceived equity

The signing of the Treaty brought to its makers another big challenge over how to position and introduce the Treaty to their respective peoples back home. This was particularly true as another issue of great magnitude was awaiting them at home: how to convince the local people and other political parties of the success of the Treaty and how to gain legitimacy and public acceptance of the deal offered by the Treaty, and therefore the maintenance of its implementation. Overcoming these challenges has proven to be easier said than done, as both the Israeli and Jordanian negotiators had been constantly trying to deflect criticism of the Treaty in the first place. The negotiators had been subjected to a great deal of criticism from their own peoples, who felt that the Treaty did not serve their respective national interests. This was partly precipitated by the difficulty in communicating and explaining the great complexity of the technical part of the Treaty and exacerbated by a legacy of negotiation secrecy. In Jordan, in particular, people feel strongly disappointed about Jordan's share allocated by the Treaty, which they believe to have favored Israel. Aside from the overwhelming perception of the unwarranted compromise of the Jordanian negotiator, people do not have a lot of trust in the normalization with Israel in general. The idea of normalizing relationships with Israel has always been viewed as an unacceptable course of action to the Arab people, especially to the Palestinians. Establishing diplomatic relations with Israel has long been viewed as a taboo subject that deterred any country from entertaining the idea of peaceful negotiation with Israel. However, as Jordan and Egypt beforehand have brokered negotiated settlements with Israel, they provided an example to other Arab countries that abandoning the option of diplomacy is likely to breed more conflict and bring atrocities and destruction to the region. These peace agreements sent a clear message that one cannot place a value on peace and as such they charted the course to a future of potential collaboration, which for the most part is no longer being viewed by many as taboo or as a compromise in and of itself. The gradual approach adopted by the Madrid Peace Process (of tackling less complicated issues first and postponing more con-tinuous matters for later stages when trust is built) has helped in changing this general view about normalization with Israel and set the stage for a long-term peace process.

In addition, the perception of the Treaty was impacted by the overall conflict resolution technique used in this process. Although the final signing of the Treaty occurred in the region, which had a positive implication on the perception of the Treaty, the initiation of the Treaty occurred in Madrid, which hindered its acceptability by the locals. In addition, the conflict resolution style used here was confined only to modern approaches. Although this approach was very effective in resolving the issues, it created problems for the implementation. It hindered the acceptability of these diplomatic efforts by the indigenous population of the Middle East, who tended to perceive these modern approaches as invasive "Western panacea" (Abukhater, 2009b). Regardless of their religious or cultural backgrounds, the indigenous people have their own techniques and processes that they use for mediating and resolving severe conflicts over water resources, which are not very different from those modern approaches. This calls for a strategy that incorporates both modern and indigenous processes of conflict resolution and reconciliation in order to boost the acceptability of these techniques and processes and, as a result, advance negotiations toward more plausible and implementable agreements (Abukhater, 2009b).

The makers of the Treaty decided to "play it safe" by keeping the details of the negotiations secretive. This was done to ensure a smooth negotiation process and to avoid political complications and resistance from those who opposed any peace-ful agreement altogether before the real fruits of the Treaty materialized. A public campaign to communicate the purpose and outcome of the Treaty was conducted after the Treaty was signed, as the process of negotiation and Treaty formation had been kept away from the eyes of the press (Haddadin, 2002a; Shamir and

Haddadin, 2003). "Playing it safe," however, did not work as intended, since it meant that the public was to be kept uninformed about the progress and stages of negotiation and, therefore, seemed alienated from the decision-making arena. The fact that the process deliberately sought no public involvement or feedback whatsoever made the general public unaware of what was transpiring behind closed doors and failed to gain public trust. This was very detrimental to gaining acceptance back home. There were many voices in Jordan and Israel who opposed and criticized the outcome of the Treaty as being a "sellout" of the Jordanian and Palestinian water rights (Haddadin, 2008). This shortcoming is related to the centralized forms of government that Israel and Jordan embrace, which resulted in difficulties in obtaining public trust of the process and support of its outcome. Regardless of whether or not the process yielded equitable outcomes, this process will still be perceived by the public as being unfair and insensitive to their needs and interests. Garnering, sharing, exchanging, and disseminating information and encouraging formal and informal public participation and dialog are essential in fostering public awareness and support of the process.

Outside of the politicians involved in the negotiation of the Treaty, there are very few people in Jordan who think that the deal provided by the Treaty has been good for their country. Phillips (2008) explains: "the fact that they [Jordanian end users] are sitting there with only $220\,m^3$ and Israel on the other side with $360\,m^3$ suggests that the deal is clearly poor for Jordan." This perception of the Treaty has adverse implications on treaty implementation, which is impacted by, and is contingent upon, the perceptions of both the negotiators and the general public. Similarly, Zeitoun (2008b) argues that satisfactory outcomes or "rightful alloca-tion" lie in the eyes of the beholder. In this sense, what might seem satisfactory to the negotiator might not appear that way to the public, the beneficiary parties, or the end users. For example, the Jordanian negotiator, who seemed most vulnerable at the time of negotiation considering the hydro-political and hydro-hegemonic situation, namely the asymmetry of power imbalance and *de facto* water use, might have perceived and promoted the deal as being satisfactory even before it was revealed to the public (Elmusa, 1995). Therefore, given these circumstances in the early 1990s, cementing a deal with Israel that opened channels for future cooperation was in itself a satisfactory outcome to the Jordanian negotiator. This does not mean, however, that the Treaty itself is equitable or perceived as such by others. Zeitoun (2008b) argues to the contrary, stating that "having no agreement is better than reaching an inequitable agreement." For Israel, however, water was not the main motive for promoting such a treaty with Jordan. Water was secondary compared with resolving the overall political conflict through cementing a peace treaty with a second Arab state (Egypt being the first to initiate and sign a peace treaty with Israel) (Elmusa, 1995). Israel's desire to foster cooperation, along with the political impetus and the good relations between the two leaders, provided the needed momentum to facilitate the conclusion of the Treaty, which was promoted by Israeli as well as Jordanian negotiators as being equitable.

The deal made regarding the Wadi Araba groundwater was also subject to criti-cism. Allowing Israel to continue using Jordanian land and natural resources that were occupied by force and farmed by Israeli farmers during a period when both countries were in a state of war was viewed by many people in Jordan, and in many Arab countries, as one of the greatest compromises made by Jordan, by giving up some of its natural resources located within its territories. This was viewed as leasing land to Israel (Haddadin, 2002a). This, according to Haddadin (2002a), was not the case as the Treaty did not include any provisions of this sort. However, the Treaty referred to the Israeli farmers as "land owners," acknowledging their "private land use rights," and further allowing them to enter and use this land for 25 years subject to automatic renewal (Elmusa, 1995). Much of this criticism has emanated from the fact that the Treaty conferred to the Israeli farmers the status of "land owners."

As the two countries succeeded in maintaining relatively successful and steady implementation of the terms of the Treaty, one would expect that this public perception of the Treaty would improve. However, this perception did not change much in the wake of the Treaty implementation. In the first year, following the ratification of the Treaty, both parties' trust in the success of the Treaty had already started to fade, as the political arena witnessed a set of adverse events. In particular, the Jordanian perception of the Treaty was negatively impacted by the lack of progress on the Palestinian, Syrian, and Lebanese tracks of negotiations and the continued Israeli confiscation policy of the Palestinian lands in East Jerusalem (Haddadin, 2002a). This political impasse, coupled with the lack of successful outcomes in almost all other fields stipulated in the Treaty (namely joint economic cooperation), gave rise to growing opposition toward normalizing relations with Israel (Haddadin, 2002a; 2002b). With the failure of the JWC in reaching an agreement regarding the delivery of the additional 50 MCM to Jordan and other ambiguous items of the Treaty, exacerbated by the growing water shortage in Amman, the Jordanian public continued to lose faith in the Treaty as being equitable. As disappointment over the Treaty in Jordan grew stronger so did the opposition to it, as was expressed by the Jordanian Parliament in 1995 (Haaretz, 1995).

The Treaty itself has been subjected to an avalanche of opposition and criticism from people on both sides of the river as well as many transboundary water experts around the world. In spite of this criticism, Dr. Haddadin,[117] akin to his Israeli counterparts, takes a hard-line defensive position, maintaining that the Treaty was a breakthrough success and a testimony to the ingenuity of Jordanian diplomacy. His argument is grounded in the circumstances under which the Treaty was reached. In that sense, he argues that, considering the situation at the time of the Treaty, the deal that the Jordanians obtained was favorable, explaining that if Jordan renegotiated the Treaty now it would not be able to secure the same amounts it had been able to secure in 1994 because of the conducive political atmosphere that dominated the early 1990s (Haddadin, 2008). He further adds, "I had to come to an agreement that satisfied the Jordanians and cannot be criticized by the Arabs at the same time. I asked for what was approved by the Arab Technical Committee in 1955 and I got it in 1994." Similarly, Shamir (2008), arguing that equity is a subjective matter, believes that the Treaty was equitable

"because this is the way it is perceived by both sides." However, few people agree. Some water experts seem skeptical in accepting the argument regarding the success of the Treaty claimed by its negotiators. Zeitoun (2008b) contends that the vulnerability of the Jordanian negotiator at the time of negotiation was one of the key reasons that led to the outcome of the Treaty being inequitable. Phillips (2008) also disagrees with the Jordanian negotiator's assessment of the Treaty as being equitable to Jordan, making reference to the current per capita water availability for Jordan and that of Israel, which represents a disproportionate allocation of water favorable to Israel. Utterly refuting any allegation of the Treaty being equitable as not being serious claims, he further asserts that the current per capita water availability in Israel and Jordan indicates that the Treaty clearly favored Israel. "It [Treaty] is not equitable, and how can it be equitable when one party [Israel] has twice as much water as the other [Jordan]?" (Phillips, 2008); Phillips adds that the reason Jordan "got a bad deal" was because "they negotiated poorly on water."

Negotiators on both sides hold a very favorable perception about the way the political and technical aspects of the Treaty negotiation were handled as well as its outcome, particularly the way it allocated water for both sides. The Jordanian negotiator in particular continues to defend the deal obtained by Jordan, a claim disputed by many others who view the Treaty as a defeat for Jordanian diplomacy and a huge success for Israeli diplomacy. Cognizant of the significance of maintaining peaceful and cooperative relationships with the other side and the high cost of conflict, the Treaty negotiators believe that brokering the Treaty was a great success on their part (Haddadin, 2008). As such, moving from a stage of rivalry to a more cooperative stage through means of diplomacy is valued by the makers of the Treaty as a strategic value worthy of attaining and maintaining at any cost to prevent possible clashes between the two countries in the future. To them, the Treaty has set an example of the role of preventive hydro-diplomacy, which cannot be overemphasized in resolving conflicts through negotiation.

Outcome equity

The Treaty governs the allocation of regional water resources (including both surface and groundwater) available for Israel and Jordan. This includes:

1 allocations for Israel and Jordan from the Yarmouk and the Jordan Rivers;
2 allocations for Israel from the Arava Valley/Wadi Araba groundwater; and
3 allocations for Jordan from unidentified sources.

Originally, the makers of the Treaty (particularly the Jordanian negotiators) set out to adopt the sharing formula provided by international law and the Unified (Johnston) Plan, which was the proposed basis for water allocation from these three sources (Haddadin, 2002a). However, this was easier said than done, as they were challenged by the opposition of their Israeli counterparts.

Numerous obstacles surfaced as the negotiators started to address these issues in meaningful detail in the first serious round of negotiation held in Aqaba (Shamir and Haddadin, 2003). At first, the negotiating teams were unable to reach a consensus on the basis for the negotiation, as the Israeli counterparts declined the adoption of the Unified (Johnston) Plan sharing formula regarding the Yarmouk. These bases constituted major points of contention that were discussed during the onset of the negotiation in Aqaba. However, as the Israeli and Jordanian negotiators continued to work their way through negotiation, it became clear that both sides came to an implicit agreement to utilize the Unified (Johnston) Plan as the basis for their negotiation over water (Haddadin, 2008). Regardless of what allocation strategy was adopted, water was allocated nonetheless. As shown in Table 4.9, which summarizes the outcome of the Treaty, the volumetric water allocation for each party from different shared water resources can be outlined as follows:

Table 4.9 Water allocations in the Israel–Jordan Peace Treaty (MCM)

Jordan	Israel	Implications
Yarmouk residual for Jordan (estimated between 60–282 MCM) *This amount is an estimate compilation based on the amounts provisioned by the terms of the water agreement.*	Jordan River residual for Israel (estimated at 600 MCM)	Undetermined and varied amounts. Greater uncertainty for Jordan as a result of the unregulated Syrian use of the Yarmouk. Israel maintained its current use of the Jordan River (particularly at its confluence with the Yarmouk and Tira't Zvi near the West Bank border)
20 MCM from Jordan River Deganya Gates (summer)	20 MCM from Yarmouk (winter)	Israel to utilize Lake Tiberias for storage of floodwater. Jordan has no storage facilities
10 MCM from desalinated water in the north (10 MCM from Deganya Gates (summer) until plant constructed)	Maintaining existing use in Araba Additional 10 MCM from Araba groundwater in the south	Israel's existing use given higher priority
50 MCM "additional water"	25 MCM from Yarmouk (12 MCM summer, 13 MCM winter)	Guaranteed amounts for Israel. Uncertain and unguaranteed 50 MCM additional water for Jordan. Ambiguous cost sharing (Jordan will cover cost)
Total (before treaty): 120–130 MCM **Total (after treaty): 90–312 MCM**	Total (before treaty): 640–690 MCM **Total (after treaty): 625–665 MCM**	Overall, Israel is allocated more water. Israel maintained its current volumetric water extraction

Data sources: adapted from Elmusa (1995); Louka (2006); and Sosland (2007).

The Yarmouk   Overall, Israel, according to this Treaty, receives an annual amount of 25 MCM from the Yarmouk, along with whatever amount of flood water is available for it to capture (Elmusa, 1995). Jordan, on the other hand, receives an undetermined amount that varies according to the variation in the annual natural flow of the river from one year to another. It is important to note that there is no precise figure for how much flow is available in the Yarmouk each year, especially when the current Syrian water use is unregulated and for the most part remains undetermined, in spite of its water agreement with Jordan.

Varying figures regarding the Yarmouk flow have been reported in different periods. For example, in the period 1927–1954, believed to be a wet period, the river yielded an annual average of about 467 MCM (Salameh and Bannayan, 1993). Other sources provided more conservative figures of about 350 MCM/year. There are strong indications that the flow of the Yarmouk has recently declined significantly in the wake of excessive groundwater extraction. Recent figures indicate a total flow of only about 245 MCM/year (Haddadin, 2006). Excluding the Syrian share of the Yarmouk water, which is estimated to be 160–170 MCM, and the allocated amount of 25 MCM for Israel, the expected annual amount of water left for Jordan could be as low as 60 MCM (based on a 245 MCM/year estimate) and as high as 282 MCM (based on the 467 MCM/year estimate registered during the period of 1927–54).

Because Jordan is the residual user on the Yarmouk, the additional Syrian usage, beyond the 90 MCM specified by the Unified Plan, diminishes Jordan's share from the River. Zeitoun (2008b) argues that the Treaty does not provide a fair share for Jordan from the Yarmouk. The fact that the level of Syrian extraction from the Yarmouk water is unknown and unrestricted, and is likely to increase in the future in keeping with their population growth, makes it hard to exactly predict how much water is left for Jordan to capture. What is certain here, however, is that Jordan is likely to receive even significantly lower amounts of water in the future than the expected estimates set out above.

The Jordan River   According to the Treaty, Jordan should be guaranteed an annual water supply of at least 30 MCM (20 MCM directly from the Jordan River and 10 MCM desalinized water from Israel), in addition to undetermined amounts it can extract from the reach between the Yarmouk confluence and the Wadi Yabis/Tira't Zvi near the border with the Palestinian West Bank (Elmusa, 1995; Beaumont, 1997). The Treaty does not permit Jordan to extract water from the Jordan River below Lake Tiberias. Instead, much of this water, totaling about 600 MCM/year, is expected to be diverted by Israel's National Water Carrier for out-of-basin water use south of the country. Maintaining its current use of the Jordan River water through its National Water Carrier, which is proven significant, constitutes a great diplomatic success for Israel.

Groundwater   The Treaty has not provided substantive details about groundwater allotment. The only relevant part of the Treaty is Article 4, which deals with the issue of groundwater in the Emek Ha'arava/Wadi Araba, namely the wells that have been used by Israel since 1948, despite their location in Jordanian territories that were seized by Israel, piecemeal, between 1968 and 1970. The Treaty permits Israel not only to continue using these water resources and assets, most of which are transferred to Israel across the border, but also to increase their use by 10 MCM/year. The only requirement made by the Treaty is that Israel must not cause harm to other Jordanian users, which seems meaningless, since there are very few Jordanian users in this area because of the proximity of these wells to Israel's military activities. Considering the estimated pre-1994 yield of these wells of about 8 MCM/year, this stipulation of the agreement allows Israel to extract 8–18 MCM/year, although Israel has no legal grounds to continue utilizing the groundwater resources of an area that was occupied by force (Elmusa, 1995).

However, this water allocation configuration might seem fair when considering the whole picture, particularly swapping land and water in the north and the south. According to Haddadin (2008), the additional amount of 10 MCM that Israel receives from the increasing extraction of groundwater in Wadi Araba/Emek Ha'arava is parallel to another amount of 10 MCM that Jordan receives in the north from "the desalination of 20 MCM of saline springs now diverted to the Jordan River" (Treaty of Peace, 1994: Annex II, 6(3)). Although neither the water–land issue nor the two water quantities were explicitly linked in the Treaty, they nonetheless seem to have balanced the exchange of land and water between the two countries (Shamir and Haddadin, 2003). However, many experts argue that the failure of the agreement to address financial compensation for water that Israel will continue to extract in these areas creates a situation of *de facto* water exchange in the north and south of the region (Elmusa, 1995).

Water quality   Water quality was given particular attention and consideration by the makers of the Treaty. The Treaty was explicit about swapping water resources that are compatible in terms of their quality to both Israel and Jordan. To that end, the Treaty stated that "the quality of water supplied from one country to the other at any given location shall be equivalent to the quality of the water used from the same location by the supplying country" (Treaty of Peace, 1994: Article 3). However, in the summer of 1998, the issue of water quality resurfaced as Jordanian users in West Amman suffered from deteriorated water quality as a result of the diversion from the King Abdallah Canal. The issue, which provoked strong opposition that led to changes in the Jordanian political and governmental arenas,[118] still occupies the minds of the Jordanian water planners and policymakers (Shamir and Haddadin, 2003).

The Treaty also prohibited the disposal of "industrial wastewater into the courses of the Yarmouk and the Jordan Rivers before they are treated to standards allowing their unrestricted agricultural use" (Treaty of Peace, 1994: Article 3). However, there were no water quality or environmental protection guidelines in the Treaty, other than advising the two parties "against any pollution, contamination, harm or unauthorized withdrawal of each other's allocations" (Treaty of Peace, 1994: Article 3). Monitoring quality of water along the joint waterway is left to the work of the Joint Water Committee.

Overall assessment of water allocation

To evaluate the outcome of the Treaty, a set of legal and practical parameters need to be considered. These include various water demand standards developed by different experts, the Unified (Johnston) Plan of 1955, and the parameters of outcome equity identified by international law as an adequate basis for assessment.

COMPARED WITH VARIOUS INTERNATIONAL STANDARDS OF WATER DEMAND

A number of different standards were put forward by different experts to address the question of water allocation equitability along the Jordan River. In this region, which represents a modern and over-populated society, there are several sectors and services that require substantial amounts of water in order to supply different human, productive, social, cultural, and aesthetic needs. A typical town in the region consumes about 73 percent of municipal water in residential uses, whereas the rest of the water goes to other uses, such as industrial and commercial, public gardens and parks, educational and public institutions, construction, and hotels and recreational uses (Water Care, 2004). There are different standards for different demand levels. A minimum subsistence level water demand for drinking must be secured to prevent the effects of dehydration and potentially life threatening diseases.[119] On average, a person consumes 3 liters of water per day (Water Care, 2004). The minimum daily amount of drinking water needed for humans to survive ranges from about 2 liters in temperate climates to about 4.5 liters in hot climates (Howard and Bartram, 2003; WBCSD, 2005). According to the World Health Organization (WHO), about 30 liters per day (5 liters for drinking and cooking and another 25 to maintain hygiene) is required for survival. However, there are various standards that recommend conflicting thresholds of water use.

Due in part to the fact that needs are easier to quantify than rights, there have been conscientious attempts to quantify human needs for water irrespective of location, meteorological conditions, or socioeconomic factors. Although some experts developed their water requirement standards based on human and various socioeconomic development needs, others, such as Phillips, advocate equal per capita allocation of water irrespective of other socioeconomic factors to determine equitable and reasonable allocation (Phillips, 2008). This way, the proposed desirable water allocation can take into consideration the relative population sizes of each co-riparian (Phillips et al., 2007b). This method of equal per capita allocation is not new. Many experts have proposed similar per capita standards based on the concept of equality.[120]

Although both Israel and Jordan enjoy optimal water access higher than 200 liters/person/day, there is still a pronounced gap between the amounts of water per capita currently available for both countries. Given these standards and the pronounced gap between the annual per capita water availability in Israel and Jordan, the majority of international water experts believe that the Treaty provided a water allocation scheme that can conservatively be characterized as being less than equitable, at best. For instance, Phillips (2008), akin to many others, advocates against attributing equity to the outcome of the Treaty and further argues against unfounded claims of its fairness. He asserts that, considering the relative current water availability, in reference to the current annual per capita water availability of 360 m³ and 220 m³ for Israel and Jordan respectively, the outcome of the Treaty seems to be in favor of Israel. To that end, he states that "the belief that anyone can think that this Treaty is equitable is beyond me" (Phillips, 2008). Although Haddadin (2008) strongly disputes this criticism and fundamentally disagrees with those claiming that the Treaty was unfair to Jordan, Phillips (2008) emphasizes the fact that "no one serious in the international water arena believes that the Treaty is equitable – they all believe that Israel did better than Jordan." Similarly, Zeitoun (2008b) explains that Jordan water needs are still not being met, considering the current per capita consumption. He also concludes that this provides conclusive evidence that the Treaty was not fair to Jordan in terms of mere volumetric water allocation.

When comparing the parties' overall *de facto* water consumption in the early 1990s, prior to ratifying the Treaty, and the current water consumption, it becomes apparent that the situation did not change much. Israel, according to the Treaty, is guaranteed 625 MCM and continues to extract from 625 MCM to 667 MCM from the Jordan River Valley, compared with 640–690 MCM prior to the Treaty. Similarly, Jordan's share continues to be minimal compared with that of Israel, even after signing the Treaty. The Treaty allows Jordan to extract an unspecified amount that varies from at least 90 MCM to 312 MCM at most from the Jordan River basin compared with 120–130 MCM in the early 1990s. Under this Treaty, Jordan was unable to guarantee its minimum *de facto* water consumption it received in the early 1990s. Needless to say, the Treaty did not change the situation for the Palestinians of the West Bank, whose access to the basin's water continues to be utterly denied.

COMPARED WITH THE UNIFIED (JOHNSTON) PLAN OF 1955

It is crucial to evaluate the outcome of the Treaty against the Johnston Plan, especially because it is the only benchmark available on a basin-wide scale and was used by the negotiators as the basis for their negotiation. Although all five co-riparians adjacent to the river have the right to access the river's water, the reality demonstrates otherwise. Israel diverts 75 percent of the river's water before it reaches the West Bank (PASSIA, 2002). Although Israel, Syria, and Jordan access the river's water, the Palestinians are denied access to the Jordan Valley. Only prior to 1967 did the Palestinians have access to their share of the water. Table 4.10 illustrates the quotas assigned by the Johnston Plan compared with the current utilization of the Jordan River's water for each country.

Table 4.11 provides a brief comparison between the water allocation scheme of the 1994 Treaty and the Unified (Johnston) Plan. The Treaty, being a bilateral agreement, tended to abandon an important feature of the Unified (Johnston) Plan by segmenting the river system rather than managing it based on an integrated

Table 4.10 Actual use versus quotas under the Johnston Plan in the Jordan River basin

Country	Quota (MCM/ year)	Percentage of total	Actual use (MCM/ year)	Percentage of total	Description
Lebanon	35	3	20	< 2	Both quotas and use from the al-Hasbani
Syria	132	10	200	17	Quotas: 90 from the Yarmouk, 22 from the Jordan, and 20 from the Banyas; actual use: all from the Yarmouk
Israel	400	31	690	60	Quotas: 375 from the Jordan, 25 from the Yarmouk; actual use: 550 from the Jordan and 70–100 from the Yarmouk
Jordan	720	56			Quotas: 100 from the Jordan, 377 from the Yarmouk, and 243 from the western and eastern side wadis
East Bank	505	39	250	22	Quotas: 297 from the Jordan and the Yarmouk and 206 from the side wadis; actual use: 130 from the Yarmouk and 120 from the side wadis
West Bank	215	17	0	0	Quotas: 180 from the Jordan and the Yarmouk and 35 from the side wadis
TOTAL	1, 287	100	1,160	100	

Note: Percentages do not add up due to rounding.
Data sources: Elmusa (1998); Shamir and Haddadin (2003).

basin-wide strategy that included all parties and key regional actors.[121] The Plan seemed to favor a just, durable, and comprehensive settlement of the Arab–Israeli dispute compared with the Treaty, which bilaterally addressed the water needs of only two riparians. For political and ecological reasons, this problematic issue could be addressed through incorporating other regional players into a multilateral agreement for the management of the river system.

Compared with the Johnston Plan of 1955, these figures provided by the Treaty seem to favor Israel's existing water use. The amounts allowed for Israel's use by the Treaty exceed those identified by the Johnston Plan. Israel, according to the Treaty, is entitled to the use of a total of at least 625 MCM, in addition to an extra amount that varies from 0 to 40 MCM. Considering the 625 MCM, which is the least amount guaranteed for Israel by the Treaty, Israel's share of water increases by 56 percent compared with the amounts identified by the Johnston Plan, which was confined to only 400 MCM.[122] Jordan, on the other hand, was assigned an annual total of about 720 MCM according to the Plan.[123] The Treaty assigns to

Table 4.11 Israel–Jordan Peace Treaty (1994) compared with the Unified Plan (1955)

Item	Negotiations outcome (1994)	Unified Plan (1955)
(1) Surface water allocation:		
The Yarmouk	Israel receives 20 MCM in winter. Israel receives 12 MCM in summer and 13 MCM in winter (total of 25 MCM); Jordan receives the residual of the flow (estimated between 60–282 MCM). Undetermined and varied amounts for Jordan. Greater uncertainty for Jordan as a result of the unregulated Syrian use of the Yarmouk	Israel receives 25 MCM and Jordan receives 377 from the Yarmouk
Jordan River	Israel receives residual of the Jordan River (estimated at 600 MCM). Israel maintained its current use of the Jordan River	Israel receives 375 from the Jordan; Jordan receives 100 from the Jordan
Other	50 MCM/year of drinkable quality and Jordan will cover cost. The source of this water is undetermined*	Jordan receives 243 from the western and eastern side wadis
(2) Groundwater and desalinated water	Israel maintains existing use in Wadi Araba and increases its use of groundwater by 10 MCM if hydrogeological conditions allow; Jordan gets 10 MCM from Israel in the form of desalinated water. Until the desalination plant is operational, Jordan gets its share from Lake Tiberias*	Israel does not get any portion of Jordanian ground water.* The Plan is concerned with only surface water
(3) Storage	Diversion/storage dam on the Yarmouk at Adassiya; storage on the Jordan River course, and storage off the Jordan River course (on the side wadis)*	Storage on the Yarmouk at Maqarin by a dam 126 meters high (300 MCM), raisable at the expense of the Arabs to 148 meters. Diversion dam at Adassiya*
(4) Lake Tiberias storage	Was not mentioned or considered	Storage of floods not impounded by Maqarin dam (about 60–70 MCM) in Lake Tiberias or any other economic site*
(5) Total water allocation	625–665 MCM for Israel and 90–315 MCM for Jordan (Israel's share of water increases by 56%)	720 MCM for Jordan; 400 MCM for Israel

Item	Negotiations outcome (1994)	Unified Plan (1955)
(6) Priority of uses	Fixed amounts for Israel** Israel's use is given higher priority**	Fixed amounts were allocated for other co-riparians. Other co-riparians were given higher priority than Israel, which was allocated the residual of the system**
(7) Cooperation	Two parties undertook to cooperate to make more water available and to alleviate water shortage. A joint committee was established*	No cooperation stipulated. An international engineering board would oversee the water releases
(8) Operation and maintenance	Except for Israel's use of the Wadi Arab groundwater, cost sharing remained ambiguous	No mention of that aspect*
(9) Political motives	Achieving viable and sustainable peace in the region and mutual recognition of equal partnership	Achieving a just and lasting peace between the Arab states, the Palestinians, and the Israelis*
(10) Other co-riparians	Not considered	All co-riparians included
(11) West Bank's share	Not accounted for (excluded from Jordan's share)	Included in Jordan's share (additional 215 MCM/year)

Data sources: *Haddadin (2000); **Elmusa (1995; 1998) (based on the work of Phillips *et al.*, 2007b).

Jordan a total amount of water that varies from only 60 MCM, at a minimum, to 282 MCM per year, at a maximum, from the Yarmouk, plus an additional 30 MCM/ year from the Jordan River. This indicates that Jordan is expected to receive an annual total ranging from 90 MCM to 312 MCM. Although Jordan was able to increase its existing total water budget by more than 25 percent, these amounts seem insignificant compared with a sizeable annual total of 720 MCM proposed by the Johnston Plan (Elmusa, 1995). To that end, Phillips (2008) explains that "even on the basis of this single benchmark, Jordan failed to land a fair deal or one that is equivalent or comparable to the deal offered by the Johnston Plan." Therefore, power structure imbalance created an uneven playing field where the powerful state was able to impact the outcome in this situation.

Moreover, there is no mention of the impact of the Treaty on, or how the Treaty could be impacted by, other parties in the region, namely Syria and the Palestinians. The Treaty completely disregarded the amounts assigned for the Palestinians of the West Bank, which, according to the Johnston Plan, was about 215 MCM. One may expect that, since the West Bank is now under the control of Israel, these amounts earmarked for the West Bank, totaling nearly 215 MCM, would be delivered by Israel. However, the Treaty failed in providing any assurance for the release of these amounts, which continue to remain undelivered to the West Bank to this day. In addition to the comparatively massive increase of water earmarked for Israel, Syria is expected to extract increasing amounts of water beyond what had been allotted in the Johnston Plan, which was only 132 MCM per year. These increasing amounts are estimated at more than 170 MCM/year. This puts Jordan in an even more precarious situation, especially considering that Syria is not obligated by this Treaty or any other agreements to reduce their water extraction to the level specified by the Johnston Plan (Beaumont, 1997). Expecting Syria to respect and abide by the amounts of water allocated by the Johnston Plan when Israel does not is an unlikely scenario. This creates uncertainty regarding the annual flow of the Yarmouk, which is completely transferred to Jordan, as Israel is allocated fixed amounts that were given higher priority, whereas Jordan takes the balance of the flow.[124] Considering that Jordan does not have the facilities in its current water system to allow for adequate storage capacity, extracting water allocated to it by the Treaty and planning for droughts seem insurmountable.[125] This suggests that Jordan will be vulnerable and will most likely suffer the consequences of any future droughts due to the high evaporation rates and the lack of water storage facilities. This poses a significant implication for Jordan, especially during times of drought, which would be handled by Jordan alone.

Although the Johnston Plan made a clear reference to Lake Tiberias, the lake was not mentioned in the Treaty because Jordan is not a riparian of the lake.[126] This created a problem in the implementation of the Treaty which surfaced in 1998, at which point a new agreement was reached specifying the lake as a storage facility (Shamir and Haddadin, 2003). Storing Jordan's water in Lake Tiberias means the *coup de grace* to the Maqarin Dam project proposed by the Johnston Plan to store water for Jordan and provide hydropower for Syria (Elmusa, 1995).

Despite the fact that the Jordan River water below Lake Tiberias is equally divided between both countries, the only amount that Jordan secures from the Jordan River, according to the Treaty, is 30 MCM/year, compared with 100 MCM/year determined by the Johnston Plan, as shown in Table 4.11. The Jordanian share beyond the established 30 MCM seems insignificant considering that the flow below Lake Tiberias is depleted by the extensive diversion of Israel's National Water Carrier, which is given priority by the Treaty. This means that Jordan compromised 70 percent of its use of the Jordan River water by signing this Treaty. Consequently, the idea of equitable water sharing that the Johnston Plan attempted to attain is not evident in the volumetric water allocation of the Treaty. Although no Jordanian groundwater was allocated to Israel in the Johnston Plan, the Treaty allows Israel to maintain its current use of the groundwater in Jordanian territories and further increase it by up to 10 MCM. The Plan seems to have provided a clear description of the amounts and sources of water for each riparian and no additional water components were left undetermined, unlike in the Treaty, which allocates an additional 50 MCM without specifying its source, as discussed before.

Although the Jordanian negotiators advocate that the Treaty offered Jordan a good deal on water, there is no evidence to support this claim. They often make reference to the Johnston Plan, which they claim was used as the basis for the negotiation and allocation. Phillips (2008), commenting on this claim, contends that "if this claim is true then they did very poorly." In short, the outcome of the Treaty failed as far as its volumetric equivalency to the Plan.

Based on the principles of equitable utilization identified by international law, three measures are used to evaluate the outcome of the Treaty. The first evaluates the extent to which each individual riparian country has received an equitable share of the basin water; the second evaluates the extent to which the relative allocation for the two riparian countries is equitable; the third evaluates the extent to which all co-riparians have received an equitable share of the basin water. Eight operational factors (hydrogeomorphology-based drivers of outcome equity) are developed based on the principles of international water law, based on geography, hydrology, and demography (as will be discussed in more detail later in the book). The results of applying these factors of equity are outlined in Table 4.12).

Utilizing this model suggests that Israel and Jordan should receive equal percentages of 22 percent, Syria should receive 32 percent, and Palestine and Lebanon should also receive equal allocations representing 12 percent of the total average annual flow of the Jordan River, which is estimated at 1340 MCM. This means that Jordan's equitable share should be equivalent to that of Israel, about 295 MCM. When comparing these amounts with the amounts allocated by the Treaty,[127] the Treaty outcome appears inequitable for two reasons. First, when considering the regional water sharing, the Treaty seems oblivious and problematic to the shares of the other co-riparians sharing the basin, namely Syria and Palestine. In this

Table 4.12 International law-based multicriteria measurement

Outcome equity drivers	Lebanon	Syria	Israel	Palestine	Jordan	Weight*
Catchment area	664	7,301	1,867	2,344	7,663	8.0%
Contribution to annual discharge	115	416	155	148	506	15.0%
Climate (rainfall)	508	508	184	361	222	14.0%
Existing utilization	0	200	690	0	250	20.0%
Economic and social needs (projected water demands)	3,850	23,555	2,800	1,290	1,760	12.0%
Total country population	3.8	18.0	6.7	3.6	5.6	10.0%
Costs of alternative water (per capita GDP)	6,200	3,900	28,900	920	5,000	8.0%
Availability of other water resources in the country (per capita water consumption)	1,780	2,830	360	100	220	8.0%

Data sources: World Bank (2000b); *Mimi and Sawalhi (2003); and Phillips *et al.* (2007b).

regard, the Treaty allocates more relative water for these two countries combined in comparison with the relative water allocated to the other co-riparian countries. This is the byproduct of excluding other co-riparians from the negotiation process, which was bilateral in nature. Second, Israel, according to the Treaty, receives a much higher amount of water than Jordan, although the model suggests that Jordan should receive an equivalent volumetric water allocation to that of Israel. Notwithstanding, these volumetric allocations, although representing significant implications for Jordan's water security, are very minuscule to Israel, considering its significant annual water budget (Elmusa, 1995).

Environmental assessment

Sustainability is an integral parameter of water allocation equity. Rendering measures to rehabilitate the ecosystem and preserve the environmental integrity of the Jordan River system is of great importance. It is believed that the allocation scheme offered by this Treaty is far from sustainable, since it tends to yield serious consequences that may harm the integrity of the environment. First and foremost, the Treaty did not take into account, or was even aware of, the unique and significant demographic and water demand differences between Israel and Jordan, or the alternative water sources options available to each nation.[128] Therefore, this water supply allotment pattern reflects a disproportional distribution relative to both populations. Second, the Treaty did not recognize, or make mention of, the rights of the environment and the health of the ecosystem. Instead, Annex II of the Treaty contends that, "in order that waste of water will be minimized, Israel and Jordan may use, downstream of point 121/Adassiya Diversion, excess flood water that is not usable and will evidently go to waste unused." This perception that unused water "will evidently go to waste" seems to propagate an environmentally unfriendly mechanism of interacting with natural water resources. The Treaty failed to perceive these earmarked amounts of water as necessary for maintaining healthy ecosystems. This perception, viewing these natural resources as a commodity that needs to be excavated, encourages the over-pumping of water for various human uses. This practice represents little regard for environmental concerns, namely in-stream flow. It is important to note that this unused, excess flood water referenced in the Treaty is not wasted water; it is crucial to the sustenance of healthy ecological flow and the integrity of the ecosystem as a whole.

Third, the Treaty did not acknowledge the rights of future generations to the use of this water. It suggested a water allocation plan that was confined to only the current population and immediate demands, without providing any further discussions of how to manage these water resources for future generations. Having said this, the Treaty seems unsustainable based on the fact that it disregards the right of the environment and future generations. According to Haddadin (2008), the makers of the Treaty should have paid more attention to environmental factors to prevent degradation to the water quality and the environment.

Conclusion: overall treaty assessment

Both Israel and Jordan seem to be cognizant of the compromises that they ought to make if they choose to walk away from negotiation and the abominable ramifications of relinquishing the option of diplomacy. Although the BATNA (Best Alternative to a Negotiated Agreement) for Israel might seem very advantageous, Israel, akin to all other countries in the region, cannot afford the price of walking away from the negotiation table. In that sense, Arabs and Israelis share the same strategic goal of achieving regional hydro-stability. Each agreement is uniquely distinctive based on its participants and context. Although the Treaty might be favorable to Israel in terms of water allocation, or the tangible products, the issue that remains of far reaching importance to both parties (regardless of who gains access to more or less water) is future cooperation and sustaining peaceful relations, or the intangible products.

In this particular context, water and security are closely intertwined. Establishing channels of diplomacy, peace, and cooperation, and putting an end to a long period of hostility and conflict are the mutual gains for both Israel and Jordan. Commenting on this intangible feat, King Hussein of Jordan contended that "water is the sole cause for war between Israel and Jordan" (Fuwa, 2003: 40). By settling the issue of water with Israel, Jordan eliminated major causes of future tension. In particular, Jordan has been trying to improve its water supply by maintaining the existing water infrastructure, namely the King Abdallah Canal, and constructing various water-related developments along the Jordan River, including dams and reservoirs. However, these Jordanian attempts were impeded by frequent shelling by Israel (Fuwa, 2003). For Jordan, it is clear that the intangible products, namely resolving the water issue which is a major factor behind the regional tension, outweighed the tangible products of this particular Treaty, given the unprecedented political stability that it offered both states. As such, when considering the overall gain of Jordan, for example, one must pay attention to the other strategic, economic, and political gains that it was able to harness as a result of this Treaty.

However, many experts argue that if water is not properly and equitably resolved for both parties by international treaties, sooner or later tension will surface over inequitable sharing of water (Zeitoun, 2008b). In this case, the consequences will not be confined to only political aspects, but also strategic and economic, as the countries continue to search for alterative water resources to meet future demands. Jordan, for example, is still suffering from a lack of adequate water supply and has gone to great lengths in acquiring additional water to alleviate its demand, such as the Red Sea and Dead Sea project which constitutes a large economic expenditure.

The link between the technical and political aspects of water is inextricable. As such, the manner in which the negotiation process was conducted, namely the pronounced divide and disconnect between the political and technical tracks, was subject to heavy criticism. A key criticism targeting the treaty formation process holds that the political negotiations were conducted almost entirely in separation from the technical negotiations on water, although water permeates into all political aspects. Unlike the preceding diplomatic attempts, the formal process of negotiation tackled issues related to both high and low politics. Water itself is an important item of low politics-related issues and also has political and technical dimensions that need to be closely managed during both tracks of negotiation. Although the negotiation benefited from combining these issues of high and low politics, water negotiations failed to adequately tie the political and technical dimensions of water. These technical and political dimensions remained separated during the water-related negotiation and water was only negotiated on the technical level, whereas its political side was overlooked. Some experts further argue that the entire Treaty, as well as the events leading up to it, was highly political in nature, which caused the political impetus to overtake the technical aspect of water (Phillips, 2008).

Additionally, the formal negotiations were merely dominated by politically charged issues, or issues of high politics, such as security, mutual recognition, land, and borders. Water was negotiated only on the technical level, which dominated the Picnic Table Talks held in the pre-negotiation stage and continued in the 1990s negotiation. These were not kept on a par with the political aspects, which continued to dominate the negotiating table throughout the negotiation. Even when there was enough political momentum to ratify a political agreement, the technical aspect of the agreement was not mature enough to be finalized in the agreement. Nonetheless, it was added on at the end of the agreement in Annex II, which was simply attached to the larger political Treaty when it was ready to be signed, with little or no reconciliation efforts between the two aspects of the negotiation. As such, a major shortcoming of the technical Annex of the Treaty relates to this pronounced disconnect between the political and the technical process, which were detached from one another until the very last moment when enough political pressure to officiate the Treaty had persisted. Because there was little linkage between these two negotiation tracks, the result was an incomplete water-related Annex. Critics believe that this produced an agreement that was less than equitable and caused Jordan to draw the short straw, in terms of water quantity and quality. However, the makers of the Treaty have argued that the outcomes of both the political and technical sides of the Treaty are inseparable from one another and must be considered in their entirety and communicated to the public as such (Shamir and Haddadin, 2003). By doing so, one can understand the intricate relationships between water and other political and strategic issues, such as land, borders, and mutual recognition, and, therefore, adequately assess the success of the Treaty as a whole.

Furthermore, as a result of the evident disconnect between the technical and political level of negotiations, negotiators on the political level had little knowledge of the whereabouts of the technical and hydrologic data except for the technical reports provided to them. Similarly, during the technical negotiations, the Israeli and Jordanian teams were able to depoliticize water and separate it from the larger political dispute. This was a double-edged sword. Although this helped achieve a timely settlement, it tended to create tension and sometimes conflict between the political and technical aspects. Another critique emphasizes the lack of interest-based negotiation on the political level. Although the technical level exemplified, for the most part, interest-based negotiations, the debate on the political level was shifted toward more position-based negotiation (Haddadin, 2008; Daoudy, 2009). According to Philips (2008), the reliance on position-based negotiation only serves the interests of the hegemon, which usually has enough influence to force the opposition into an acceptance of the deal it offers. Conversely, the least powerful states are more likely to take an interest-based approach to negotiation as a way to delimit the influence of power structure imbalance.

Nonetheless, the Treaty offers a successful example of the role of preventive *hydro-diplomacy* in treaty formation and a framework for a negotiated solution as well as future negotiations. Aside from the gap between the political and technical aspects, the Treaty was an instrument for peace and diplomacy. This, however, did not guarantee a successful equitable outcome or, worse yet, a successful perception of the Treaty as being equitable. To that end, there is ample evidence from the aforementioned discussion that the outcome of the Treaty suffers from a lack of equity, which also contributed to the general perception of the agreement as being less than equitable. Nonetheless, the Jordanians had to bite the bullet and live with this deal.

5 The Lesotho Highlands Water Project Treaty of 1986

> Our people thirst for progress. Our land thirsts for water.
> (Pik Botha, South African Minister of Foreign Affairs,
> at the signing of the LHWP Treaty)

This chapter casts light on the 1986 water Treaty between Lesotho and South Africa over the shared Orange River, which illustrates how process equity components can strongly impact the perception of the Treaty as being equitable. Highlighting the importance of perceived equity and the impact of process equity on the implementation of water-related agreements, this chapter provides a brief background on the nature of the water crisis between the two riparian states and presents a detailed analysis of the Treaty formation, implementation, and outcome. The logic model is used as a guiding tool for the assessment of the Treaty, which gives an overview of its provisions, process of negotiation, and outcome, and discusses its prospects and pitfalls in order to draw useful lessons and conclusions for water-related negotiations.

Achieving cooperation over water allocation and swapping water (which is needed in South Africa) for economic assistance (which is needed in Lesotho) are key issues that warranted and incentivized the negotiation. Issues related to water benefit sharing (quid pro quo), economic incentives, third-party involvement, environmental degradation, monitoring of project operation, flexibility of implementation to new changes, and the establishment of a basin-wide water regime are influential factors in determining the actual outcome of the agreement and the way it is perceived. Understanding the profound characteristics of integrating the political, economic, and social dimensions in a single agreement, which makes this Treaty unique in its ability to highlight water-related, as well as non-water-related, benefits for both basin states in the negotiation and implementation of the project, helps unravel these important lessons for water negotiators (Grover, 2007).

This chapter includes two major sections, the first of which introduces essential background information and water availability and use data; the second section provides the water equity analysis of the Treaty and its various implications.

Background: basin spatial hydrology and geomorphology

This section presents the basin spatial hydrology and geomorphology with respect to its geographical, climatic, and hydrological aspects, the spatial patterns of current and previous water utilization trends and dynamics, conflict and cooperation over water, and other human-induced events that influence the basin water characteristics and inter-basin relations among riparian states. The importance of this section to the case analysis is due to the factual information that it provides regarding current and future population growth and associated water demand and requirements. This helps in understanding water stress imposed by population growth and evaluating available water resources to meet this increasing demand. This section also introduces the cause of the water conflict and previous agreements in order to provide a brief background of how the conflict evolved (or became "ripe") for negotiation over the past years.

Geographic, demographic, and hydro-meteorological context

Geography

There are about 59 international river basins located in Africa, as shown in Figure 5.1 (Wolf, 2002). A considerable number of international water treaties have been officiated in many of these river basins to manage conflict and move toward cooperative management of these shared water resources. These treaties are negotiated and signed over the disputed water of 19 river basins, including Congo/Zaire, Corubal, Gambia, Gash, Incomati, Juba–Shibeli, Kunene, Lake Chad, Limpopo, Maputo, Niger, Nile, Okavango, Orange, Ruvuma, Senegal, Umbeluzi, Volta, and Zambezi (Wolf, 2002).

The Orange River basin is the largest basin in South Africa (Fullalove, 1997). Hosting a variety of water transfer projects for municipal, commercial, subsistence, agricultural, and industrial water supply needs, the basin is also considered the most developed transboundary river basin in the region. According to the Department of Water Affairs and Forestry (DWAF, 2005), the Orange River, alternatively referred to as the Senqu River in Lesotho, where it originates, was named in honor of the Dutch House of Orange in 1779 by Colonel Robert Gordon, the commander of the garrison of the Dutch East India Company (Cape Town). Prior to its current internationally recognized name, the river was known as Gariep by the earliest precolonial indigenous population (Earle *et al.*, 2005).

As shown in Figure 5.2, the Orange River rises in the eastern part of the continent about 193 km (120 miles) west of the Indian Ocean, namely in the Drakensberg mountains along the border between South Africa and Lesotho. As the river penetrates further into the African interior, it flows westwards making its way through the heart of South Africa for about 2,200 km, collecting several large tributaries until reaching the Atlantic Ocean, where it eventually drains into the Alexander Bay (Heyns, 2004). The Vaal and Senqu Rivers are the two main tributaries of the Orange River, as shown in Figure 5.2.

The river is considered a boundary river (or international river), as it delineates significant parts of the international borders between South Africa and Namibia, South Africa and Lesotho, and South Africa and Botswana. It also defines several provincial boundaries within South Africa itself, namely the southwestern border of the Free State province. The geographic configuration of the region and the

Figure 5.1 International river basins in Africa, as delineated by the Transboundary Freshwater Dispute Database project, Oregon State University (2001).[1]

river path suggests that Lesotho is the upstream riparian, while South Africa is the downstream riparian.

The Treaty is situated within a unique multinational geographic context located at the far southern tip of the African continent. The uniqueness of this geographic context stems primarily from the spatial configuration of South Africa in relation to Lesotho. Geographically, Lesotho is a landlocked country completely surrounded by, and of a much smaller size than, South Africa (Davis and Hirji, 2003). This unique spatial configuration of the basin makes this case interesting for analysis and enhances its potential to offer useful lessons and parallels. It has the potential to provide an example where water is used to eliminate hostility and overcome unusual spatial and geographic configuration and socioeconomic adversity. Within this unique geographic context, four basin states, including

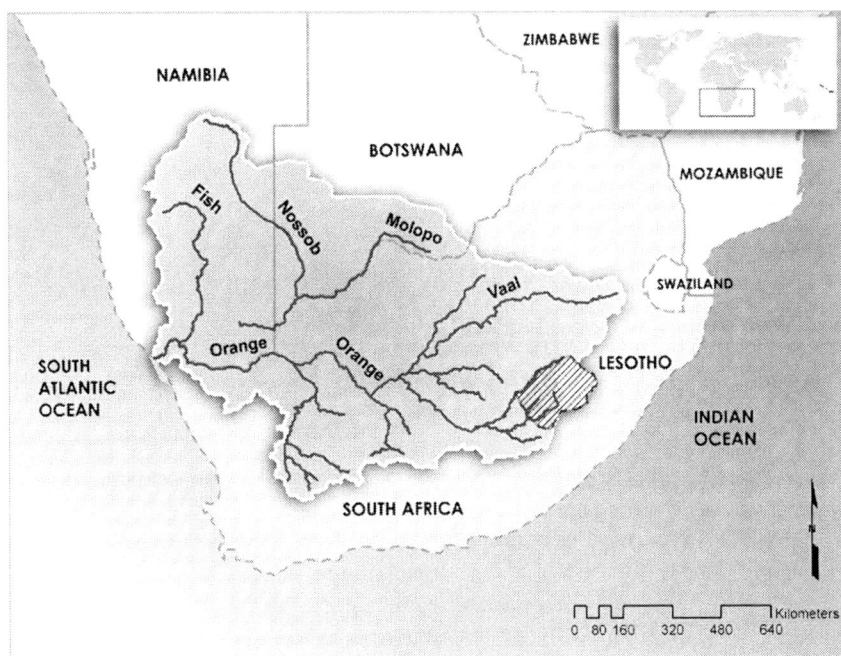

Figure 5.2 The Orange River basin and its major tributaries.[2]

South Africa, Namibia, Botswana, and Lesotho, share access to the Orange River. Although it is difficult to precisely determine the effective catchment area of the river basin, the Orange River basin is considered the largest river basin in the southern African region, covering a total catchment area of approximately 945,500 km[2] (Wolf, 2002; DWAF, 2005). With such a significant catchment area, the basin is also deemed the most developed river basin in the region, incorporating a variety of water transfer projects for domestic, commercial, irrigation, and industrial uses (Fullalove, 1997). Table 5.1 shows the geographic configuration of the basin and the countries it houses. The largest land area of the basin, 563,900 km[2] (more than 59 percent of the total basin area), is located in South Africa, followed by 240,200 km[2] (about 25.4 percent of the total basin area) in Namibia, 121,400 km[2] (approximately 12.9 percent of the total basin area) in Botswana, and the smallest land area of the basin, 19,900 km[2] (representing only 2.10 percent), falls in the boundary of Lesotho (Wolf, 2002).

Demographic characteristics

The basin population is increasingly and steadily becoming more urbanized, a transition that started in the early 1990s (Earle *et al.*, 2005). Historically, people started to move to the more industrialized parts of the basin in search of job opportunities, namely in the mining industry. These demographic changes have placed

Table 5.1 Orange River basin

Countries	Area of basin in country	
	km²	%
South Africa	563,900	59.65
Namibia	240,200	25.40
Botswana	121,400	12.85
Lesotho	19,900	2.10
Total area	**945,500**	

Data source: adapted from the Atlas of International Freshwater Agreements (Wolf, 2002).[3]

greater pressure on existing water resources and imposed greater demand for more water in these urbanized areas of the basin.

Table 5.2 shows that the basin population exceeds 19 million people, with an average population density of about 10 people per square kilometer. The population is predominantly concentrated in the upper reaches of the industrialized eastern part of the basin, with population density exceeding 500 people per square kilometer in some areas, compared with a much lower population density in the dry and comparatively less developed areas of the western part of the basin. This is because the eastern part of the basin is more urbanized, whereas the western part of the basin includes heavily irrigated agricultural land owing, in part, to its flat terrain.

Arguably, it was the arrival of the European settlers that triggered the establishment of the country of Lesotho, formerly known as Basutoland.[4] In 1966, Lesotho attained full independence from British rule that had lasted almost a century. Currently, the vast majority of the Lesotho population is Basotho, a homogeneous

Table 5.2 Major features of the Orange River basin

Feature	Description
Total basin area	896,368 km²
Area rainfall (mm/year)	Average: 330; range >2000 to <50
Average discharge	Vaal River, 4.27 km³/year; Senqu River, 4.73 km³/year; estuary, 11.5 km³/year
Water demand	Total = 6,500 km³/year: agriculture 64%, urban supply 23%, rural supply 6%, mining and other 7%
Population	19 million (year 2002)
Water availability	<1000m³ per capita

Note: 1 km³/year = 1,000,000,000 m³/y = 1,000 M m³/year.
Data sources: Earle *et al.* (2005).

rural population, concentrated in small mountain towns predominantly dependent on farming. According to the 2006 Census data, with an estimated land area of about 30,355 km^2 and a population density of about 59 people per square kilometer, Lesotho has a total population of 1.881 million people, of which 23.8 percent is urban and 76.2 percent is rural (Bureau of Statistics, 2007). With an annual growth rate of 1.4 percent, the demographic ethnic composition of the country is predominantly Basotho, at 99.7 percent (Bureau of African Affairs, 2008). This indicates that the homogeneous Basotho population dominates the demographic and anthropogenic landscape of Lesotho.

Except for the settlement of Orania, near Hopetown, South Africa exhibits a more diverse population in terms of ethnicity and language. With a surface area of about 1,219,090 km^2 and a population density exceeding 39 people per square kilometer, South Africa is the fourth largest country in Africa, with a total population exceeding 47 million (UNESCO, 2006). As estimated by the Bureau of African Affairs in 2008, population growth in South Africa is –0.501 percent. The demographic ethnic composition of the country is predominantly black African, constituting 79 percent, followed by white, 9.6 percent, colored, 8.9 percent, and Indian/Asian, 2.5 percent (Bureau of African Affairs, 2008). These numbers indicate that the population of South Africa is composed of a variety of ethnic groups.

Hydro-meteorological setting

Although the Orange River basin is located, for the most part, in an arid region, its climate varies as the river flows from east to west transforming from cold, temperate, to dry climactic conditions. Overall, the average annual flow of the Orange River is estimated to be in the region of 12,000 million cubic meters (MCM) (FAO, 1997; DWAF, 2005). Table 5.3 shows the availability and distribution of water resources in the Orange River basin.

Table 5.3 Water resources of the Orange River basin

Description	Catchment area (km^2)	Natural run-off (million m^3/year)
Vaal River basin	196,290	4,300
Senqu/Orange to South Africa/Lesotho border	24 680	4,010
Caledon River basin to Welbedacht Dam	15,270	1,240
Remainder of Orange River upstream of Orange/Vaal confluence	~59,400	1,300
Remainder of lower Orange basin excluding Fish River basin	~670,000	420
Fish River basin	76,000	480
Total	~1,000,000	11,750

Data source: DWAF (2005).

The two main tributaries that contribute to much of the Orange River water are the Vaal and Senqu Rivers. The Vaal River, which rises from a number of streams and wetlands in northeast South Africa, meets the Orange River at the western part of the Free State, delineating a significant portion of the boundary of the province. It is one of its main tributaries that contribute significant amounts of water into the main flow of the river. While the average annual run-off of the Orange River basin, excluding the Vaal, is about 7,450 MCM, of which 5,760 MCM is exploitable, it is estimated that the average annual run-off in the Vaal basin is about 4,300 MCM, of which 2,150 MCM is exploitable (FAO, 1997). Flowing southwards, the Vaal supplies most of the industrial and municipal water to South Africa's industrial and commercial powerhouse, Guateng province, which includes Johannesburg and Pretoria (Poivey, 1997; Earle *et al.*, 2005). The Vaal hosts and provides water to 48 percent of the population of South Africa. With such a highly urbanized and industrialized catchment area, the Vaal is becoming increasingly polluted (Heyns, 2004). The other main tributary is the Senqu, which originates in the mountainous region of Lesotho at an altitude of about 3,000 m above sea level (Heyns, 2004; Earle *et al.*, 2005). Because of the substantial volumes of water that the Senqu River contributes into the main flow of the Orange River, the latter is often alternatively referred to as the Orange–Senqu River. However, the river also receives trivial amounts of water from smaller tributaries as it makes its way westward in the last 800 km.

The various range of ecological and climatic regions through which the river meanders, results in various amounts of annual precipitation and run-off (Figure 5.3). The mean annual precipitation (MAP) for the Orange River basin is estimated at about 325 mm (FAO, 1997; DWAF, 2005). The eastern part of the basin receives notably more precipitation than the western part, as shown in Table 5.4. This great variation in the amount and in the spatial and temporal distribution of rainfall causes great variation in the run-off in each sub-basin. The high precipitation and low evaporation rate, together with the relatively shallow soil cover, causes significant run-off in the Lesotho Highlands.[5] Although the average

Table 5.4 Orange River basin rainfall by country[6]

Area	Average annual rainfall in the basin area (mm)		
	min.	*max.*	*mean*
Botswana	165	520	295
Namibia	35	415	185
Lesotho	575	1040	755
South Africa	35	1035	365
Orange basin	35	1040	325

Data source: FAO (1997).

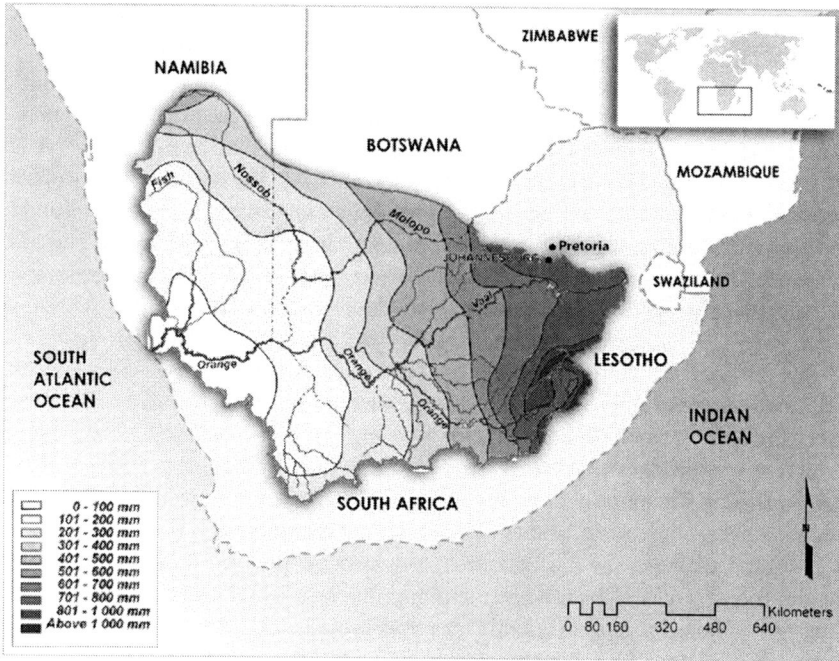

Figure 5.3 Annual precipitation in the Orange River basin. Adapted from the Department of Water and Forestry, Republic of South Africa (DWAF, 2005).

annual run-off that Lesotho and South Africa receive is far in excess of their water requirements, much of this run-off is not exploitable, and is therefore not usable.

In short, the basin climate is characterized by temperature increase, associated with increased rainfall variability and a higher frequency of extreme hydro-events, namely flooding and droughts, which in turn lead to declining yields in rain-fed systems and increased volatility of agricultural production (DWAF, 2005). This, associated with the limitation of available resources, leads to greater vulnerability to such drastic climate changes and restricts farmers' adaptive ability to cope with such extreme hydro-events (Faurès and Santin, 2008).

The course of the river, which is characterized by its high level of fragmentation, includes a significant number of dams for irrigation and hydropower purposes. As estimated by Earthtrends (2008), there are a total of 37 dams ranging in height from 15 to 150 m, with a total reservoir capacity of 142,000 MCM (FAO-Aquastat, 2007). The basin houses 24 large dams, the largest of which are the Gariep Dam[7] near Colesberg and the Vanderkloof Dam[8] (Turton, 2003). This intensive river development pattern has tended to cause environmental consequences and has impacted the view of the outcome of the water agreements, which will be discussed later in this chapter.

Cultural hydrology: an ever growing water demand

The relationship between man and water, and between riparian disputants, greatly influences the way people view, value and consume water to meet their dramatically varying needs, on the one hand, and the way they tend to lead cooperative or conflictive relationships, on the other. Water is needed not only for essential human needs but also for economic development. Although essential human needs are, for the most part, comparable within the same geographic and climatic contexts, resulting in relatively similar water consumption trends, economic development water needs are highly variant depending on the degree of social and economic development of each nation. Conflict over water allocation, however, is not confined to water allocation for economic development. Rather, conflict and disagreements over water often arise over essential human water needs. This troubling side of water conflicts raises a vexing question regarding water access, which is a byproduct of power structure imbalance, rather than needs per se. Given this, it is imperative to understand the dynamics of power structures and regional hydro-hegemony, and the historical and existing water uses in the basin.

Mindful of these pivotal dynamics, this section presents essential elements of the basin's cultural hydrology, including different water consumption trends, hydro-dependency, and economic development, water conflict, and previous water agreements. This section aims to provide a better understanding of the underlying issues impacting equity in water sharing and distribution among different riparians. This gives the reader background information essential to the understanding of the dynamics of water-sharing equity in the basin.

Water consumption trends, hydro-dependency, and economic development

WATER RESOURCES IN THE ORANGE RIVER BASIN

Although the basin is generally on the brink of water scarcity, the higher amounts of precipitation that Lesotho receives makes it a water-abundant country when compared with South Africa, which suffers from severe water shortages for its industrial-based economic development. This provided incentive for swapping water for economic assistance, which will be discussed in more detail later in this chapter. The relatively greater amount of precipitation that Lesotho receives is largely due to high precipitation in the northeastern part, where rivers start to form as a result. Based on the hydrologic characteristics of the basin, the annual supply of surface water is estimated at about 4.73 km^3 per year (FAO-Aquastat, 1995). This amount of available surface water greatly exceeds the water demand of Lesotho. Therefore, the LHWP was initiated to take advantage of this excess water, in response to the disparity between water availability and actual water needs.

With such a high precipitation rate and abundant surface water, one should expect that adequate ubiquitous water access is secured and homogeneously distributed throughout the community. However, this is not the case. Because of the variation of weather across Lesotho, certain areas receive higher amounts of water, whereas others suffer as a result of the lack of adequate water access. The irony is related to the gap between local consumption of water and national policymaking. For the local Basotho, access to water for crop production is evidently very limited, even with these substantial amounts of water available at the national level in Lesotho (Leboela and Turner, 2003). This created a gap between local consumption trends of water and the national decision-making authorities, which regard water as an exceedingly available resource for everyone in Lesotho.

Nonetheless, during the past 20 years, Lesotho has made significant progress in providing adequate and clean water to its population. In the early 1990s, over 76 percent of households in Lesotho lacked access to a clean and reliable source of potable water (Sechaba Consultants, 2000). This situation has drastically changed since then. According to the Lesotho Demographic Survey of 2001, progress in water allocation was confirmed as 75 percent of households had access to reliable and clean running water. However, there are still some mountain areas experiencing limited access to clean water.[9] Overall, the percentage of households with access to clean water has been steadily rising from 52 percent in 1990, to 64 percent in 1993, to 73 percent in 1999 (Sechaba Consultants, 2000).

Considering the regional demographic and hydrologic configuration of the basin, South Africa has the largest basin area of about 60 percent, and contributes a comparable amount of run-off. Although Namibia covers a relatively large portion of the basin area, it contributes only less than 4 percent of the basin annual run-off. Constituting only 5 percent of the total basin area, Lesotho contributes over 40 percent of the basin annual run-off. Although Botswana constitutes a considerable portion of the basin area, it hardly contributes to its run-off. The interesting spatial configuration of Botswana provides the country with a unique and influential political position and "diplomatic maneuverability," in the sense that it can form cooperative relations with other riparians, which might enhance its leverage on other important issues (Turton, 2003). South Africa emerges as the highest water consumer, using more than 82 percent of the total annual water use in the basin.

The nomadic lifestyle of the indigenous inhabitants of the basin kept the overall water use at a minimum in the past. This low water use changed dramatically with the arrival of the European settlers. To date, it is estimated that the per capita amount of renewable water resources potentially available in South Africa is about 1,103 m^3/year (FAO-Aquastat, 2007). With nearly a quarter of this amount being extracted for human consumption, the overall per capita water withdrawal is estimated at around 275 m^3/year (FAO-Aquastat, 2007). Of the total water withdrawal, approximately 63 percent goes to agricultural use, catering to nearly 1.5 million hectares of irrigated land largely used for large-scale commercial farming (FAO-Aquastat, 2007). In comparison, the per capita amount of potentially available renewable water resources in Lesotho is estimated at about 1,681 m^3/year, which exceeds both the average per capita amount of renewable water resources in the basin and in South Africa in general (FAO-Aquastat, 2007). However, because of the hydrogeomorphologic nature of the country, only

1.65 percent of this amount is being captured by Lesotho. With such a low rate of water extraction, the overall per capita water withdrawal is estimated at around 28 m^3/year, equivalent to one-tenth of that of South Africa (FAO-Aquastat, 2007). This disparity in water withdrawal per inhabitant between the two countries connotes the disproportional allocation of resources, due to technologic advancement, varying access to water, and other political and natural processes. Of the total water withdrawal, only 20 percent goes to agriculture, serving 2,640 hectares of irrigated land (less than 1 percent of this irrigated land is cultivated) (FAO-Aquastat, 2007).

HYDRO-DEPENDENCY AND ECONOMIC DEVELOPMENT

Economically, Lesotho, reflecting its geographic configuration, is almost entirely integrated within South Africa. With a work force of 704,000 people, a gross domestic product (GDP) totaling USD 1.43 billion (per capita GDP of USD 850), and an average inflation rate of 12 percent, Lesotho is heavily dependent on foreign aid. In 2007, it received economic aid totaling USD 350 million from various donors, including the United States, the World Bank, Ireland, the United Kingdom, the European Union, Germany, and the People's Republic of China (Bureau of African Affairs, 2008). The country's economy is based on water and electricity sold to South Africa, manufacturing, earnings from the Southern African Customs Union (SACU), agriculture, livestock, and export of labor to South Africa (particularly in the mining industry). Industry (including apparel, food, beverages, handicrafts, construction, and ecotourism) and agriculture (including corn, wheat, sorghum, barley, peas, beans, asparagus, wool, mohair, and livestock) are the major sectors of the economy, constituting 47 percent and 17 percent of the total GDP respectively (CIA, 2008).

South Africa is much stronger in terms of its economic base than impoverished neighboring Lesotho (Mckenzie, 2009). Although South Africa's economy rivals other developed countries, it still exhibits some characteristics of developing countries, with uneven distribution of resources and wealth. South Africa, which shares four river basins with other, less developed, riparian states, is economically the most advanced country in the Southern African Development Community (SADC) region (Turton, 2003).[10] With an unemployment rate of about 23 percent, South Africa's GDP is estimated at USD 283 billion (about USD 5,900 per capita GDP) (Bureau of African Affairs, 2008). Agriculture and mining is a primary sector in South Africa's economy, comprising only 8 percent of its GDP. The industrial sector (including mining, machinery, textiles, chemicals, fertilizers, information technology, electronics, and agro-processing) accounts for 22 percent of its GDP, and services for 70 percent of its GDP (Bureau of African Affairs, 2008).

Water shortages and the uneven distribution of freshwater across the basin are the two main factors that hinder the economic development of the region (Heyns, 2004). The river has a significant economic role in the region, as it delivers much-needed water for irrigation, commercial and industrial development, mining, and hydro-electric power. Water is used for different purposes in different areas of the basin. While the Orange River supplies water for industrial and municipal uses in its upper reaches, it also sustains agricultural production in the middle to lower reaches in support of the livelihood of people in the most arid part of the basin. The annual total water use in the Orange River basin is estimated at 6.5 km^3, with agriculture being the major consumer of water (Earle *et al.*, 2005). The role of agricultural production in the regional economy has been reduced since the mid-1980s, which in turn increased the reliance on food imports. The region is increasingly moving toward production of higher value crops, such as table grapes, olives, and nuts (Earle *et al.*, 2005).

Nearly half of Lesotho's population depends on crop cultivation and agriculture located in the western lowlands, which accounts for almost half of the country's income (FAO-Aquastat, 1995). Both commercial and small-scale agriculture are practiced in Lesotho. In addition, livestock are kept in the drier parts of the basin, while grapes and vegetables are farmed in a narrow strip with intensive irrigation schemes (Earle *et al.*, 2005). As an essential element for sustainable economic development, water is found in nature in many forms, namely precipitation that directly feeds crops, groundwater for domestic uses, and rivers for larger transfer schemes for irrigation (Turner, 2003). The livelihoods of the Basotho reflect a traditional lifestyle dependent primarily on crop and livestock production. As in many drought-prone areas, Lesotho faces the challenge of how to sustain crop production, in order to rejuvenate its economic development potential, in the face of extreme climatic events (Turner, 2003). In rural Lesotho, the lack of clean and reliable water supplies is a major cause of poverty among the Basotho population (Sechaba Consultants, 2000). Chakela (1999), points out that "Lesotho's climatic conditions are optimal for the annual cultivation of most Temperate Zone crops, including maize, sorghum, wheat, beans, peas, vegetables and fruits." However, such crop production has been drastically impacted by the dramatic instability in the climate. The issue of an unclean water supply, compounded by poor maintenance of reticulation systems, is considered a crisis of a national magnitude in Lesotho (Moeti *et al.*, 2003). For the local people, the domestic water supply remains a limiting feature during times of drought, as well as a key potential component in the improvement of their livelihood (Turner, 2003).

For South Africa, water is essential not only for domestic uses, but also in large-scale irrigation schemes and industrial development, namely in Pretoria and Gauteng provinces (previously Johannesburg). It is estimated that South Africa will fully utilize its available water resources by the year 2020, assuming the continuation of the current development trends (Turton, 2003). Thus, South Africa emerges as one of the driest and most dependent countries in the whole basin (Turton, 2003).[11]

An estimated 20 percent of the Orange River flow (including the Vaal as its main tributary) drains into South Africa alone (Turton, 2003). With its highly elaborate and developed infrastructure, the Vaal supplies water to more than 12 million people in support of municipal uses and economic development (DWAF, 2005).[12] In particular, urban and industrial uses in Gauteng, Vereeniging, Vanderbijlpark,

Table 5.5 Irrigated areas, water availability, water requirements and irrigation potential in the Orange basin in South Africa

Sub-basin	Irrigated area (ha)	Actual water use (km³/ year)	Water available (km³/year)	Irrigation water requirement (m³/ha/year)	Irrigation potential (ha)
Vaal	160,000	1.366	1.770	14,000	126,400
Orange	140,000	1.413	2.488	11,000	226,100
Total for Orange basin in South Africa	300,000	2.779	4.258	25,000	352,500

Source: FAO (1997).

and Potchefstroom, along with other mining operations, are the greatest consumers of water in the eastern parts of the Vaal basin, totaling 77 percent of the overall water consumption, while agricultural uses dominate the western part of the river basin (Earle *et al.*, 2005). In such an agricultural region, irrigation potential is intertwined with economic development. Table 5.5 summarizes irrigated areas, water availability, water requirements and irrigation potential in the Orange basin in South Africa. The data reveal that much of the water resources available in the Orange River basin are underutilized and can rather be used to a higher capacity to foster economic revitalization.

Water conflict and hydro-politics along the Orange River

On the whole, South Africa, Namibia, Botswana, and Lesotho share access to the Orange River, which is of great economic significance to all four co-riparian states. In particular, the degree to which Namibia and South Africa achieve their future economic development goals is primarily contingent upon the degree to which they utilize the available water resources of the river. While Lesotho receives royalty payments from South Africa for water exports via the LHWP, Botswana itself has a stake in keeping this option of acquiring water through the LHWP alive in the future (Earle *et al.*, 2005). For Botswana, the river provides strategic leverage to foster its position in other river basins of greater importance (Turton, 2003; Earle *et al.*, 2005).

Tension regarding accessing and distributing the river waters has created a long-standing conflict between Lesotho and South Africa that continued for decades. The impact of water sharing is not confined to water demand and supply issues, but is also a part of a larger, more intricate set of political and socioeconomic factors that, while feeding into the tension over water allocation, constitute a major challenge to the region at large. In this region, water sharing has been a major issue that played a significant role in the formation of the region in general and the nation building of Lesotho in particular. Since the early 1930s, there have been various desultory efforts to attain cooperation over the sharing of the river waters and many proposals have been formulated and considered. However, none of these plans produced a feasible and implementable solution given the political and economic realities at the time of these proposals. The protracted inter-governmental negotiations over the sharing of the Orange water were conducted as early as the 1950s, but were often handicapped by the political tension that poisoned the negotiation atmosphere and caused delays, or worse yet, an impasse. This was particularly related to the issue of the apartheid system, which was most hated and criticized in Lesotho, and many other countries in the region and around the world. This racial segregation system posed a continuous reminder of the ethnocentric nature and intractability of the conflict. Nevertheless, the Orange River basin as it stands today exemplifies a case of cooperation, rather than conflict, over shared water resources (Turton, 2003; Earle *et al.*, 2005).

Previous water agreements

The water of the Orange River basin has been a major point of contention among riparian states throughout the past century. Conflict has occurred; so have negotiation and cooperation. Table 5.6 illustrates significant historical hydro-political and geopolitical events related to the Orange River water resources allocation and conflict resolution that led to the most recent hydro-diplomatic events. Feasibility

Table 5.6 Hydro-political events, Treaty development, and phases

Year	Event
1930–1977	Feasibility studies and surveying of water potential in Lesotho
1978	Joint Preliminary Feasibility Study carried out by consultants from South Africa and Lesotho
1983–1985	Joint Detailed Feasibility Study
1986	LHWP Treaty signed by the Government of Lesotho and the Republic of South Africa
	Establishment of Joint Permanent Technical Commission to represent two governments
1990	End of apartheid era, South Africa
	Construction begins on Phase I
1996	Workers protest at the LHWP Site in Butha Buthe: several workers killed, many wounded
1998	Phase IA completed
	First water supply from Lesotho to South Africa
2004	Phase IB completed
	Present Phase II feasibility study being conducted bi-nationally with 50/50 input and cost sharing between Lesotho and South Africa

Source: Newton and Wolf (2008).[13]

studies started as early as 1930 and continued through the 1970s and the 1980s. Most recently, there have been many serious attempts to address the water conflict. A number of water-related agreements have been negotiated and signed in the past few decades to manage the allotment of the Orange River water.

As shown in Table 5.6, the signing of the 1986 Treaty between Lesotho and South Africa and the consequent adoption of the LHWP set the stage for a new era of diplomacy and cooperation. Negotiation, however, started much earlier in the 1950s, and again in the 1960s, to investigate the establishment of a project to transfer water from the Senqu River to South Africa, which was never implemented because of the disagreement over water payment (Newton and Wolf, 2008). The two governments appointed a joint technical team for the same purpose in 1978. The result was two further feasibility studies, the first of which suggested that the two countries share the cost of a project to transfer 35 m³/s, involving the construction of four dams, 100 km of transfer tunnel, and a hydropower component. The second feasibility study, which led to the 1986 LHWP Treaty, recommended doubling the amount of transferred water to 70 m³/s, as will be further discussed in this chapter. This Treaty, which has maintained implementation until the present day with no significant changes, was followed by a number of protocols and ancillaries between the two countries in 1991, 1992, and 1999 (Table 5.7) (Wolf,

Table 5.7 Previous treaties on the Orange River basin

Date	Treaty basin	Signatories	Treaty name
October 24, 1986	Orange/ Senqu	Lesotho, Kingdom of; South Africa, Republic of	Treaty on the LHWP between the government of the Republic of South Africa and the government of the Kingdom of Lesotho
November 19, 1991	Orange	Lesotho; South Africa	Protocol IV to the Treaty on the LHWP: supplementary arrangements regarding phase IA
August 31, 1992	Orange	Lesotho Highlands Development Authority; South Africa, Republic of	Ancillary agreement to the deed of undertaking and relevant agreements entered into between the Lesotho Highlands Development Authority and the government of the Republic of South Africa
September 14, 1992	Frontier or shared waters	Namibia, Republic of; South Africa, Republic of	Agreement between the government of the Republic of Namibia and the government of the Republic of South Africa on the establishment of a permanent water commission
January 1, 1999	Orange	Lesotho; South Africa	Protocol VI to the Treaty on the LHWP: supplementary arrangements regarding the system of governance for the project

Data source: adapted from Atlas of International Freshwater Agreements (Wolf, 2002).[14]

2002). In 1992, an agreement between Namibia and South Africa was signed to establish a permanent water commission.

Parallel to the implementation of the LHWP, there have been many attempts to govern the water allocation and sharing of the Orange River basin among the different co-riparians. The first water regime was established in 1978 between South Africa and Lesotho, known as the Joint Technical Committee (JTC). These attempts to govern the basin continued until the ratification of the LHWP Treaty in 1986, when the two countries decided to change the name from JTC to JPTC (Joint Permanent Technical Commission), to reflect these new political changes. Most recently these efforts culminated in 2000 in the establishment of the Orange–Senqu River Commission (ORACOM), which provided a multilateral basin-wide tool to govern the Orange River basin, as will be discussed in more detail later in this chapter.

Analysis of the Lesotho Highlands Water Project Treaty

> A giant economic baby is in the process of being born.
> (M.M. Lebotsa, Minister of Lesotho Highlands Water and Energy Affairs)

In 1986, the LHWP was memorialized as a treaty, between the Government of the Kingdom of Lesotho and the Government of the Republic of South Africa. The Treaty was signed at Maseru on October 24, 1986. The negotiation teams were comprised of legal experts, senior government officials, specialist consultants, and engineers from both countries. The final product was a handbook for the implementation of the largest binational construction project on the African continent. This section provides an analysis of the LHWP Treaty in terms of its process, as well as outcome. Although both of these aspects related to treaty formation will be addressed in adequate detail, the following analysis will pay significant attention to *outcome equity* and various issues that arise throughout the implementation of the Treaty.

Intent of the Treaty

As outlined in the Treaty Preamble, the parties, cognizant of the value of water in achieving cooperation and a higher quality of life, signed the Treaty for the intention of enhancing cooperation regarding the regional development of "mutual water resources [that] can significantly contribute towards the peace and prosperity of the Southern African region and the welfare of its peoples" (LHWP Treaty, 1986: Preamble). Both the mutual benefits and equitable sharing of the water resources of the Orange/Senqu River and its tributaries were considered by the parties to "promote the traditions of good neighbourly relations and peaceful cooperation." As such, the intention of the 1986 Treaty between the Government of Lesotho and South Africa, as outlined in Article 3, "shall be to provide for the establishment, implementation, operation and maintenance of the Project."

The purpose of the LHWP, as identified by Article 4, was to enhance the utilization of the Orange/Senqu River water by "storing, regulating, diverting and controlling the flow of the Senqu/Orange River and its affluents." This joint project is intended to transfer water from the Highlands of Lesotho, where water is available and abundant, to the Vaal River catchment in South Africa, where water is needed the most, to augment water supply in the South African province of Gauteng, particularly the industrialized regions of the Pretoria Witwatersrand Vereeniging (PWV) triangle, and to generate hydropower for Lesotho (LHDA, 2003). Furthermore, Article 4 also permits both parties to

> undertake ancillary developments in its territory including: (i) the provision of water for irrigation, potable water supply and other uses; (ii) the development of other projects to generate hydro-electric power, and (iii) the development of tourism, fisheries and other projects for economic and social development.
>
> (LHWP Treaty, 1986: Article 4)

The overall *raison d'être* for this project can be summed up as (1) redirecting water flow into areas with high water demand in South Africa, (2) generating hydropower and enhancing the water irrigation for Lesotho, and (3) fostering the regional socioeconomic situation (Fullalove, 1997).

Process equity and treaty formation

The Treaty was cemented following a miasma of events that shaped the future of the region of Southern Africa. In this region, with such a significant river development, the LHWP emerged as the largest water development project to be officiated and implemented between the governments of Lesotho and South Africa and one of the largest of its kind in the world. In particular, as part of the LHWP, the Katse Dam system, incorporating five large dams constructed on the Malibamatso River in remote rural areas in Lesotho, was deemed Africa's largest dam project. With such a staggering magnitude, the project revealed a blend of ingenuity in engineering and innovation in ADR techniques and negotiation strategies.

The conflict over water resources, in this context, was a result of a combination of water scarcity and mismanagement of water resources, which led to negative sentiment and rivalry, and eventually caused regional instability that was nothing short of cataclysmic (Mirumachi, 2007). This mismanagement of water resources was a direct result of poor water management policies and weak institutional structure that together initiated and continued to exacerbate the problem (Grover, 2007).

The problem of freshwater scarcity emanates from insufficient availability of both surface water and groundwater to meet the rising demand of both urban and rural populations in the Southern African region. For Lesotho, freshwater was abundant in great volumes, yet it was underutilized because of lack of adequate infrastructure to extract and transfer it. In particular, Lesotho suffered from the lack of natural resources and investment capital, which together limited the county's economic growth and development (Newton and Wolf, 2008). Water, as such, acted as Lesotho's political arrow in the quiver to promote the transfer pro-ject and reap its benefits. Conversely, with its significant industrial development and population growth, South Africa lacked water, which was the only abundant resource in Lesotho. Consequently, South Africa's national policy demonstrated long-standing interest in searching for alternative water resources. This included utilizing water from the Lesotho Highlands to meet an exponentially growing demand for water for South Africa's industrial heartland, stretching from Pretoria to Witwatersrand, which consumed most of its local water resources (Poivey, 1997; Newton and Wolf, 2008).

In the post-apartheid era, both countries realized the importance of remedying these problems and addressing their various needs. After 30 years of negotiations, a treaty between the government of South Africa and its much smaller neighbor, Lesotho, with the help of the World Bank as a natural and mutually agreed upon third party, was negotiated and officiated in 1986, allowing the execution of a multi-phased project (the LHWP) (Rothert, 1999; Mirumachi, 2007). The Treaty between the two countries embraced five phases, but committed the countries only to the first phase, involving the LHWP (Davis and Hirji, 2003). The LHWP is a major transnational project involving a two-component water supply and hydropower project resulting in a win–win situation for the variant, yet mutual, benefit of both countries.

Aiming to negotiate the technical and financial details of this water transfer scheme, the Treaty addressed the major issues from which both parties suffered, which included a water deficit in the South African industrial hub and a lack of hydropower for Lesotho's internal consumption and general development. Considering the compensation for transferred water and energy resources, the Treaty provided a strategy for South Africa to buy water from Lesotho and finance water diversion, on the one hand, and for Lesotho to utilize these payments in exchange for water and development aid for hydropower generation and general development, on the other hand (Priscoli and Wolf, 2009). These payments, along with international aid from the World Bank and other donors, helped Lesotho finance the cost of the hydropower and development components of the project.

The pre-negotiation and negotiation process

Labor-intensive development and large-scale irrigation projects along the Vaal and Orange, which today form more than 50 percent of South Africa's water demand, posed a continuous reminder of the significance of inter-basin transfer projects (Conley and van Niekerk, 1998). The idea of a regional water transfer project is not new. In 1928, the South African Irrigation Department proposed a tunnel to transfer water from the Orange River to quench the thirst for water in the Eastern Cape (Turton, 2003). This idea grew into what was known as the Orange River Project (ORP) in 1968, with the slogan of "taming a river giant" (Turton, 2003). Because of South Africa's growing interest in Lesotho's waters, two feasibility studies were conducted in the 1950s and 1960s (Fullalove, 1997; Hassan, 2002). However, these attempts were not successful in reaching agreement on payment for water exports. In the early years of the 1950s and 1960s, these plans to tap into the available water resources of Lesotho were undergoing a thorough assessment, most of which suggested the adoption of the LHWP as the premiere option for augmenting and redistributing the regional transboundary water supply.

The two key project proposals that were under consideration by officials and engineers from the 1950s until the 1980s were the Shand's Oxbow Scheme and the Malibamatso Scheme (Whann, 1995; Mirumachi, 2007). At that time, these projects were mostly concerned with transferring water at an "attractive price," and the idea of swapping water for hydropower was not completely mature. Upon the request of the High Commissioner to Lesotho, Sir Evelyn Baring, a survey of the available water resources of the region was administrated to determine the best course of action to alleviate water stress in South Africa (LHWP, 2009). Although it was considered ambitious, a five-phase project was proposed in 1956 by the South African engineer Ninham Shand, which aimed at securing enough water for the development of South Africa's industrial region (Stevens, 1967; Turton, 2003; LHWP, 2009). This water would be exported from the Orange/Senqu River in Lesotho to South Africa's industrial heartland to supply the gold mines through the construction of the "Oxbow Scheme," an element of Shand's project of 1956 (Meissner and Turton, 2003; Mirumachi, 2007). This project proposal, which was estimated at a cost of L 15 million, included a high-altitude dam, a hydro-electric power station, and a tunnel through the Maluti Mountains utilizing the Elands River. It was envisioned to generate more than 350 million kilowatt-hours of energy each year, while transferring about 380,000 m^3 of water each day (Stevens, 1967; Mirumachi, 2007). However, the idea of buying water and electricity from Lesotho was not favored by South Africa, which seemed sceptical, at the time, about giving much attention to such plans (Mirumachi, 2007).

Another major challenge that posed a high economic risk for South Africa in considering these plans was the lack of developed infrastructure capacity in Lesotho, which could potentially require substantial economic investment by South Africa (Meissner and Turton, 2003). Although this scheme was rejected by South Africa at first, the drought of the mid-1960s compelled the country to reconsider it as a viable option (Turton, 2003; Mirumachi, 2007). However, unwillingness to compromise its sovereignty by relying on Lesotho's resources to support its economic development prevented South Africa from giving this plan full consideration (Meissner and Turton, 2003). On the other hand, state officials in Lesotho were divided with regard to the scope of the project. Some preferred the internal utilization of Lesotho's water resources on a less extensive scale, whereas others preferred the option of negotiation in order to transfer this water to South Africa on a large scale.

Ongoing negotiation between the two countries was terminated in the early 1970s, as disagreement between the countries ensued over royalty payments for the water transfer project, and this caused a political impasse on the hydro-diplomatic front until 1976 (Whann, 1995; Turton, 2003). Meanwhile, the demand for water in South Africa increased as drought conditions intensified, which compelled South Africa to pursue the Tugela–Vaal transfer project instead (Turton,

2003). However, this project, which was established in 1974, was not large enough to meet the demand for more water. By 1983, this project was yielding 1 MCM of water that was transferred to South Africa's most industrialized region (the PWV region at the time) daily, with the expectation that this amount would increase to 3 billion m³ per year, which is equivalent to two-thirds of the Tugela River's annual discharge (Robbins, 1983). Although negotiation between the two countries was resumed in 1975, no agreement was reached regarding the technical or financial aspects of a transfer project (Turton, 2003). The situation of severe drought compelled South Africa to seriously consider the feasibility of other plans that addressed both the water needs of South Africa and the economic development needs of Lesotho (Whann, 1995).

Investigations regarding the feasibility of the LHWP continued unilaterally and intermittently. The goal was to identify and evaluate the feasibility of a dual-purpose project, capable of delivering much-needed water from the upper Orange and Malibamatso Rivers to the Pretoria–Witwatersrand area, on one hand, while ensuring Lesotho's self-sufficiency in terms of power supply, on the other. To determine the viability of such a project, British and South African teams, in collaboration with the World Bank, conducted preliminary feasibility studies in which technical teams from South Africa started to collect rock samples from different sites in Lesotho during the mid-1970s (Mirumachi, 2007; LHWP, 2009).

In addition to water scarcity and resource needs, political issues related to security and lack of mutual trust that dominated the political landscape for many years because of the apartheid system, directly influenced the course of negotiation between the two countries. Since its independence in 1966, Lesotho had been actively searching for a source of revenue to support its economic development needs, and water was potentially the answer. On the other hand, the Basotho people had long been struggling to maintain their identity, national integrity, and political independence against their very powerful neighbor's apartheid policies that infringed on their political sovereignty (Weisfelder, 1979; Scott, 1985). Mounting domestic pressure in Lesotho led the government to publicly criticize the apartheid system in the 1970s, which caused political tension between the two countries (Mirumachi, 2007). This, and the unsettling political upheaval in South Africa, created a situation of mistrust causing a political impasse, hindering the ability, once again, of the two countries to negotiate between 1976 and 1978 (Turton, 2003; Turton et al., 2004). Because of the political situation at that time, which was less than ideal for the two countries to operate jointly with regards to water schemes, separate unilateral efforts had been conducted in 1967 and 1968, including preliminary feasibility studies. This changed in 1978 when a JTC that included experts from both countries was established and started its joint work to conduct another feasibility study for the project (Mirumachi, 2007; LHWP, 2009). The initial findings of these studies were indicative of a possible water transfer project similar to the Shand's proposal in the 1950s. As negotiation went on, additional joint feasibility studies were warranted in 1979 and 1983, and both countries further agreed to conduct a final feasibility study in 1986, which recommended a 70 m³/s water transfer project with a hydropower generation component, reflecting Lesotho's desire to generate electricity (Poivey, 1997; Turton, 2003). Although both countries were in agreement regarding conducting these feasibility studies, donors were skeptical to support a project that might serve South Africa's apartheid system, or might be viewed as such (Whann, 1995). Because of this, negotiators agreed to a final arrangement whereby both countries would conduct two separate feasibility studies that would be reviewed by the other party and would eventually be combined into one final report. To that end, the Lahmeyer MacDonald Consortium worked on behalf of Lesotho, while the Olivier Shand Consortium worked on behalf of South Africa.

Although many proposals had been developed, considered, and mooted by the negotiation teams since the 1950s, progress on the diplomatic front was sluggish. This was because the development and progress in the negotiation of the Treaty reflected the overall political situation and the nature of the state officials' relationships. In this regard, it was not until the political relationship between the leaders of the two countries improved that significant progress was attained in the negotiation of the Treaty. Conversely, any deterioration in the political relationships between the two countries' leaders had a dramatically adverse impact on the progress of negotiation. However, the ingenuity of the makers of the Treaty was demonstrated as a result of their ability to separate the overall political conflict and turmoil between the two countries from the progress attained on water-related negotiations. Despite the anti-apartheid stance of many of Lesotho's leaders in the late 1970s, namely Chief Leabua, the diplomatic front took a proactive turn as the negotiators of the Treaty agreed to continue their efforts to conduct the feasibility studies (Whann, 1995). These political aspects of the conflict, which had led to an impasse in the past, had been side-stepped and were kept separate from the overall progress of the Treaty formation, which in turn made it possible for the two countries to reach agreeable solutions that ultimately served the interests of both sides.

With these feasibility studies under way, the particularly long drought of the 1980s and the tighter control over water use at the local level rejuvenated interest in tapping into Lesotho's water to address recurrent water shortages.[15] As the need to find a long-term solution became apparent, the national focus of both countries was shifted to one in agreement with proceeding with such a large-scale scheme, which could potentially generate cooperation and development (Mirumachi, 2007; Lang, 2008). In 1983, the two countries mutually agreed to conduct a final feasibility study and a more detailed layout of the project (Poivey, 1997). These studies underwent several stages and aimed to evaluate two project alternatives: (1) the LHWP, a three-phase project with a transfer capacity of 50 m^3/s, and (2) the Orange–Vaal Transfer Scheme (OVTS), with a transfer capacity of 70 m^3/s (Whann, 1995; Mirmachi, 2007). The latter scheme included a 500 km aqueduct that would transfer water from the southwest of Lesotho to the Vaal Dam, placed on a higher altitude than the intake (Wallis, 1996). A third alternative that was also considered suggested another layout that combined the best elements of both projects to maximize efficiency and minimize cost. Because of the Vaal Dam's higher altitude, water needed to be pumped and the OVTS alternative was proven to be more costly (Wallis, 1996; Mirumachi, 2007). About two years prior to the signing of the Treaty, both teams realized that the LHWP was the best single alternative that would provide the needed efficiency with minimum cost. Although the OVTS was not the most favorable project alternative, it was nonetheless regarded as the benchmark against which the cost and benefit of the LHWP could be determined, namely the amount of royalties that South Africa would pay to Lesotho for the price of transferred water (Whann, 1995). A very detailed and lengthy feasibility study was completed and mutually agreed upon by both parties in December 1985, and was revealed to the public shortly after, in April 1986 (Whann, 1995). The completion of these more detailed studies in April 1986 indicated a strong preference and potential for the project, and signified the possibility of cementing a diplomatic agreement (LHWP, 2009).

Historically, South Africa had been actively searching for alternative water resources to supplement its limited available water resources and cater to its growing industrial and domestic water needs. South Africa did not have many feasible alternatives to supplementing its annual water budget, which was insufficient to meet the demand. To augment its water supply, a number of options for water allocation and transfer plans had been proposed and long considered. According to Whann (1995) these include (1) the Tugela River project in KwaZulu and Natal Provence; (2) the Zambezi River project, on the border of Zimbabwe and Zambia; (3) the Okavango Swamp project in Botswana; and (4) the LHWP. Except for the LHWP, none of these alternatives seemed favorable to South Africa at the time. Although the first alternative, the Tugela River project, was already in place, it suffered practical and institutional difficulties which hindered its implementation. In addition to these implementation-related challenges and political disputes, the project was not capable of accommodating the substantial magnitude of the water need of South Africa's industrial uses because of its relatively small size. The Zambezi River project was also politically problematic because of the tension between South Africa and the governments of Zambia, Zimbabwe, and Botswana, caused by the South African apartheid policies (Whann, 1995). Similarly, the economic and environmental costs of the Okavango Swamp project were significant, in part because of its geographic location and the long distance between the origin of the water and its intended destination, the PWV triangle. Although representing an appealing and effective approach to South Africa's water supply augmentation, desalination is a very costly alternative that, given the size and scale of water demand, might seem prohibitively unfeasible, in addition to the difficulties in maintaining such systems. Although these projects provided a variety of water supply options, none had so far proven to be politically, economically, or practically feasible. With these difficulties surrounding the available options for South Africa, the LHWP emerged as the most feasible project, given its magnitude and political and technological prospects. For South Africa, the only other intrinsic alternative to cooperation with Lesotho was to wait and collect water as it naturally flowed downstream and then attempt to pump it back up into the higher elevations, which would have been a costly and environmentally disastrous alternative (Mckenzie, 2009). In contradistinction, the LHWP operated by utilizing natural gravity, with no pumping required to move water into South Africa (Mckenzie, 2009).

The aforementioned reasons demonstrated that the LHWP was the most suitable alternative for all parties, considering the political and socioeconomic factors. What motivated the two countries to pursue the LHWP was the economic advantage that the project offered Lesotho, the large size of the project, and its feasibility compared with other options for water transfer available to South Africa, as discussed above. The two countries, as a result, realized the value in undertaking this project and entering into a diplomatic agreement that would govern the construction phases of the project and maintain cooperative and diplomatic relations between them. Although the project provided a technically feasible and economically viable option, political tension and security concerns prevented the two countries from implementing the project, which was greatly politicized as a result. In particular, in 1982, South Africa, viewing Lesotho as an "extremist state," accused the latter of housing the then banned African National Congress (ANC) armed wing, Mkonto we Sizwe (MK), and was further concerned about the communist influence in Lesotho (Turton, 2003; Mirumachi, 2007). Similarly, Lesotho was also concerned about South Africa's role in destabilizing the country by supporting the anti-government Lesotho National Liberation Army (LNLA) (Mirumachi, 2007). As a result, South Africa blocked the border with Lesotho and further demanded that Lesotho sign a security agreement as a prerequisite for any future negotiations over the LHWP, which in turn was viewed by Lesotho as a superfluous and unfounded request (Turton *et al.*, 2004; Mirumachi, 2007). To that end, Lesotho did not make a connection between lack of security and progress on the LHWP negotiation and insisted on separating these two major issues to advance negotiation to a final settlement (Turton *et al.*, 2004). With the national security of both countries at stake and the diplomatic relations in shambles as a result of the growing local and international opposition to the apartheid system, little progress was achieved on the hydro-political front (Mirumachi, 2007).

The decisive breakthrough that influenced the signing of the Treaty and the implementation of the LHWP was the regime change in Lesotho (Mirumachi, 2007), particularly the military coup of January 1986, in which Major-General Justin Metsing Lekhanya overthrew Chief Leabua Jonathan (Esterhuysen, 1992; Turton, 2003). Unlike the previous regime, the new government in Lesotho was more inclined toward achieving cooperative arrangement with its counterpart South African government, despite the apartheid regime (Mckenzie, 2009). This helped in smoothing the way and creating a "window of opportunity" for a more serious negotiation and consideration of the LHWP (Meissner and Turton, 2003), which had been informally mooted for many years, but to no avail. With this new government, which was no longer viewed by South Africa as openly anti-apartheid and therefore unfriendly, along with the domestic support for strategic economic investment through the utilization of water resources in Lesotho, the prospect for reaching a reasonable and acceptable deal over the Orange River waters was closer than ever. Their diplomatic and technical efforts to cement a lasting and meaningful agreement, which started in the 1950s and lasted for 30 years, concluded on October 24, 1986, with the signing of the LHWP Treaty by leaders of the two countries: the South African Minister of Foreign Affairs, R. F. Botha, and Colonel T. Letsie of the Military Council in Lesotho (Poivey, 1997; Turton, 2003; Mirumachi, 2007). Both the drafting and signing of the Treaty was facilitated by the World Bank, which financed 4 percent of the project and, more importantly, was the catalyst for acquiring international finance for the project (Hassan, 2002).

Although the first draft of the text of the Treaty was developed eight months before its signature, further consideration of various issues surrounding the financial and hydrological aspects of the project delayed the ratification of the Treaty until it was ready. To that end, a South African civil servant (Department of Foreign Affairs) contended that "although we retained a number of basic features and principles of this draft, the document itself was completely re-written during the eight month drafting period immediately prior to the signing thereof" (cited in Whann, 1995). The makers of the Treaty regarded these changes made to the draft as minimal technical details, although many people in Lesotho provide a different account of these changes. They contend that, because of the political tension and deterioration in the relationship between the two governments in 1985, the draft agreement was rejected by Leabua's government (Whann, 1995). The opponents' account further holds that it was not until the coup of January 20, 1986, which brought a more accommodationist government to power with the support and help of the South African government, that the Treaty was signed after it was significantly revised (Whann, 1995; Rothert, 1999). These revisions were, according to this account, significant and detrimental to the interests of Lesotho (Whann, 1995). However, the negotiators of the Treaty still hold a very firm stance, affirming that the final draft of the Treaty was not much different from the first draft and the only differences were not more than minor technical changes.

Post-negotiation, Treaty implementation and the project's technical aspects

INSTITUTIONAL ARRANGEMENT AND WATER REGIME CREATION

The Treaty produced a massive inter-basin water transfer project that was to be developed over multiple stages, spanning across the period from 1990 to 2017 (Poivey, 1997). The estimated cost of the construction of such a project, with six sizable dams and tunnels, ranged from USD 6 billion to 8 billion (Lang, 2008). To facilitate implementation, collaboration between the states, both on the technical and political levels, was warranted and the need to create formalized institutional arrangements to channel this collaboration seemed evident (Fullalove, 1997). To that end, the Treaty resulted in the establishment of three new project authorities. The JPTC, based in Maseru in Lesotho, is a bilateral organization that was established as per the terms of the Treaty on the LHWP (Wallis, 1996; Turton, 2003). The project is also jointly implemented by two public sector organisations: the

Lesotho Highlands Development Authority (LHDA) in Lesotho and the Trans-Caledon Tunnel Authority (TCTA) in South Africa, both of which fall under the JPTC structure (Whann, 1995; EIB, 2002; LHDA, 2003). Representatives from the LHDA and the TCTA participate in the JPTC, which oversees and mentors the development of the project implementation in Lesotho. Its primary functions include coordinating all project activities in both countries, monitoring the work of both the LHDA and TCTA, and ensuring the relevance and feasibility of proposed activities (Fullalove, 1997). Because, in part, of the different interests of both countries in the project, namely Lesotho's interest in maximizing its economic gain and South Africa's interest in mitigating the cost of obtaining access to water (Mckenzie, 2009), the JPTC also acts to facilitate the activities of both the LHDA and TCTA, and further mediate and resolve conflict of interests between them (Fullalove, 1997; Turton, 2003).

The JPTC was later changed into the Lesotho Highlands Water Commission (LHWC) and still maintains its regular biweekly meetings (Turton, 2003). Reporting to the LHWC on all aspects related to the project, the LHDA was established to manage the implementation of the project construction, operations and maintenance located within Lesotho's borders (LHDA, 2003). It is the largest of the three institutions, given that the largest portion of the project is constructed in Lesotho. It incorporates four departments, including engineering and construction, operation, finance and administration, and environment and public relations (Fullalove, 1997). The TCTA, which is also responsible for managing the projects inside Lesotho, was created by the Department of South African Water Affairs and Forestry and oversees the construction, operations, and maintenance of the delivery tunnel, transporting water from Lesotho to South Africa (Wallis, 1996). These agencies are also responsible for securing financial support by raising funds toward the completion of the project.

As indicated in Article 2 of the Treaty, the designated governmental authorities for the implementation of specific provisions of the Treaty include the South African Department of Water Affairs and the Lesotho Ministry of Water, Energy and Mining, required to ensure the delivery of water and the enhancement of social and economic prosperity of the peoples of South Africa and Lesotho respectively (Wallis, 1996). The design and construction of the master plan of the project was conducted by four major consulting companies, including the Lesotho Highlands Consultants, the Lahmeyer MacDonald Consortium, the Lesotho Highlands Tunnel Partnership, and the Highlands Delivery Tunnel Consultants. The project infrastructure was developed under separate contracts with the Highlands Infrastructure Consultants, the Highlands Water Venture, the Lesotho Highlands Project Contractors, and the HMC Tunneling Venture. To ensure sustainable and cost-effective design, planning, and implementation, cutting-edge technological and management practices were employed, with the help of the World Bank. This was done to ensure meeting international standards and to maximize stability, accessibility, safety, productivity, and efficiency (Wallis, 1996; Fullalove, 1997; Hassan, 2002).

TREATY IMPLEMENTATION AND THE PROJECT'S TECHNICAL ASPECTS

The LHWP is one of the most ambitious and intricate schemes of its kind in the world currently under way (Villiers *et al.*, 1996). The LHWP is an ongoing transnational water transfer project that diverts water from the Senqu River in Lesotho northward into the Vaal River in South Africa, utilizing natural gravity to store water in dams and transfer it without the need for pumping (Poivey, 1997; Mirumachi, 2007). The impact of the project is nothing short of substantial, as when completed it will double the water resources available to the Gauteng region, the economic and industrial hub of South Africa, and enhance the socioeconomic conditions of the impoverished country of Lesotho (Mckenzie, 2009).

Constructed in large part in Lesotho, the project incorporates substantial tunnel system development and extensive infrastructure construction consisting of a system of five dams and underground tunnels (more than 200 km) and a hydropower plant at Muela (Wallis, 2000). These infrastructure components were constructed to store and transfer excess water from Lesotho to South Africa's Gauteng area (previously Johannesburg). Because, in part, of its technical intricacy and large size (in addition to the boycott of international aid for the then apartheid South Africa), the project is managed, financed, and implemented in five phases, starting with the higher altitude region in Lesotho, as shown in Figure 5.4 (Villiers *et al.*, 1996; EIB, 2002).

The first phase of the project is being implemented in two steps that are divided geographically into two portions (IA and IB) (Villiers *et al.*, 1996; Wallis, 1996). The first portion (roughly three-quarters of the first phase and about 95 percent of the total required construction) is located in Lesotho, whereas the other portion (about 22 km of the delivery tunnel) is located in South Africa (Wallis, 1996). As shown in Table 5.8, Phase IA of the project (which was completed in 1998 at a cost of USD 2.4 billion) comprises the 185 m high concrete Katse Dam on the Malibamatso, a 48 km water transfer tunnel to the 55 m high Muela Dam, and a 72 MW hydropower station (Villiers *et al.*, 1996; Rothert, 1999; EIB, 2002; Mirumachi, 2007). This phase is constructed on the upper portion of the Orange River, namely the Senque and Malibamatso, and utilizes the altitude difference to transfer about 18 m^3/s (a total annual yield of 567 MCM) into the Vaal Dam, located 70 km south of Johannesburg (Villiers *et al.*, 1996; Wallis, 1996; Fullalove, 1997).

Phase IB of the project consists of the 145 m high Mohale Dam at another tributary of the Senqu, the 38 km Mohale transfer tunnel and the Matsoku tunnel and weir (Davis and Hirji, 2003). This phase was completed in 2002 at a cost of USD 1.5 billion (Newton and Wolf, 2008). The size of the project infrastructure gives us a glimpse of the engineering feat and sheer magnitude of this project. For example, the Katse Dam was constructed with a maximum height of 185 m, crest length of 700 m and a reservoir capacity of 1,950 MCM. Over 2 MCM of concrete was used for the construction of this sizable dam. Because of its large size, the Katse Dam was divided into several building blocks for implementation (Wallis, 1996). Further, the tunnel system was designed in the first phase to be

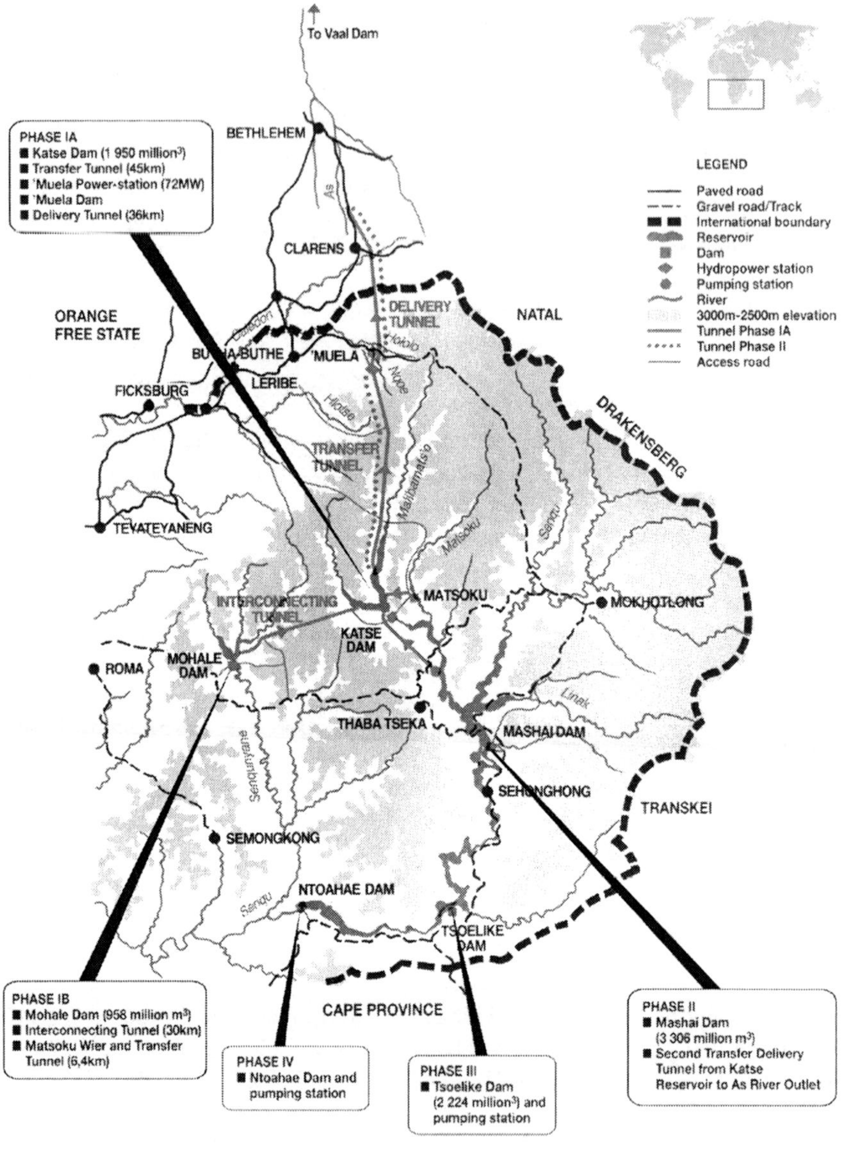

Figure 5.4 Phases of the Lesotho Highlands Water Project. Source: adapted from Wallis (1996).

large enough to handle the combined flow from both the Katse and Mohale Dams (Villiers *et al.*, 1996).

Monitored by the World Bank and a multidisciplinary team of experts, Environmental Action Plans (EAPs) for Phase IA of the project were carried out during the construction of this phase (Fullalove, 1997; Hassan, 2002). EAPs for

Table 5.8 Phases of the Lesotho Highlands Water Project

Phase	Yield	Facilities
IA completed 1998	18 m³/s flow 72 MW hydropower	185 m Katse Dam (1,950 MCM) on the Malibamatso River 55 m Muela Dam on the Nqoe River Muela hydropower station (72 MW) Transfer tunnel (45 km) from Katse Reservoir to Muela hydropower station Delivery tunnels (36 km) from Muela Dam to Axle River outlet Transmission lines to Maseru Access roads and other infrastructure
IB completed 2002	9 m³/s flow 2 m³/s flow	145 m Mohale Dam (958 MCM) on the Senqunyane River Transfer tunnel (30 km) from Mohale to Katse Reservoir 20 m Matsoku Weir on the Matsoku River Delivery tunnels to Katse Reservoir Increased capacity of the Muela power station (110 MW) Access roads and other infrastructure
II III IV V	Total estimated yield for all phases (I–V): 70 m³/s flow	Mashai Dam (3,306 MCM) on the Senqu River Pumping station connecting Mashai and Katse Reservoirs Tsoelike Dam (2,224 MCM) on the Senqu River Pumping stations and transfer tunnels Ntoahae Dam on the Senqu River Malatsi Dam on the Senqunyane River

Source: adapted from Davis and Hirji (2003).

Phase IB were conducted in 1994, leading to an instream flow requirement policy, which was implemented in 2002 (Davis and Hirji, 2003). Because of the socioeconomic and demographic changes and environmental challenges imposed by the construction of the project, the feasibility of the second phase was commissioned for communities downstream of the project in 2004 (Newton and Wolf, 2008). This study was also originally scheduled to be completed by December 2007 for its widely anticipated presentation before the parliaments of the two countries. However, this study has been delayed (Lang, 2008).

Upon the completion of Phase IA in 1998, it was initially expected that South Africa would receive 18 m³/s of additional water (enough to meet its needs until 2003), while Lesotho would enjoy an additional 72 MW of electricity per hour (Villiers *et al.*, 1996; Poivey, 1997; Davis and Hirji, 2003). It was also estimated that Phase IB would yield an additional 9.7 m³/s of flow, which would satisfy South Africa's water needs until 2007 (Villiers *et al.*, 1996; Davis and Hirji, 2003). In the wake of the implementation of the first phase of the project, industrial development and irrigation in South Africa had received a tremendous boost from the additional water transfer, which is estimated at 30.2 m³/s at the time of this writing (Wallis, 2000). In addition, hydro-electric power stations operating at Muela Dam to generate and transmit hydro-electric power to Lesotho have increased power supply to much of the country. The project also provided water to the vast downstream regions, transforming thousands of hectares of this arid area into a highly productive agricultural land.

The second phase will result in the construction of an embankment dam (the Mashai Dam) on the Senqu, and a transfer tunnel. Its height of 197 m makes the Mashai Dam the largest dam in the project yet, and one of the largest of its kind in the world (Villiers *et al.*, 1996). It was expected that this phase would increase the project yield by 55 m³/s, which was deemed sufficient to meet the water demand until 2007 (Villiers *et al.*, 1996; Fullalove, 1997). However, project implementation to date suggests that a rather smaller dam without a second transfer tunnel is being implemented instead (Mckenzie, 2009). In the third phase, the construction of the sixth embankment dam on the Senqu at Tsoelike (90 km downstream of the Mashai Dam) and associated pumping stations and tunnels will materialize, generating enough water (about 6.1 m³/s) until 2021 (Villiers *et al.*, 1996). The fourth phase involves the construction of a 125 m high concrete dam at Ntoahae and associated pumping stations and tunnels (Fullalove, 1997), while the Malatsi Dam and associated infrastructure will be constructed on the Senqu as part of the fifth phase of the project (Davis and Hirji, 2003).

Although plans for subsequent phases were put forth by the 1986 Treaty, implementation of these phases has not been carried out yet and further negotiations, studies, and analyses are needed in the future for the completion of the project (EIB, 2002). Upon the completion of all phases in 2020, the project is expected to include four dams and a network of interconnected tunnels with a transfer capacity of 70 m³/s of water (Fullalove, 1997; Poivey, 1997; Mirumachi, 2007). Until the project is fully implemented, the Treaty (in its Annexure II) identified minimum amounts of water to be transferred to South Africa, as illustrated in Table 5.9.

Overall, the project is running behind in terms of its implementation schedule

Table 5.9 Minimum quantities for water delivery identified by the Lesotho Highlands Water Project Treaty

Calendar year	MCM	Calendar year	MCM	Calendar year	MCM
1995	57	2004	695	2013	1452
1996	123	2005	772	2014	1545
1997	190	2006	852	2015	1640
1998	258	2007	932	2016	1736
1999	327	2008	1014	2017	1835
2000	398	2009	1098	2018	1934
2001	470	2010	1183	2019	2036
2002	543	2011	1271	2020	2139
2003	618	2012	1361	After 2020	2208

Source: LHWP Treaty (1986: Annexure II).

and financial arrangement, despite the optimism of the Chief Executive of the LHDA, Masuoha Sole, who asserted the success and financial feasibility of the project implementation. Although the project was scheduled to deliver water by 1995, this portion of the project was not completed until 1998 (Rothert, 1999). As these five phases have not been implemented to date, many experts question not only the value of their implementation, but also their impact on the environment. Currently, because of environmental and financial constraints, another plan which utilizes the existing tunnels is being considered without the construction of additional costly transfer tunnels (Mckenzie, 2009). In addition to the delay in its implementation, the project was also over budget (Whann, 1995). The delay in the project implementation was partly blamed on financial constraints, particularly the ability to acquire needed funds from donors to cover the cost of construction, and labor violence, which erupted frequently. Labor violence between the imported skilled workers and unemployed Basotho caused the construction of the project to be delayed for several weeks (Whann, 1995). These troubling aspects in the implementation were the result of employing and importing non-Basotho workers, which triggered anger and resentment in Lesotho. To that end, many people believe that the project could have been more beneficial to the general public of Lesotho had it considered employing only Basotho workers, rather than distributing the benefit among Lesotho's political elites.

By providing South Africa with much-needed water for industrial and economic development in the central Gauteng province on the one hand, and Lesotho with a source of income and hydro-electric power in exchange for this water on the other hand, the Treaty not only appropriately and astutely addressed each country's needs, but also provided a practical example of how water can be viewed as an economic commodity with a monetary value that can be justly compensated for. The fact that the Treaty makers took into account both countries' specific needs, as different as they were, rather than allegations of entitlement and sovereignty, boosted a strong trust-building process between the disputing parties and facilitated a positive-sum situation, in which both countries seem to generally hold a favorable perception of the Treaty as being equitable and fair in terms of its process and outcome alike.

Collaborative collection of exact water availability data in Lesotho was not conducted until after the Treaty was signed. While the methodology for determining royalty payments was established by the Treaty, there was a general understanding that there would be agreed upon accurate hydrological data in the future, based on which exact figures of royalty payments would be calculated (Mckenzie, 2009). Accordingly, two separate teams, one from South Africa and the other appointed by Lesotho, were put in charge of the assessment of water availability in Lesotho and determining how much water can be delivered to South Africa based on available hydrologic data. These teams occasionally meet to reach an agreement on the hydrological data. According to Mckenzie (2009), there was pronounced disagreement on the factual basis for more than a decade, which required an ongoing data mediation and verification on both sides. The figures reached by the South African team suggested 20 percent less water than what was suggested by the team appointed on behalf of Lesotho. One might speculate that Lesotho's tendency of overestimating its water availability was primarily driven by its interest in maximizing the royalties received from South Africa in exchange for water.

The implementation of the Treaty and operation of the project was proven fixable in the face of changing realties and new circumstances. Over time, several changes occurred with regards to both the political and technical aspects of the project. Particularly, there was mounting skepticism of the ability of the two countries to resume smooth implementation of the project in the wake of a miasma of events, namely the 1990 coup against Lekhanya, the 1993 reinstatement of civilian rule, and the 1995 restoration of King Moshoeshoe II in Lesotho (Whann, 1995). The state officials and diplomats, as well as many experts, were all attesting to the growing jitters in Lesotho and South Africa of possible failure of their cooperation in the wake of such events. However, these skepticisms were proven anachronistic, as the Treaty has been, and is still, adhered to by the two countries notwithstanding these changes. This is because the compartmentalization and segregation of the high politics issues helped in depoliticizing the negotiation and its outcome. This depoliticization and removal of the project from external political influence produced and maintained steady implementation.

Flexibility was evident in the incorporation of several protocols as part of the Treaty and the creation of new water regimes to govern and facilitate the implementation and operation of the project. First, although the Treaty has not been subject to any major changes since its signature in 1986, a number of additions were included in the Treaty. According to the LHWP website, (LHWP, 2009), the following six protocols, some of which were initiated and referred to by the Treaty, were added:

Protocol I: includes the royalty manual, which outlines a detailed mechanism for calculating the royalties due to Lesotho.

Protocol II: addresses the impact of the project on the Southern African Customs Union (SACU)[16] revenue due to Lesotho.

Protocol III: determines cost apportionment between Lesotho and South Africa.

Protocol IV: includes states' supplementary arrangements regarding cost-related payments, concessionary finance, and insurance. It also details when royalty payments to Lesotho would start. These payments were agreed to start when the water level in the Katse Dam reached 1,993 m above sea level, which was achieved in September 1996.

Protocol V: addresses how taxation in Lesotho can apply to the project.

Protocol VI: changed the name and responsibilities of the JPTC to the LHWC, which became necessary during the course of implementation. It also changed the role of TCTA beyond the auspices of the LHWP.

According to Woodhouse (2008), these revised protocols, while responding to new changes, were primarily based on what was accomplished in 1986. This is reflective of an incremental approach which first attempts to cement successful bilateral agreements before reaching a basin-wide multilateral settlement. This is because bilateral agreements, although they can be narrowly focused on specific issues unique to the two riparian states, are generally more efficient and easier to accomplish compared with multilateral agreements, which require the involvement of all parties. As the situation becomes more conducive for multinational cooperation, this approach may later proceed with a multilateral strategy that ties all of these binational agreements into a unified and comprehensive multinational framework involving all stakeholders within the basin.

Second, resilience in implementing the Treaty may also call for certain changes in established water regimes and may further necessitate the initiation of new ones reflecting the fluctuating political, economic, and social landscape. The establishment of the ORACOM in 2000 was the first step toward the creation of a multilateral basin-wide regime that overlaps with existing bilateral regimes without superseding them (Turton, 2003; Earle *et al.*, 2005). According to Woodhouse (2008), ORACOM provided a regional instrument for the management of transboundary water resources involving all states in the South African region. Based on the defining principles of international water law, namely the Helsinki Rules of 1966 and the 1997 United Nations Convention on the Non-Navigational Uses of International Watercourses, the ORACOM adopts an equitable vision of reasonable and equitable utilization and inflicting no significant harm in managing the shared water resources and exchange of information regarding these resources.[17] The establishment of ORACOM helped in fostering regional cooperation and a shared vision among all riparian states regarding the development of the river basin (Pallett *et al.*, 1997). For the ORACOM to succeed in achieving its goals, compromising a certain level of sovereignty by riparian states over their water resources is both warranted and feasible (Turton, 2003). Figure 5.5 shows a diagram that depicts the changes over time in the water regime structure governing the river basin. It is clear that the basin is moving toward a multilateral water regime creation and management, despite the history of bilateral agreements.[18]

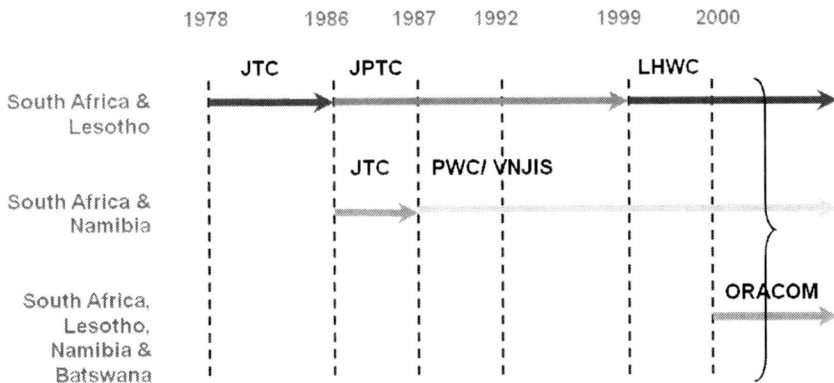

Figure 5.5 Historic overview of regime creation in the Orange River basin. Sources: adapted from Turton (2003); and Earle *et al.* (2005).

For example, the creation of the National Water Resource Strategy in 2012, which aims at achieving equitable water management, provides a framework for co-riparians to manage their regional water resources through the enhancement of multilateral cooperation. Several strategies for achieving equitable growth and sustainable development are put forth, such as implementing equity policies, making water the centerpiece of integrated development planning and a decision-making process that involves all parties, optimizing available water resources, and implementing sustainable business practices, to name a few (Department of Water Affairs, Republic of South Africa, 2012).

Content analysis

The Treaty, which is an internationally recognized bilateral agreement between the governments of Lesotho and the Republic of South Africa, provided significant technical, financial, and legislative details regarding the construction and operation of the LHWP. Many experts regard the text of the Treaty as highly technical in nature, as it was developed by a number of technocrats, including civil engineers, financial experts, and legal advisors. The Treaty text includes 18 articles in addition to a preamble and three annexures, including a project description, minimum quantities of water delivery, and privileges and immunities. Article 1 of the Treaty provides ample explanation and definitions of the terms and acronyms used in the body of the Treaty text to eliminate any misunderstanding and misinterpretation of its terms. While Article 2 outlines the designated governmental authorities for the implementation of specific provisions of the Treaty, Articles 3 and 4 outline the purpose of the Treaty and of the project, which is discussed earlier in this chapter. According to Article 4, both countries are allowed to undertake unilateral river development projects related to the LHWP from both sides of the border without harming the interests of the other side specified by the Treaty. Given this, any harm to either Lesotho's economic base or South Africa's water transfer is not permitted under this Treaty, as it provides protection for both countries' interests.

Article 5 of the Treaty includes provisions pertaining to project implementation. Making reference to project phases and technical details included in Annexure I, this article specifies 70 m³/s as the ultimate water delivery goal of the project. These minimum water delivery amounts listed in Annexure II are characterized by the Treaty as fixable and can be modified as the parties jointly wish. It further indicates that "water deliveries to South Africa from Sub-phase IA of the Project shall be due to commence in the year 1995 and water deliveries to South Africa from Sub-phase IB of the Project shall be due to commence in the year 2002" (LHWP Treaty, 1986: Article 5, paragraph 2). However, these deadlines were not met as the project implementation suffered delays due to financial and political constraints, as discussed earlier in the implementation section of this chapter. Although this part of the Treaty gives protection to South Africa in terms of its water transfer rights from Lesotho, many South Africans who opposed the project believed that the source of this water is not secured as it would be under the control of the Lesotho government, which at the time was not viewed as a friendly partner (Whann, 1995). To that end, the Treaty explains that

> the conveyance systems for the discharge from the most downstream hydro-electric power station in the Kingdom of Lesotho shall be designed, built, operated and maintained in such a manner that neither Party shall be in a position to interfere unilaterally with the flow of water to the Designated Delivery Point.
>
> (LHWP Treaty, 1986: Article 5, paragraph 3)

Similarly, the Article also provided protection for Lesotho's hydro-electric power generation, which would not be affected notwithstanding any dysfunctionality in the water delivery system.

Article 6, which describes the general duties regarding the project, states that implementation of the project phases subsequent to the first phase "shall be subject to the consent of each Party prior to such implementation" and that refusal of implementation by any party necessitates compensating the other party "for any wasted Project implementation costs reasonably expended by such other Party in anticipation of the implementation of such subsequent phase." This Article also lays out the overall responsibility for the parts of the project that fall within each of the countries, stating that Lesotho, akin to South Africa, is responsible for the part located in its territory. As such, both the benefits from the LHWP and the financial responsibilities of each country are outlined. In this regard, the Treaty stipulates that Lesotho would bear the cost of the hydro-electric power component of the project, from which it benefits, whereas South Africa would cover the cost of the water transfer component of the project, including construction, operation, and maintenance in addition to the price of the water it acquires from Lesotho (Poivey, 1997; LHWP, 2009). Accordingly, Lesotho has no financial obligation for the cost of the water transfer component of the project, which is entirely the responsibility of South Africa (Poivey, 1997).

As for the technical aspects of the project, the Treaty (in Articles 6, 7, and 8) stipulates the establishment of the Lesotho Highlands Development Authority (LHDA) in Lesotho, and the TCTA in South Africa, along with the Joint Permanent Technical Committee (JPTC) to oversee project implementation, operation, and maintenance. This includes developing implementation and operation plans and monitoring the quantity and quality of water transferred to South Africa while ensuring the wellbeing and economic vitality of Lesotho's communities adjacent to the project (Whann, 1995). Article 6, paragraph 16, further stipulates that "tendering for the Project Works shall be by competitive bidding without discrimination as to the nationality of any tenderer." It also prevents the parties from imposing any restrictions on "goods, materials, plant, equipment or services . . . required for the implementation, operation or maintenance of the project or any phase thereof" (LHWP Treaty, 1986: Article 6, paragraph 18). The Treaty also calls for Lesotho to comply with the minimum rate of water delivery specified in Article 7.

Articles 7 and 8 provide more details about the nature and responsibility of these agencies, namely the LHDA and the TCTA, including (but not limited to) preparing and compiling a long-term cost plan, a detailed cost plan covering the ensuing Financial Year, short-term and long-term funding plans, a schedule of the repayment of all loans, a long-term cash flow forecast, and a detailed cash flow forecast for the ensuing Financial Year. Article 9 delineates the nature, duties, and legal rights and responsibilities of the JPTC, most notably that it is "composed of two delegations, one from each Party" and that "each Party shall nominate three representatives constituting its delegation, as well as an alternate for each of the nominated residents in Maseru." Any decision of the JPTC has to be mutually approved by these two delegations to be valid. The JPTC has "monitoring and advisory powers relating to the activities" of the LHDA and the TCTA that impact the generation of hydro-electric power in Lesotho and the delivery of water to South Africa respectively (LHWP Treaty, 1986: Article 9, paragraphs 8 and 9).

Article 10 outlines the methodology and mechanism for determining cost-related payments. To that end, it is stipulated that the costs incurred subsequent to the entry into force of the Treaty include all costs related to

> the implementation, operation, maintenance, and other recurring costs of the Project . . . ; the establishment, administration and operation of the Lesotho Highlands Development Authority and the Trans-Caledon Tunnel Authority . . . ; measures necessary for catchment conservation as well as the prevention of pollution . . . ; the taking out of insurance . . . ; annual audits . . . ; land or any interest in land acquired for the purpose of the implementation, operation and maintenance of the Project . . . ; the measures in order to ensure that members of local communities in the Kingdom of Lesotho affected by Project related causes shall be enabled to maintain a standard of living not inferior to that obtaining at the time of first disturbance as well as compensation for loss to such members as a result of such causes not met by such measures; and interest payments.
>
> (LHWP Treaty, 1986: Article 10, paragraph 3)

These costs due by each Party as cost-related payments would be paid to the LHDA or the TCTA. The deliberate use of the term "not inferior to," as opposed to "better" or "superior to" the conditions prior to the implementation of the project, indicates that the makers of the Treaty were primarily concerned not with improving the overall quality of life of the Basotho population, but rather with maintaining the *same* quality of life notwithstanding its original inferior standards of living. Although the LHDA championed the project on the basis of its economic benefits to the Basotho, this provides conclusive evidence (through word choice) that this issue was not adequately addressed in the Treaty. It remains unclear whether this was a calculated strategic decision on the part of South Africa or simply a matter of linguistic efficacy. The Treaty also established a compensation strategy for impacted entities, both individual and communal. According to the Treaty, compensation will be provided "for any loss to such member as a result of such Project related causes, not adequately met by such measures" (LHWP Treaty, 1986: Article 7, paragraph 18). Based on a participatory process that involved impacted communities and stakeholders in addition to governmental institutions, an adaptive and comprehensive Compensation Policy[19] was drafted as early as 1989, and was revisited and reviewed in 1997 at the beginning of Phase IB (EIB, 2002; LHDA, 2003).

Article 11 details the financing arrangements of both the LHDA and the TCTA. In this regard, the two agencies are responsible for raising funds

> by way of loans, credit facilities or other borrowings, in such amounts as may be required for the implementation, operation and maintenance of the Project or any phase thereof or which may otherwise be required to meet the obligations of such Authority or to perform its functions.
>
> (LHWP Treaty, 1986: Article 11, paragraph 1)

Article 12 addresses issues related to royalty payments, which are not clearly outlined as precisely as the cost of the project itself. Making frequent reference to the Royalty Manual, this Article delineates a strategy for sharing the benefits and calculating the royalties to be paid to Lesotho for the water transferred to South Africa once Phase IA of the project is completed (LHWP, 2009). These royalties, which are paid monthly, are proportional to the amount of water to be transferred to South Africa (including capital expenditure and water cost), whereas the price of water is identified to be comparable to the alternative water-only OVTS project, discussed earlier in this chapter. These payments, which may vary based on the variation in the amount of water supplied to South Africa, are to continue until December 31, 2044. The secretive nature of the amounts of royalties to be paid to Lesotho undermined the Treaty's acceptance by the general public. In particular, these amounts were outlined in a secret annex of the Treaty that was not shared with the public and was only accessible to the senior members of the negotiating teams and government officials (Whann, 1995). It is estimated, however, that these royalty amounts for the transferred water are as high as USD 60 million per year based on the 1983 prices, which in 1998 was equivalent to 14 percent of Lesotho's overall GDP (Hassan, 2002; Hitchcock *et al.*, 2007; Mirumachi, 2007).

The Treaty provides adequate resilience to uncertainty in future circumstances by permitting the parties to adjust the amount of water transfer and recompute the royalties accordingly. To that end, the Treaty states that "in the event of adjustments to the minimum annual quantities of water" or "in the event of the remaining part of the Project being cancelled at a phase subsequent to Phase I . . . , the net benefit shall be recomputed" (LHWP Treaty, 1986: Article 12, paragraphs 6 and 7). However, the Treaty gave South Africa the option to unilaterally cancel the future phases of the project "at a phase subsequent to Phase I" and recalculate the royalties to be paid to Lesotho (LHWP Treaty, 1986: Article 12, paragraph 8). This provision imposes great uncertainty and poses threat to Lesotho's economic and financial stability. This is because the compensation for transferred water that Lesotho will receive will always be tied to the amount of water that South Africa decides to acquire, which is mainly contingent upon its annual water needs that may fluctuate from one year to the next. This clause, while providing flexibility to future conditions, tends to exacerbate the dependency of Lesotho's local and national economy on South Africa. In addition, this particular clause was also subject to criticism on the basis that it gave South Africa the upper hand in determining the price of water, as the Treaty permitted the reallocation of Lesotho's royalties, which in turn were reduced from 56 percent to only 50 percent (Whann, 1995).

Article 13 provides details regarding the amounts and due dates of payments of excess water, downstream releases, and water abstractions. Both water releases to (at the request of) South Africa and water abstractions by (at the request of) Lesotho are subject to compensation by South Africa and Lesotho respectively. Article 14 defines emergency clauses, joint action and cooperative procedures in cases of *force majeure*, which is defined as "any substantial impairment of the implementation of this Treaty caused by . . . an extreme hydrological or other natural event . . . ; the use of force by the states; armed insurrection and other forms of civil strife; and episodes of sabotage" (LHWP Treaty, 1986: Article 14, paragraphs 1 and 2). Composed of only one paragraph, Article 15 lays out the social and environmental considerations by asserting the "protection of the existing quality of the environment" and the issuance of compensation to ensure "the welfare of persons and communities immediately affected by the Project" (LHWP Treaty, 1986: Article 15). To that end, promoting and protecting the environmental and social aspects calls for developing environmentally and socially sound implementation and mitigation processes, along with feasible and efficient technical and engineering requirements of the project. Studies regarding the social and environmental impacts were conducted for both phases (IA and IB) of the project, and the EAP was developed to mitigate the social and environmental ramifications of the implementation of the project (LHDA, 2003).

Article 16, which outlines conflict prevention and resolution mechanisms, identifies the JPTC as the supreme entity to conduct investigation upon the request of any party within 14 calendar days and recommend necessary measures to mitigate issues of conflict to the LHDA and TCTA. Persisting conflict, which cannot be resolved by direct negotiation between the parties, will be subject to arbitration by a joint, definitive and binding Arbitral Tribunal (LHWP Treaty, 1986: Article 16). "Savings clauses" are included in Article 17, which validates and asserts the compatibility of the terms of the Treaty with previous bilateral and multilateral agreements involving the two parties, namely the "Customs Union Agreement, the Trilateral Monetary Agreement between the Governments of Lesotho, Swaziland, and South Africa of 18 April 1986, and the Bilateral Monetary Agreement between the Governments of Lesotho and South Africa of 18 April 1986" (LHWP Treaty, 1986: Article 17). According to Article 18, the Treaty can be reviewed and amended every 12 years, or by the request and agreement of the parties. This offers the parties more flexibility in implementing the Treaty and more control over its validity and resiliency in the face of ever-changing realities. It also provides an opportunity for the parties to continue collaboration during the implementation stage to evaluate and revise the terms of the Treaty to keep up with these changes.

Perceived equity

The perception of the Treaty played an important role in its acceptance and level of implementation. The construction of the LHWP in 1984 drew tremendous controversy (Lang, 2008). At first, because of the political tension and the impact of apartheid, the success of the political and economic costs and benefits of the project were in question (Whann, 1995). There was considerable public opposition to the prohibitively high construction cost and the political, economic, and environmental impacts of the project. Public opinion and the general perception of the Treaty in both countries were shaped by a number of factors. First, the perception of the Treaty as inequitable in terms of water pricing, and of the project as a South African panacea to extend its control over Lesotho's most precious natural resource by undermining its sovereignty, appears to be due, in large part, to the lack of public involvement in the Treaty formation, on one hand, and the ambiguity surrounding project cost-sharing and the secrecy in estimating royalties, on the other. Much of the public opposition appeared to stem from the general view of the miscalculation of the economic cost of the project and the fact that such transfer projects regard water as merely a commodity that can be bought and compensated for. In estimating the cost of the project, surplus water was regarded as free rather than a marketable resource subject to compensation. In addition, the cost of irrigation and power generation was ignored, which resulted in this miscalculation in the economic cost of the project (Viessman and Welty, 1985). According to Whann (1995), many opponents saw some of its provisions as "inimical" to the sovereignty and interest of Lesotho, as they tended to reduce the royalties received by Lesotho by stipulating the option of recalculating them. This means that the amounts of these royalties can be changed at any time.

Second, the negotiation of the Treaty and financial arrangement of the LHWP were not the only issues that the Treaty makers had to contend with. Public opinion and attitude in both countries of the Treaty and of each other had a great impact on the acceptability and success of the agreement. Opposition in South Africa was mainly concerned with the fact that this source of water may not be guaranteed as the "faucet" would be controlled by the Lesotho government, which at the time was viewed as unstable and anti-apartheid (Whann, 1995). This attitude, however, changed as the project started to yield water to the designated delivery points in South Africa. The public attitude in Lesotho about the whereabouts of the revised final text of the Treaty and the perception that it had been manipulated for the purpose of serving the interests of the South African government and its allies in Lesotho created an atmosphere of suspicion and uncertainty regarding the intention and benefit of the Treaty. The Treaty, as such, was viewed not as a byproduct of the ongoing cooperation between the two governments, but rather as a tool of control and imposition to solidify the grip of the leaders of the new regime in Lesotho, who were viewed as "collaborators," rather than servants of public interests. This created public distrust in the process and its outcome and made it hard for the Treaty makers to gain public support for the agreement. This shaped, and was shaped by, the view of the previous leader of the country, Leabua, as a heroic leader who resisted the influence and interference of South Africa in the internal affairs of Lesotho. Although the state officials, who were in favor of the Treaty, spoke warmly about its outcome, they failed to address public concerns of this nature. By refusing to speak to or engage the public in a meaningful discussion, Lesotho state officials were not prepared to gain public acceptance of the economic benefit that the Treaty offers the country. This unpreparedness of state officials sparked tremendous opposition that remained unaddressed for a long time.

Third, although generally welcomed on the governmental level in both countries, skepticism and criticisms of the Treaty surfaced at the local level. In both countries, some people regarded the Treaty as a "sell-out" to the other party. The Treaty was particularly opposed and challenged in Lesotho, as much of this opposition was related to the relationship between the Lesotho military regime and South Africa (Rothert, 1999). People who opposed the regime in Lesotho viewed the Treaty as yet another governmental scheme to impose control over the population of Lesotho and its natural resources, whereas supporters of the regime were in favor of such diplomatic arrangements and saw the Treaty as a venue for future collaboration and economic stability.[20]

The state officials in Lesotho were under tremendous political pressure and were particularly criticized by skeptical and hostile citizens who viewed the Treaty as illegitimate and a huge political and economic miscalculation. Although the Treaty and the proposed LHWP aimed at enhancing the economic conditions of the Basotho, gaining public acceptance of the deal offered by the Treaty posed a tremendous challenge to the makers of the Treaty. This was partially due to the lack of public involvement in the Treaty formation, which created confusion, uncertainty, and therefore skepticism. The general public was not very aware of the negotiation process and the way the cost of the project was estimated and its future impact on their communities. For many years, this project was subject to harsh criticism and opposition in Lesotho. Many people in Lesotho perceived the project as a South African attempt to squander the only natural resource in Lesotho. Some people in Lesotho, particularly those who were in favor of the Leabua's government, went as far as referring to the LHWP project as "annexation of Lesotho without compensation" and "the final act of eliminating the nationhood of a people," demanding the renegotiation of the Treaty (Whann, 1995).

Fourth, in spite of the futile LHDA's public outreach efforts, secrecy surrounding many aspects of the Treaty negotiations and project implementation made it difficult for the public to be informed and involved. It is argued that the LHDA's attempts to share information with the public through its public meetings and publications (LHDA, 2003), namely the *Mohloli* magazine, the *Ka Metsing* (in the water) newsletter and two public radio programs, were not enough to increase public awareness of the project. Because of the difficulty in explaining the complicated technical and engineering aspects of the project to the non-technical public, the project was viewed in some circles as an apartheid project supplying water to South Africa while providing minimal benefits to the people of Lesotho (Rothert, 1999; Lang, 2008). In Lesotho in particular, people have frequently expressed concerns regarding the potential economic benefit that the Treaty is expected to bring to their lives. Many people did not understand the technical aspects of the project, the pricing system of water, in what ways it would be invested in Lesotho, and how this would eventually impact their livelihood. This general attitude was rendered stronger by the lack of information sharing about the formation and operation of the Treaty, particularly the complicated technical aspects of the project, which was claimed to be simply too complicated to explain (Lang, 2008). Furthermore, the previous government in Lesotho, which was the leading opposition party, was particularly critical of the Treaty, and sought a revision of its provisions. For example, at a meeting held in October of 1988 at the National University of Lesotho to present the LHWP, the Lesotho Highlands Development Authority (LHDA) spokesman was harshly criticized and openly challenged by some members of the audience who supported the Leabua Jonathan's government's approach. Particularly, Joel Moitse, a former cabinet minister in Leabua Jonathan's government, was openly vocal about his views of the Treaty as being out of touch with reality by failing to account for public feedback (Whann, 1995). Their criticism of the Treaty was centered on the fact that the makers of the Treaty negotiated behind closed doors without attempting to address the different public concerns and various interests and that the deal was cemented without any public involvement. By alienating the public and excluding any kind of public debate, the Treaty was viewed as inequitable by many who opposed the new military regime. Although the LHDA continued to emphasize the job-creation opportunities that the Treaty opened for the Basotho, the project was not entirely well received by many who originally opposed it on the basis of lack of public engagement.

Fifth, many experts believe that the environmental impacts of transferring these great quantities of water on the water quality and the deterioration of the ecosystem are overlooked. This perception is shaped by the general view that the Treaty paid little attention to environmental factors and the consequent deterioration of water quality of the river system. This also contributed to the negative perception of the project as being a non-environmentally friendly undertaking. Collectively, those factors tended to sow the seeds of suspicion and distrust in the Treaty in general. Because of the impact of these aforementioned factors, and the long-standing hostility and tension created by South Africa's apartheid policies, there was a general predisposition to criticize the Treaty. It is worth pointing out that much of the opposition to the project was a direct byproduct of the general opposition to the apartheid system, rather than a result of a general dissatisfaction in the project outcome per se (Whann, 1995).

However, as the project reached its fruition, the view of both state officials and the general public changed, namely when the project started to demonstrate its positive impact on the political economy of the two countries and the region at large. Opportunities for tourism and commerce ensued, creating many jobs for the Basotho communities. Thus, the overall economic base of the region was considerably improved in the wake of the project implementation. Furthermore, the end of the apartheid system made it easier for Lesotho to pursue economic prosperity through effective utilization of the project revenues in more socially conducive and economically efficient and effective ways, and in turn change the adverse public attitude of the Treaty. The recent change in the current civilian regime in Lesotho and its democratic government created a more favorable political climate and increased states' interest in maintaining continuous, open channels of public feedback and attaining equitable sharing of water resources.

Outcome equity

Cooperation over water resources entails certain benefits as well as costs. Benefits can be either water-related benefits or non-water-related benefits, while costs include both direct and indirect costs of the negotiation of the Treaty itself and its consequent implementation. Both the negotiation process and financial arrangements produced by the Treaty demonstrate how the two countries negotiated costs and benefits for the construction and implementation of the LHWP (Mirumachi, 2007). The defined goals and general benefits of such a large-scale development project, as summed up by Turton (2003), include

> a gravity feed of water in the Vaal Dam; sharing the cost-saving between both countries; avoidance of air pollution that would result from the use of coal-fired electricity to pump water . . . ; development of a degree of economic self-sufficiency in Lesotho; acceleration of socially and environmentally appropriate development in Lesotho; and meeting the needs of rapidly growing populations.

As shown in Table 5.10, Lesotho and South Africa developed a mutually acceptable cost and benefit-sharing mechanism. Direct benefits to Lesotho included economic and infrastructure development, hydropower and electricity generation, annual royalties for water transferred, and a solid foundation for nation-building capacity. In turn, Lesotho was responsible for the cost of the hydropower component of the project. Direct benefits to South Africa included augmentation of a secure and cheap water supply for its industrial and economic development, and political and strategic gains. Direct costs to South Africa included the cost of construction, operation, and maintenance of the project (except for the hydropower component), annual royalties payments to Lesotho, and compensation and environmental mitigation costs.

Quid pro quo (water benefit sharing)

The Treaty produced a unique outcome, based on the principle of benefit sharing, which is difficult to quantify and measure in numerical terms. Yet this outcome is frequently referred to as one of the most appealing and successful models in modern water-related negotiation. Despite its arable and degraded landscape and

Table 5.10 Costs and benefits of the Lesotho Highlands Water Project

	Lesotho	South Africa
Benefits	Royalties for water transferred: • Receiving an annual revenue of USD 55 million from South Africa for 50 years • No financial risk for the water transfer component Infrastructure and economic development: • Roads and telecommunications to improve health, education, and trade services • Job creation and employment for local community • Additional enhancement of GDP Hydropower and electricity generation Solid foundation for nation building	Cheap water supply for its industrial and economic development (about 40% of Lesotho's "white gold") • Augmentation of water supply for industrial and economic growth • Lower water prices for end users Political and strategic gains: • Reduction of political stress and hostility • Settling the political dispute • Establishing cooperative relationship with its neighbor
Direct costs	Covering the hydropower component of the project	Covering the cost of construction, operation, and maintenance of the project, except for the hydropower component Annual royalties payment to Lesotho of USD 55 million Compensation and environmental mitigation costs Socioeconomic programs

the lack of major natural resources, Lesotho has a large surplus of water, which is much more than what is needed for Lesotho's domestic, agricultural and industrial uses. Only 6 percent of the available water resources are utilized in Lesotho (EIB, 2002). The majesty of these water resources, which remained largely unused until the Treaty materialized, was described during both the precolonial and colonial eras (Whann, 1995). The Treaty proposed a plan to take advantage of these underutilized water resources for heavy industrial uses in South Africa. Although these rivers do eventually drain into South Africa, they do not flow into the industrialized parts of the country that need water the most. Dam construction has been suggested, and favorably viewed by many experts since the colonial era, as an effective tool to capture and transfer this unused water for more productive purposes (Whann, 1995). By relying on dam construction and tying Lesotho's economy to that of South Africa, the Treaty seems to have produced an outcome that contributes to a better, mutually beneficial management of these shared transboundary water resources.

The Treaty provided an example of a quid pro quo (or "this for that") negotiation strategy. Under the auspices of the notion of water-sharing benefits, the two countries adopted a unique approach to negotiation over water resources, whereby South Africa would receive water and Lesotho, in turn, would receive hydropower and compensation. Swapping water for hydropower and royalties has proven successful in catering to both parties' needs and interests (Mckenzie, 2009). Water was particularly needed in South Africa but was abundantly available in Lesotho. Hydropower and economic development were particularly needed in Lesotho and could be provided by the much stronger economic base of South Africa. Supporters of the project often emphasize the mutual benefits that it has brought to both countries.

For South Africa, the most notable benefits are the acquisition of a cheap water supply for its industrial and economic development, on one hand, and the reduction of apartheid-centered political stress and hostility, on the other. The LHWP provides water (about 40 percent of Lesotho's "white gold") for commercial, industrial, and domestic water uses in Gauteng province, a heavily industrialized region that includes large cities, namely Johannesburg and Pretoria (Rothert, 1999; Mirumachi, 2007). Prior to this project, South Africa, particularly in Gauteng, suffered from an acute water shortage and was lacking adequate access to safe water and sanitation, because, in part, of its arid landscape, very limited water supply, and frequent droughts (EIB, 2002). This led to significant dam construction along the Vaal, Ash, and Wilge rivers to augment water supply. As the Gauteng region accommodates more than 42 percent of South Africa's population (Wallis, 1996) and houses 70 percent of its workforce and 40 percent of its overall GDP, the lack of water supply prior to the implementation of the LHWP constituted a great threat to the country's economic development and future growth (TCTA–LHDA, 2003). It is reasonable to deduce, therefore, the importance of the additional water supply, which the country now receives at a minimal cost, in providing the much-needed relief to assuage such acute water pressure (Mirumachi, 2007).

For Lesotho, benefits from the LHWP are for the most part non-water-related benefits, and include royalty earnings from the sale of water transferred to South Africa, massive infrastructure construction, power generation, and nation building. These economic incentives, namely the royalty earnings that are bundled with the adoption and implementation of the LHWP, were the primary motive for Lesotho to engage in diplomatic talks with South Africa (Mirumachi, 2007). These royalties, which are deposited directly into the Lesotho government's treasury by the South African government, are estimated at nearly USD 55 million to 60 million each year (Poivey, 1997; Mirumachi, 2007).

Table 5.11 shows the volumetric water transferred and the amounts of royalties, which both vary depending on the data source. It is worth mentioning that, in determining these volumetric water transfers to South Arica, there was no consideration for demand-side management, as the per capita water availability in Lesotho is much higher than the per capita demand (Mckenzie, 2009). In addition, this excess water will naturally drain into South Africa if not collected and transferred. The economic impact of these royalties, which are regarded as a major benefit for Lesotho, is significant in rejuvenating the overall economic base of the country (Poivey, 1997). Because water is one of the few commodities that Lesotho can sell to South Africa, Lesotho has been very keen about implementing future

Table 5.11 Scheduled water deliveries and associated royalty payments

Year	Planned deliveries (MCM)		Actual deliveries (MCM)		Royalty payments	
					Million Maluti,	Million USD,
	LHDA	LHWC	LHDA	LHWC	LHDA	LHWC
1999/2000	538	327	540	539.6	146.93	22.18
2000/2001	573	398	574	570.3	158.05	24.31
2001/2002	591	470	584	436.8	182.95	28.15
2002/2003	615	543	585	554.2	205.91	32.38
2003/2004	95	618	687	692.6	207.85	31.49

Sources: LHDA (2003); Tromp (2006).

phases of the project in order to receive more economic assistance in return for the delivered water (Mckenzie, 2009).

The realization of such a project resulted in direct benefits, such as enormous economic, capital, and infrastructure development for Lesotho, which is reportedly the ninth poorest nation in the world (Villiers et al., 1996; Wallis, 1996). In the past, Lesotho's economy was, in large part, handicapped by lack of job opportunities and severe infrastructure problems. Additionally, because of the lack of essential infrastructure development, including sufficient road networks, it was hard for local communities, investors, and visitors to navigate the country's mountainous terrain (Wallis, 1996; Fullalove, 1997). These aspects were acknowledged by the Treaty makers and adequately addressed by the Treaty. The substantial infrastructure development that occurred in Lesotho has drastically transformed the country and provided a strong basis for substantial economic development on the macro scale (Hassan, 2002). Because of its economic and job-creating nature, the project created a significant number of new jobs for the Basotho, many of whom worked in the mining industry in South Africa (the largest employer of Basotho). By 1992, between 3,000 and 4,000 new jobs were created because of the project, generating approximately USD 44 million per year (Whann, 1995; Rothert, 1999).

Many development plans were initiated as a tool to create jobs and improve the overall economy, and to provide a long-term indirect compensation plan for communities impacted by the project in terms of loss of land, loss of income, or resettlement (Fullalove, 1997; Hassan, 2002). This involved providing monetary compensation or replacement homes to affected communities or individuals. One of the most remarkable projects is the rural development plan, which was scheduled to start in 1991 and continue for 10 years, providing training and infrastructure development (Whann, 1995; Fullalove, 1997; EIB, 2002). This project involved the construction of 200 km of tunnels, over 1,000 km of road and highway networks, bridges, telecommunication networks, improved electricity and water supply, and health centers and sanitation facilities in the rural areas of Lesotho (Fullalove, 1997; Rothert, 1999; LHDA, 2003). New roads were constructed, while old ones were improved and upgraded (EIB, 2002).[21] In addition, public outreach programs and water awareness and education campaigns were also carried out to educate the local population about the economic opportunities that the project could bring to their communities (Fullalove, 1997). The LHDA was also actively involved in a training campaign to prepare the local population and enhance their social welfare and employment in the project construction (EIB, 2002).

Previously, Lesotho was largely dependent on its neighbor, South Africa, for most of its electric power acquisition,[22] namely from South Africa's Electricity Supply Commission (Poivey, 1997). Now, Lesotho can produce its own electricity at the Muela Hydropower Station to meet its domestic demand for electricity. Lesotho appears to be almost completely self-sufficient in its electricity production and distribution in the wake of the first phase of the LHWP. It has generated approximately USD 24 million annually from the production of electricity and water to South Africa (Bureau of African Affairs, 2008). Not only was Lesotho able to mitigate its dependency on South Africa for electricity generation, but it also was able to sell some of its domestically generated electricity to South Africa through the Lesotho Electric Corporation (Villiers *et al.*, 1996; Fullalove, 1997; Mirumachi, 2007). These direct material benefits that the LHWP brought to Lesotho gave the new government much-needed legitimacy and stronger domestic support, on one hand, and moved the country closer toward a state of "economic self-reliance," on the other (Ciccolo, 1992; Mirumachi, 2007).

Other indirect benefits include the advancement in the tourism and recreation sector (Wallis, 1996), while further political gains include the ability to maintain a mutually cooperative relationship between the two countries and foster future conflict prevention. It became evident to both countries, namely after the establishment of the SADC, that the benefits from the LHWP were not confined to only isolated development projects, but also provided an impetus to a prominent integrated regional economic development future (Mirumachi, 2007). To that end, the project provided a platform for fostering effective socioeconomic development and cordial relations and remedying the long-standing animosity triggered by the apartheid and self-interest visions that dominated the region for many decades. However, the impact of the project, namely the displacement of local communities, triggered mixed feelings and reactions. Although many amplify the aforementioned benefits that the project brought to both countries, others, however, question the success of the project in boosting the socioeconomic base of the Basotho at the local scale. To be able to construct these massive infrastructure projects, Lesotho needed to acquire massive areas of arable land, which meant evacuating and destroying populated areas and agricultural lands, and displacing many Basotho communities at a great cost (Fullalove, 1997; Rothert, 1999). It is estimated that, in the implementation of Phases IA and IB, approximately 2,300 hectares of arable land and 3,400 hectares of grazing land was lost and more than 30,000 people were directly impacted (Rothert, 1999). However, only 71 households were displaced as a result of Phase IA, which is considered insignificantly minuscule when compared with other projects of a similar size and magnitude (Fullalove, 1997).

Although those who were relocated because of Phases IA and IB received resettlement compensation through the implementation of the rural development plan, many regard the allocation of benefits on the local level to be problematic. In particular, some locals and non-governmental organizations (NGOs) believe that the project benefited only those who are in power and little, if any, benefit reached the local populations. This is due in large part to the corruption that plagued the project construction, implementation, and operation,[23] along with its social and environmental impacts. A local Basotho villager, relocated by the project, sums up this unresolved local problem that impacts nearly 20,000 Basotho people by stating, "I realize that we will be plagued by hunger because life here is mainly by the thing called pokotho, that is money. Now, we find that we are lost because the money we had been promised has not been given to us. Even the money we were paid for our compensation has been reduced drastically. We found that we will end up living much more poorly than where we came from. There are some painful things about resettlement" (Simpson, 2008).

Nonetheless, none of the NGOs working on the LHWP or the impacted local communities wish that the project did not exist. Rather, there is a widespread acceptance of the project and its outcome, despite the aforementioned criticism that the project was subject to. Once this massive infrastructure achieves its cost recovery, which might take several years, the economic base of the country is expected to flourish in the wake of the project and the long-term benefits will become more evident (Mirumachi, 2007). Given this, the Treaty makers were successful in reaching a satisfactory solution and producing a bilateral project that benefited the two countries in different, yet considerable ways. Rather than relying on mere volumetric allocation of water alone, the Treaty, considering these interests, emerged as a pivotal tool in managing and resolving the *sharing* of transboundary water resources rather than *dividing* them. Both parties, as such, feel that their needs are met in the wake of the Treaty as they were able to negotiate a deal that took into account both costs and benefits of cooperation along volatile waterways.

Financial arrangement and non-water-related benefits

One of the greatest challenges the negotiators faced was the financial arrangement of the costs of the project itself and the price of water. Considering the high cost of the project, which was estimated at between USD 6 billion and 8 billion, dividing the cost between the two countries constituted a momentous task. As such, allocating the financial responsibilities of the project construction, including both the water transfer and hydro-electric components, between Lesotho and South Africa posed a big concern to both countries during the negotiation process. The ridge was created because of the fact that the South African officials were not willing to absorb the cost of the hydropower generation project construction and installation (which was relatively high compared with the water transfer portion of the project). To account for the difference in cost between the two components of the project (and therefore establish the amount of royalties), an alternative scheme, which provided only water to South Africa without the hydropower portion of the project, was considered[24] (Fullalove, 1997; Mckenzie, 2009).

This new alternative, known as the "water-only scheme," provided a template to measure the difference in the cost between the two projects (one with only water transfer and the other with both water transfer and the hydropower project), which was to be absorbed by Lesotho (Whann, 1995; Fullalove, 1997). The royalties were primarily determined in the Treaty based on the differential cost between these two schemes. This is reflective of an interest-based approach to negotiation, where the parties negotiate their interests rather than positions. The overall cost of the hydropower generation component of the project alone has been estimated to be about USD 307 million, of which USD 107 million is to fund the hydropower elements used to facilitate water transfer (Wallis, 1996). It is expected that Lesotho would shoulder a total cost of about USD 236.4 million, which would be invested toward the project construction and operation (Whann, 1995).

As explained earlier in this chapter, determining the price of water, the amounts of royalty payments, and the financial arrangement was also a challenging and problematic task. However, the Treaty outlined the financial responsibility of each country. Although the Treaty stipulated that each country would be responsible for the cost of the project located in its territory, South Africa will eventually shoulder the cost of the water transfer elements of the project, whereas Lesotho will be responsible for the cost of the hydropower generation component of the project. Although the LHDA is in charge of raising required finance for the agreed upon elements of the project in Lesotho, South Africa is also required to secure assurance of the finance of water transfer facilities irrespective of their location in Lesotho (Fullalove, 1997). One of the foremost implications of such a financial arrangement is the risk and compromises that both countries have had to make. Signing the Treaty and committing to the project implementation meant great financial risk to both countries. The risk to South Africa emanates from the fact that parts of the water transfer component of the project is located in Lesotho, whereas the risk to Lesotho lies in the incremental financial arrangement and associated ambiguities pertaining to the hydropower generation component (Fullalove, 1997).

In addition to the Lesotho government, which is responsible for financing the hydropower generation component of the project,[25] and the South African government, which supports the water transfer component, the project was financed by international donors, most notably the World Bank, the European Development Fund, the European Investment Bank, the African Development Bank, the United Nations Development Programme, and other international financial institutions (Whann, 1995; Wallis, 1996; Poivey, 1997). In addition, a *pari-passu* strategy[26] was employed to ensure overall equitable allocation of project funds and that no lender was given a favorable status over others, including the World Bank (Fullalove, 1997).

Many of these international donors were initially skeptical about being involved in the project finance because of the international opposition and stigma surrounding the apartheid policies of the then political outcast: South Africa (Fullalove, 1997). Notwithstanding the sensitivity of the issue, the LHDA was able to raise foreign funds by way of loans to support the construction and operation of the project (LHWP, 2009), with the support of the Development Bank of South Africa (DBSA), which was created by South Africa to support its apartheid policies. Thus, many aspects of the project construction and operation were backed up by the financial support of South Africa. The involvement of the World Bank was beneficial in balancing the negotiation table and ensuring sound management and distribution of the project finances and benefits. In addition to managing the tendering and contracting procedures, the bank's financial commitment of USD 110 million, along with its involvement as a coordinator for the fund mobilization program, was very successful in fostering project legitimacy and credibility and gaining international donors' trust and cooperation without the fear of violating international sanctions imposed on the apartheid regime (Fullalove, 1997; Rothert, 1999; LHWP, 2009). Building such confidence in the project contributed to the advancement of the overall international perception of the project as a viable and worthwhile investment (Fullalove, 1997).

In addition to the political benefits and regional stability, economic and infrastructure development are major benefits that both countries are expected to harness in both the short and the long term. Lesotho receives fixed and variable royalties. The fixed royalties were calculated based on the cost saved by South Africa, representing the difference in cost between the LHWP and the next cheapest water transfer scheme, known as the Orange–Vaal Transfer Scheme (OVTS), whereas the variable royalties are proportionate to the amount of transferred water (Mirumachi, 2007). Additionally, any further revenues generated from the project will be invested in the construction of future phases and the ongoing royalty payments to Lesotho (Wallis, 1996). It is estimated that the cost of the LHWP is 56 percent of the OVTS project (Mirumachi, 2007). Because of the latter's high cost of coal-fueled pumping and operation, and long distances of transfer, in addition to its lower water quality, the LHWP was a preferred option (Mohamed, 2003). Accordingly, the countries agreed that 56 percent of the difference in cost would constitute the fixed amount of royalties paid to Lesotho, which has been paid since 1995 and will continue until the cost of all components of the project have been met and Lesotho is able to receive its full benefits (Wallis, 1996; 2000). These cost–benefit saving differences between the two schemes include the difference in the cost of construction, projected operation and maintenance, and the electricity required for water pumping in the OVTS project (Wallis, 1996). According to the Treaty, 56 percent of the benefits and savings (approximately USD 60 million/year from Phases IA and IB) from the implementation of the LHWP is allocated to Lesotho in the form of royalties (EIB, 2002; Hitchcock *et al.*, 2007). However, even fixed royalties are a moving target to be negotiated during the remaining period of the Treaty. Therefore, royalty payments are flexible and can vary from one year to another based on the ongoing savings between the two schemes in terms of operation and maintenance and the varying amounts of water that South Africa requires (Wallis, 1996).

From the beginning of the operation of the Katse Dam in 1994 until 2006, Lesotho received over USD 230 million in royalty payments[27] (Tromp, 2006). Similarly, it is estimated that South Africa's benefits and savings from implementing Phases IA and IB (compared with alternative proposals) is about USD 30 million per year (EIB, 2002). In September 1991, the World Bank provided USD 110 million of concessional loans to Lesotho to finance the first phase of the project (Wallis, 1996; Poivey, 1997). An estimated USD 55 million to 60 million/year in water royalties will be paid by South Africa to Lesotho upon the completion of the project (Poivey, 1997).

Environmental assessment

With the intensity of infrastructure development and other human activities sustained in the Orange River basin as part of the LHWP, a number of environmental and economic challenges continue to mount. As shown in Figure 5.6, the large-scale industrial water use around Johannesburg and Pretoria (the most industrially developed parts of South Africa), farming practices and overpopulation in the eastern part of the basin, changes in flow regime and silt load occurring at the mouth of the river, and commercial agribusiness caused many challenges ranging from industrial and mining pollution, soil erosion, deterioration of the estuary's

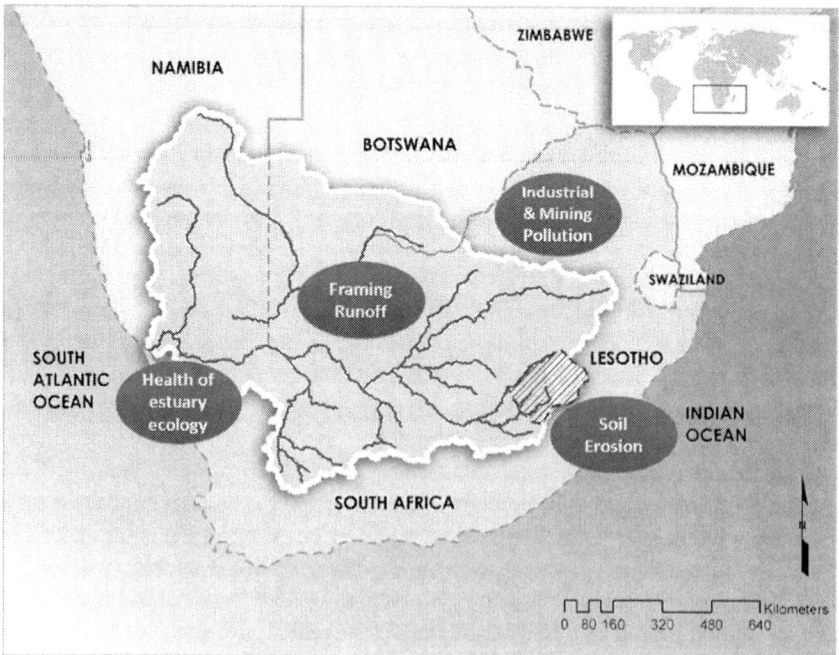

Figure 5.6 Environmental and economic challenges in the Orange River basin. Adapted from Earle and colleagues (2005).

ecological health, and farming run-off (Earle *et al.*, 2005). Most notably, diversion of nearly half of the river's flow has potentially serious impacts on endangered species, such as the Maluti minnow, rock catfish, and bearded vulture, to name a few (EBC, 2007).

Protecting the environment and mitigating pollution are key goals of the LHWP (Turton, 2003). In this regard, one of the most fundamental principles promoted by the project authorities is the maintenance of comparable living conditions. However, with the lack of comprehensive legislation in Lesotho, the LHDA approached this by adopting international environmental guidelines and incorporating them into the project construction contracts (Fullalove, 1997). In addition, the LHDA and TCTA carried out a series of environmental and social impact studies for the purpose of mitigating the adverse impacts of the project (Fullalove, 1997). Since there was no comprehensive assessment done at the time of the Treaty ratification, the five phases of the project were proposed at higher volumes of water transfer than was originally envisioned. However, according to Mckenzie (2009), these environmental studies funded by the World Bank to determine the sustainable yield of Lesotho's water resources that can be trans-ferred to South Africa without harming the environment suggested diverting less water than was originally planned as a result of environmental considerations.

Similar environmental action plans for Phases IA and IB of the project were also completed in 1991, at which point major components of the project construc-tion were already in place.[28] These studies were conducted to mitigate environ-mental harm on both downstream ecosystems and the livelihoods of communities located alongside the river and resulted in an in-stream flow requirement policy, which was later implemented in 2002 (Fullalove, 1997; Davis and Hirji, 2003; Brown, 2008). A study of in-stream flow requirements was conducted during the implementation of the first phase of the project to identify and mitigate potential ramifications of the reductions in river flow on aquatic ecosystems (EIB, 2002; LHDA, 2002). According to these environmental studies, the impact of changing the way the water is stored and released in the first phase of the project indicates a "severe level" impact in several locations downstream of the LHWP and further imposes a "critically severe" impact downstream of the Katse Dam (Brown, 2008). These studies resulted in the 2003 In-stream Flow Requirement Policy, based on which higher water was allocated to preserve and maintain the *status quo* of the ecological functioning of these sensitive reaches (Turton, 2003; Brown, 2008).

Environmental ramifications of such a large-scale transfer project, namely overlooked downstream impacts, can result from ecological and engineering miscalculations (Hildyard, 2000). Removing significant amounts of water from the river has many harmful effects on the environment and the balance of the ecosystem. This was exacerbated by the inaccuracy in estimating water levels and availability in the river system and the miscalculation and, in turn, disagree-ment pertaining to the optimum tunnel design of the project (Whann, 1995). In addressing the social and environmental considerations of the project, the Treaty in Article 15 states that:

the Parties agree to take all reasonable measures to ensure that the implementation operation and maintenance of the Project are compatible with the
protection of the existing quality of the environment and, in particular, shall
pay due regard to the maintenance of the welfare of persons and communities
immediately affected by the Project.

Other than a cursory mentioning of the need to promote the protection of
the existing quality of the environment, the Treaty seemed oblivious to the disastrous environmental ramifications of such a large project. Over-extracting the
river's water in such gargantuan proportions can potentially cause the quality of
in-stream water to deteriorate drastically. There is ample evidence suggesting that
the impact of dam buildings on the in-stream flow requirements was overlooked
(Rothert, 1999). Fostering sound environmental protection requires developing
environmental impact assessments and the consideration of all mitigation alternatives and processes. Although some studies regarding the social and environmental impacts were conducted for both Phases IA and IB of the project (LHDA,
2003), these impacts on the health of the ecosystem were not considered in a
serious manner and were further disregarded by the Treaty.

There seem to be some unexpected positive environmental impacts of the
LHWP that materialized upon the project implementation and operation.
The flow of clean water from the LHWP into the reservoir located in Gauteng
province reduced the level of salinity in the water and rejuvenated aquatic life.
Nonetheless, there is a general view that the benefits from such a project outweigh
its drawbacks, which are excepted and unavoidable given the size and nature of
the project (Fullalove, 1997).

Conclusion: overall treaty assessment

The process and outcome of such a Treaty cannot be neatly separated, as they are
intimately intertwined through a vision of mutual benefit. Understanding that
bringing about mutual gain for the social and economic benefit of both countries
requires genuine collaborative effort was key to the success of the Treaty formation
and implementation (Newton and Wolf, 2008). The Treaty offered a formalized
water-sharing agreement that governed the implementation and operation of such
a massive project with the level of political and technical sophistication that it
represented (Turton, 2003). The signing of this Treaty personified the initial deto-
nation for the launching of one of the most remarkable engineering projects in the
world. With the advancement of this project, each phase has witnessed astound-
ing new developments in translating textual account into physical reality. This
stimulus came from the Treaty itself and its ability to adapt to new circumstances
as they unfold.

On the whole, the Treaty was successful in boosting stability and economic
vitality through the implementation of the world's largest water transfer project,
the LHWP, which is well into its stride (Wallis, 1996). It worked reasonably well
in terms of its negotiation and outcome, except for limited slip-ups in the initial

perception of the Treaty as a "sellout" as a result of the lack of public involvement. In addition, issues of high politics, such as national security and sovereignty and the apartheid policies which could have stymied the progress on the diplomatic front, emerged as fundamental factors in influencing the overall climate of negotiations (Mirumachi, 2007). The fact that the source of the water needed by South Africa was in Lesotho raised concerns amnog both state officials and the public in South Africa. This myopia that surfaced and dominated the public sphere in South Africa was equally matched by myopia of another kind in Lesotho. Although South Africa was concerned about the source of its water being located in territories outside of its constitutional control, many people in Lesotho viewed the project as yet another South African attempt to undermine the national sovereignty of Lesotho. As such, it was viewed as a threat to both countries' sovereignty and stability. Another key question that raised doubts about the success of the Treaty was whether the gap between what Lesotho might regard as reasonable and what South Africa might regard as feasible in terms of project implementation and operation could possibly be bridged through cooperative arrangement.

However, the impact of such issues of high politics was kept to a nadir during and after negotiation. By alienating these issues of high politics and keeping negotiation separated from political distractions, the two countries were able to capitalize on the socioeconomic opportunities that the LHWP offers. In spite of the many political impasses, water (as an issue of low politics) acted as a source of cooperation rather than conflict. It advanced the talks between the two countries toward a more cooperative and plausible situation, where both are satisfied with the outcome of the agreement, at least among the public administrations of each country. In addition, the involvement of a neutral and well-respected third party, the World Bank, proved successful in eliminating negative stigmas about the project, attracting donors and contractors, and facilitating understanding and cooperation between the two countries during times of escalated tension and a state of war between South Africa and the African National Congress (ANC) (Turton, 2003). In addition, allowing for an ongoing reiterative renegotiation and revision process of the terms of the Treaty provided a particularly high level of resilience to cope with political and economic changes.

Although the Treaty makers laid out a solid negotiation strategy that resulted in a relatively flexible implementation process, one of the most notable pitfalls of the Treaty has been its lack of stricter and more potent environmental protection clauses. This lack of environmental consideration and impact studies in the early stages of the project was also evident in some of the serious environmental ramifications that the implementation has brought to the health of the ecosystem, which some argue to be an expected consequence of such a large infrastructure undertaking (Brown, 2008; Newton and Wolf, 2008).

The dynamics of the relationship between Lesotho and South Africa have been predominantly impacted by the question of sharing the Orange River waters. Although it now stands resolved by the Treaty of 1986, the tension between the two countries cannot be said to have completely dissipated. Although the dispute regarding the sharing of the Orange waters has been settled, the issue of equitable allocation of water and its respective benefits, namely on the local level, still occupies the minds of many scholars. In this regard, equitable distribution of economic benefits was of great importance in state-level politics, though the issue was not adequately and fully addressed on the local level. According to the UN Human Development Report in 2006, "Access to water was one of the defining racial divides in apartheid South Africa. Since apartheid was brought to an end, a rights-based legislative framework and public policies aimed at extending access to water have empowered local communities and reduced inequalities. The task is not yet complete – but there are important lessons for other countries" (UNHDR, 2006).

Although opponents of the Treaty, namely supporters of the previous Lekhanya's government, demanded the revision of the Treaty terms, the Treaty remained largely unchanged and no major adjustments to its terms have been made since its ratification in 1986, nor have the parties felt the need to resort to any dispute arbitration mechanisms identified by the Treaty. This provides an ample indication of the success of the Treaty in adequately addressing the social and economic needs of the two countries. This is because considering benefits from water is key to the success of any treaty. It is argued that benefit from water, rather than water transfer per se, is a key factor to resolving water issues of high contention and producing a treaty where a win–win situation is engendered. Swapping water for economic assistance was proven successful as South Africa needed water, whereas Lesotho needed financial aid. Under this benefit-sharing notion, there were a set of pivotal factors and other alternatives to be considered in the negotiation, such as virtual water, gray water, and green water. By the same token, the process of negotiation and elaborate technical and financial arrangements that led to the Lesotho–South Africa Water Treaty and the construction of the LHWP provide ample evidence of how states negotiate costs and benefits of cooperation, and meanwhile consider both water and non-water linkages (Mirumachi, 2007). To that end, the benefits and interests of each riparian state were established as the basis for negotiation, where the negotiators acknowledged and accepted the other party's needs and moved beyond the questions of entitlement to exploring ways to satisfy these respective needs. In that sense, the mutual gains of both parties, or a diverse "basket" of benefits (Wolf, 1999), are considered an integrative and comprehensive arrangement (Newton and Wolf, 2008).

In short, the LHWP is a real life testimony of how riparian states can capitalize on the economic opportunity that such projects offer to attain and maintain regional socioeconomic prosperity while moving toward regional hydro-political stability by assuaging regional tension and reasons for future conflict. Integrating both water-related benefits as well as non-water-related benefits, such as political and economic gains, in a single hydro-diplomatic resolution is proven successful in providing a platform for satisfactory, win–win outcomes (known, alternatively, as a positive-sum situation), and, in turn, cordial relations among riparian states. Hence, the overall benefit of the LHWP in preventing conflict and ameliorating regional hydro-stability cannot be overstated.

6 Comparative cross-case analysis, hydro-political implications, and lessons learned

> Only free men can negotiate. Prisoners cannot enter into contracts.
>
> (Nelson Mandela)

The *raison d'être* for this chapter is to provide a comparative case study analysis in order to unearth significant process equity parameters that are believed to be associated with more equitable outcomes and further outline lessons and parallels for the theory and practice of equitable concepts in water negotiation in general. This chapter also outlines a comprehensive conflict resolution strategy that can transform water negotiation from a non-cooperative to a cooperative state, where riparians in dispute can engender equitable hydro-diplomacy and satisfactory agreements. This approach, which is outlined in detail in this chapter, is referred to as a "transformative approach to conflict resolution."

Comparative cross-case analysis

Figure 2.1 in Chapter 2 showed the preliminary logic model prior to conducting case analysis. Literature suggests that important elements of the process equity parameters fall into three major categories: (1) negotiation process (treaty formation); (2) planning analysis; and (3) structure of the agreement. The model initially included 21 essential parameters of process equity as a result of rigorous research and examination of related cases and literature. However, as the research unfolded, a number of interesting, contextual anomalies and issues arose. These anomalies suggest that there are other important elements that are an integral part of the process equity parameters. These include depoliticizing water (focusing on low-politics issues); considering average, wet, and drought conditions; socioeconomic considerations; language of equity; sharing not dividing; benefit from water use; flexibility and resilience to political changes and extreme hydro-events; and inclusion of contingency planning clauses in the agreement. Similarly, some of the previous components were consolidated as they are now included as parts of other parameters, such as the notion of "time-based not volumetric-based allocation," which is encapsulated in a more sophisticated way within the notion of "sharing not dividing," as will be discussed in more detail later in this chapter. Based on the original logic model, Tables 6.1 and 6.2 show a revised list of process equity parameters incorporating the newly added components that were discovered to be important elements in the process of treaty negotiation and formation as a result

Table 6.1 Use and impact of process equity parameters on outcome in the Israel–Jordan Peace Treaty of 1994

	Process equity parameters	Use	Impact on outcome
Negotiation process	All stakeholders	No	Negative
	Trust building	No	Negative
	Interest-based (not position-based)	Yes	Positive
	Modern and indigenous	No	Negative
	Incremental approach (not precipitous)	No	Negative
	Facilitation/mediation (3rd party)	No	Negative
	Public involvement	No	Negative
	Depoliticizing water (focusing on low politics issues)	Yes	Positive
	Considering average, wet, and drought conditions	No	Negative
	Socioeconomic considerations	Yes	Positive
	Reliance on international law	No	Negative
Planning analysis	Growth rate	No	Negative
	Annual water budget	No	Negative
	Per capita consumption rate	No	Neutral
	Consumption by sector	No	Neutral
	Water demand model	No	Neutral
	Needs-based (not rights-based)	No	Negative
	Data mediation/joint fact finding (JFF)	Yes	Positive
Structure of the agreement	Language of equity	Yes	Positive
	Quantity/quality	No	Negative
	Prioritization of uses and users	No	Negative
	Sharing not dividing	Yes	Positive
	Benefit from water use	No	Neutral
	Flexibility/Resilience	No	Negative
	Adaptability to drought (extreme hydro-event)	No	Negative
	Environmental and non-environmental	No	Negative
	Contingency planning clauses	No	Negative
	Enforcement mechanism	No	Neutral
	Monitoring	No	Negative

of case analysis. The revised logic model now includes 28 refined process equity parameters.

In light of the analysis of the two case studies, a number of these parameters were found to be important in influencing the outcome and perception of the water

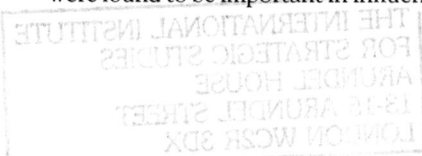

Table 6.2 Use and impact of process equity parameters on outcome in the Lesotho Highlands Water Project Treaty of 1986

	Process equity parameters	Use	Impact on outcome
Negotiation process	All stakeholders	No	Negative
	Trust building	No	Negative
	Interest-based (not position-based)	Yes	Positive
	Modern and indigenous	No	Neutral
	Incremental approach (not precipitous)	Yes	Positive
	Facilitation/mediation (3rd party)	Yes	Positive
	Public involvement	No	Negative
	Depoliticizing water (focusing on low politics issues)	Yes	Positive
	Considering average, wet, and drought conditions	No	Neutral
	Socioeconomic considerations	Yes	Positive
	Reliance on international law	Yes	Positive
Planning analysis	Growth rate	No	Neutral
	Annual water budget	No	Neutral
	Per capita consumption rate	No	Neutral
	Consumption by sector	No	Neutral
	Water demand model	No	Neutral
	Needs-based (not rights-based)	Yes	Positive
	Data mediation/joint fact finding (JFF)	No	Negative
Structure of the agreement	Language of equity	Yes	Positive
	Quantity/quality	No	Negative
	Prioritization of uses and users	No	Neutral
	Sharing not dividing	Yes	Positive
	Benefit from water use	Yes	Positive
	Flexibility/Resilience	Yes	Positive
	Adaptability to drought (extreme hydro-event)	Yes	Positive
	Environmental and non-environmental	No	Negative
	Contingency planning clauses	No	Negative
	Enforcement mechanism	No	Neutral
	Monitoring	Yes	Positive

agreement. Similarly, factors that were originally thought of as significant appeared insignificant (or neutral in some cases) in advancing negotiation to plausible and acceptable agreements, such as considering average, wet, and drought conditions, growth rate, annual water budget, per capita consumption rate, consumption by

sector, water demand model, benefit from water use, and enforcement mechanism. More discussion will follow to identify these factors and further elaborate on
how important they appeared to be in this regard. The following tables illustrate a
cross-case comparison of process equity parameters in terms of use and impact on
outcomes for both the Israel–Jordan and the LHWP water treaties, and are used
to summarize the analysis and guide the following discussion.

These tables were developed on the basis of the logic model shown in Figure
2.1 in Chapter 2, and revised in the light of the analysis of the two case studies.
These tables are structured chronologically, starting with the early negotiation
process, then planning analysis and then the structure of the final signed agreement, reflecting the natural progression of the negotiation steps, starting with
the negotiation process (and pre-negotiation) all the way to the signing of the
agreement (and post-negotiation). The two main qualitative ratings of the importance of each process equity parameter as outlined in the methodological design
chapter are: (1) whether each of these factors have been used in the negotiation
of each case; and (2) what impact each factor has had on the actual outcome of
the agreement. The highlighted factors in Tables 6.1 and 6.2 are the ones that
are corroborated by both the case study analysis presented in this book and the
larger literature. It is crucial to recognize that these treaties are not anomalies
based on the process of selection and the crucial characteristics they represent
that are universal. This provides an indication of the validity of these inferences
and the possibility of their applicability to other cases and geographic contexts.
Ten parameters in the Israel–Jordan Treaty and nine in the LHWP Treaty were
found to be verified by both the literature and the research findings.

The following analysis looks at common characteristics (which are deemed
constant across the two cases), accidental characteristics (which are elements to
rule out), and unique characteristics (which are used to illuminate and explain differences between the cases). Here we will be concerned with developing process
equity parameters for international water negotiation and sharing. The cross-case
analysis of these process equity components is guided by the logic model and is
broken into three elements. This includes the negotiation process (treaty formation), planning analysis, and structure of the agreement and is further discussed
below.

Negotiation process (treaty formation)

It goes without saying that involving all key stakeholders in the negotiation process
is critical to its success. Although neither treaty involved all stakeholders at the
beginning of the negotiation process, the LHWP Treaty was successful in bringing
all impacted and interested stakeholders together in a multinational cooperation effort to collaboratively govern a comprehensive allocation and management of the basin water resources. Conversely, the Israel–Jordan Treaty failed to
involve the other important parties and regional players, namely Syria, Lebanon,
and Palestine, in any meaningful discussion about a basin-wide management
of the shared water resources. Many people associate this lack of multilateral

cooperation with Israel's consistent unwillingness to recognize or involve the Palestinians, which in turn carries significant implications for the regional peace process. Zeitoun (2008), for example, argues that Israel has two bilateral agreements in place with Jordan and Palestine; both are in favor of Israel. By asking the question of who wins and who loses in these bilateral agreements, one can address the political causes for the lack of multilateral agreements. If all impacted co-riparians and interested stakeholders were brought to the table in a multinational negotiation process, according to Zeitoun, Israel would lose its advantage by having to compromise its disproportionally high share of water. By so doing, all involved parties would be able to reach a satisfactory settlement through a comprehensive agreement by promoting the notion of integrated basin-wide management, rather than relying on a segregated series of politically motivated bilateral agreements that tend to isolate the parties and further complicate the issue (Elmusa, 1995).

However, other water experts argue that prematurely involving other parties in these negotiations between Israel and Jordan, namely the Palestinians, will tend to complicate the matter and produce more tension. Within the context of the Jordan River, it is often argued that it is easier to negotiate bilateral treaties than to attempt to involve all parties at once, as multilateral treaties may seem to be difficult to achieve, although not entirely impossible. However, multilateral arrangements are often portrayed as more unwieldy than they actually are. There are many ample examples of successful multilateral treaties where all key parties are involved in the negotiation and formation of a final settlement. Aside from these conflicting points of views, one thing seemed certain in both cases: this lack of involvement of all stakeholders negatively influenced perceptions of the treaties' outcomes.

According to Woodhouse (2008), establishing trust between the parties prior to negotiation is very important in enhancing equitable cooperative agreements. During the pre-negotiation stage, the parties are encouraged to work together to build credibility with each other in order to boost confidence in their negotiating partner and the negotiation process itself. This can be attained by creating a relatively close, comfortable relationship with all parties. In the case of the Israel–Jordan Peace Treaty of 1994, many experts argue that the parties attempted to promote trust and confidence building measures (CBMs) and create an atmosphere of mutual respect conducive to working relations (Shamir and Haddadin, 2003). However, it is clear that trust was still lacking in the pre-negotiation stage. Both Israel and Jordan had a high level of hostility toward each other at the time they entered into negotiation. Distrust and resentment hindered the acceptability of the outcome of the Treaty.

This is also true in the case of the LHWP Treaty, where the parties were in a state of war during the negotiation process, and, even after the arrival of the new government in Lesotho and the end of the apartheid system, they still did not have much trust in one another. This was evident in the disagreement over basic hydrologic and water availability data (which were either exaggerated or underestimated to serve the interests of Lesotho in gaining more royalties and

South Africa in reducing payment amounts respectively). In such a climate, lack of trust negatively impacted the outcomes of both treaties.

Both treaties are characterized as interest-based negotiation by focusing on and recognizing the parties' needs and interests, and to a great extent the purpose of using water. This was particularly the case in the LHWP Treaty. The Israel–Jordan negotiations benefitted from the water–land (border) and the water–environment nexus, which helped the parties mutually consider issues of historic utilization of water and land as well as water quality and energy that impacted both sides. The LHWP Treaty was successful in considering both countries' (but particularly South Africa's) need for water and Lesotho's need for economic assistance. This helped in reaching agreeable solutions that served the interests of both sides and promoted not only equitable outcomes, but also outcomes that are viewed as such. These aspects of the water-related negotiation process reflected interest-based negotiation in the sense that the parties focused on allocation of water based on use rather than entitlement and considered other related factors such as land, border, energy, and economic needs. In this interest-based negotiation process, the parties searched for mutual gains and identified common interests, which helped provide maximum mutual benefits and produce agreement amicably. The bottom line was that the parties in both cases seemed to be cognizant of the ramifications of choosing to walk away from the negotiation process, as they both recognized that their BATNA was equally deplorable. Furthermore, they seemed to share the same strategic goal and water was the vehicle that delivered cooperation and development.

The absence of indigenous conflict resolution mechanisms was not particularly problematic in the LHWP case, but was to some extent detrimental to the acceptability of the Israel–Jordan Peace Treaty outcome. This is because people in the Middle East were not very open to the idea of negotiating with Israel and the negotiators relied on exclusively modern conflict resolution techniques. This created more suspicions about the intent of the Treaty and distrust in its outcome as a Western panacea to strengthen Israel's security and control in the region.

The Israel–Jordan Treaty suffered because of the lack of an incremental, developmental process of both negotiation and implementation. During negotiation, the parties attempted to tackle several key issues at once, which resulted in disagreement and impasses during various stages of the negotiation. Similarly, during implementation there was no clear incremental strategy for phasing project implementation and managing its operation. The lack of an incremental approach adversely impacted the outcome and implementation of the Treaty and caused the relationships to further corrode. An incremental approach entails a step-by-step strategy of negotiation that tends to tackle pressing issues piecemeal rather than *en masse*. This aspect of treaty negotiation and implementation was adequately addressed in the LHWP Treaty. The negotiation was conducted in several stages and for several purposes and the implementation and operation of the project was planned and executed in several phases. This strategy was successful in facilitating smooth negotiation and effective implementation, particularly to accommodate the large size of the project.

The early involvement of powerful mediators, such as the United States and the Soviet Union, in the onset of the negotiation between Jordan and Israel provided the needed framework for negotiation and set the stage for the two countries to focus on the issues at hand that concerned them the most. This was very helpful in bringing the two parties to the negotiation table and solidifying their commitment to the peace process at its early stage. With the help and support of the United States and the Soviet Union, both countries were invested in pursuing serious negotiation and finding workable solutions that could be implemented to resolve the disputed issues and normalize relationships. However, as the two countries progressed into their direct negotiation, the role of third-party mediation diminished, causing substantial disagreement and delay in negotiation, and, in some cases, damage to their relationship. From this case, it is easy to surmise that the absence of a neutral, equidistant, and even-handed third-party mediator resulted in a situation where the hegemon dominated and eventually won the negotiation.

Conversely, the early and consistent involvement of the World Bank as a third-party mediator and also as a donor and sponsor of the LHWP project proved useful in many ways. First, the World Bank was a major sponsor of the project, as it provided financial aid in support of the implementation and operation of its different phases. The bank was also involved in conducting technical, financial, and environmental studies. Second, this proactive involvement of the bank was beneficial in removing the stigma surrounding the project due to South Africa's hated apartheid policies. Many donors and international contractors, who originally saw the project as serving the interest of the apartheid South Africa and were less than willing to participate in it at first, became more optimistic about the outcome and less skeptical of its motives upon observing the involvement of the World Bank. This involvement provided an incentive to many countries and international organizations to actively participate in financing the project and moving it forward toward its full implementation.

Although public involvement is critical in achieving equity in water negotiation, little was done to involve the public in both cases. In the Israel–Jordan Peace Treaty of 1994, it was clear that both countries were not keen to inform and involve their populations in the making of the Treaty, mainly because of security reasons. The negotiations were held secretly between the countries for fear of politically related pressure and stalemate. Similarly, the efforts of the LHDA to promote public support of the LHWP initiative were not enough to inform and involve the general public. In both cases, the lack of public involvement invoked a state of public distrust and, in many cases, opposition. This is particularly true in the Israel–Jordan Peace Treaty, where those living on opposite sides of the Jordan River are deeply divided on many issues related to the Treaty negotiation and implementation.

The negotiators of both treaties were, for the most part, able to depoliticize water by separating issues of "high politics" from issues of "low politics" (i.e., water-related negotiation) and, therefore, were successful in averting political distractions and pressures that could have been extremely detrimental to the progress of water negotiation. Although in both cases the parties involved in the negotiation process were officially in a state of war, they were nonetheless concerned about advancing their negotiation over water-related matters to more plausible solutions. Depoliticizing water by separating high politics from low politics was very helpful in reaching implantable and acceptable agreements. However, many experts argue that it is imperative to tackle issues of high and low politics jointly, and to separate them only if they appear to clash or cause harm (Daoudy, 2009; Priscoli and Wolf, 2009). Arguably, conducting negotiations that are concerned only with issues of low politics that are entirely separate from the larger political discourse may also cause negotiation failure. This is because solving issues of "low politics," including water, may help in facilitating negotiation over issues of "high politics." This was evident in both cases, as the negotiation teams tackled these issues with equal footing while preventing the detrimental impacts of high political issues from adversely affecting the negotiation over water. With this in mind, the relationship between the larger political agreement and the water-related elements of the treaty is very pivotal in shaping its outcome.

In the case of the 1994 Treaty between Israel and Jordan, the magnitude of the political issues provided enough impetus for the parties to collaborate. Come hell or high water, the parties needed to reach an agreement to settle their dispute because of the severity of the political dispute and the serious ramifications that may ensue. This does not necessarily mean that they agreed on everything or that all their interests converged or have been met by the agreement. It simply means that they were able to achieve an agreement to resolve a particular set of issues under dispute, which, in a way, sets the stage for more issues to be resolved in future negotiations. Although Jordan and Israel had axes to grind with each other, they had to bury the hatchet and put their political disagreements aside to achieve a lasting deal that would guarantee the end of hostility between them. It is crucial, however, to point out the problematic nature of the Water Annex of the 1994 Treaty between Israel and Jordan, which was a direct result of the interaction between the political and technical aspects of negotiation, or issues of high and low politics. Although the negotiators were successful in depoliticizing water, the impact of issues of high politics was evident in the development of the technical components of the Treaty. The Water Annex appeared to be a subordinate add-on to the larger Peace Treaty, which has been characterized as being politically laden and merely driven by political momentum to broker a peace treaty in order to settle the dispute and ultimately normalize relations between the two co-riparians. The water issue seems very crucial to both parties, evidenced by the fact that water was added at the end of such a remarkable Treaty, particularly after the articles pertaining to issues of international boundaries and security in the Preamble (Elmusa, 1995; Phillips, 2008). Water is important to each country, but for different reasons. Water is vastly important to Israel as a strategic element of its own. To that end, Israel views water as a matter of national security. Water has been an element of national importance in Israel's strategic agenda ever since its establishment in 1948. For Jordan, water is an irreplaceable essential element in support of the socioeconomic wellbeing of its population. However, despite the importance of water to both parties, many experts argue that the technical water-related part of the agreement was nothing more than an afterthought at best. It was inevitable for both countries to negotiate and include water as part of the agreement, but its inclusion seems to be an *ad hoc* element to the larger Peace Treaty to reinforce the *status quo ante* and create a *de facto* reality on the ground for the benefit of one party or the other. The water agreement, therefore, was more of an afterthought and should have been given a more central role in the Treaty and its negotiation, given its importance.

Crucially, the 1994 Treaty between Israel and Jordan did not establish adequate mechanisms to deal with times of drought. Instead, water allocation schemes were based on times of normal water supply. The lack of consideration of average, wet, and drought conditions of the river system had a detrimental impact on the outcome of the Israel–Jordan Treaty, as both countries were fundamentally concerned with acquiring sufficient volumetric water for their needs. However, in southern Africa, the issue of river capacity fluctuation was irrelevant to (and therefore neutral to the outcome of) the LHWP, as water is abundant in Lesotho. The 1999 drought event and consequent political drama that Israel and Jordan experienced showed the danger of considering only the steady state (the average carrying capacity of the river) and disregarding times of drought (which are the norm in arid and semi-arid regions).

In both cases, offering good faith socioeconomic assistance helped the countries establish good relationships prior to negotiation. During their negotiation, Israel and Jordan were cognizant of the value of stabilizing the regional economy and mitigating poverty, which is often a key reason for conflict. This gave the parties enough motivation to work together toward a common goal and helped them in reaching acceptable outcomes and maintaining successful implementation afterwards. This was particularly true in the case of the LHWP Treaty, where the project brought significant infrastructure development and job creation to impoverished Lesotho. The economic assistance that Lesotho received in the wake of the project in the form of royalties paid by South Africa for the amounts of transferred water helped in enhancing the regional economic base in general, and the nation building effort of Lesotho in particular. As such, economic assistance is identified as a pivotal component of an equitable negotiation strategy.

There appears to be a lack of clear ground rules for water allocation and negotiation in the case of the 1994 Treaty between Israel and Jordan. During the pre-negotiation stage, the parties failed to work together to agree on or utilize appropriate strategies and acceptable ground rules to achieve resolution. There is also no indication of mutually agreed upon ground rules that utilized principles of equity as a template against which to judge the *process* and its *outcome*. The *process* seemed to be lacking this important step, as the parties relied only on practical and pragmatic solutions, rather than principles of equity introduced by international law.

Although, according to negotiators, the Treaty was not in violation of international law, many argue that the issue of applying international law, namely the 1997 UN Resolution, was not favored by Israel and, further, was not incorporated in the Treaty text or its implementation. Israel's refusal of the applicability of international law is believed to have enabled it to maintain its position and give it the upper hand, while putting its adversarial weaker parties at a greater disadvantage. Israel's views of international law and jurisprudence application were adopted by its negotiators, who forcefully argued that the applicability of international law was disputable and irrelevant when compared with following a pragmatic approach to negotiation. Although being pragmatic is a far reaching quality in negotiation, many argue that abiding by international law and being pragmatic are not inherently antithetical. The one-sidedness of the Treaty outcome can be attributed to the utter lack of reliance on the principles of international law. In the case of the LHWP, the application of international law was not disputed and was further promoted by the two parties as a key component of the negotiation process. There has been no evidence of any dispute regarding the applicability of international law at any point of the negotiation between Lesotho and South Africa, as both sides were able to acknowledge its importance. This, along with the involvement of the World Bank, which is an internationally acclaimed organization, had positive implications on the outcome and perception of the agreement and therefore its implementation. As such, the reliance on international law emerges as a crucial element in achieving equitable outcomes and favorable perceptions.

Planning analysis

Planning analysis parameters remained the same and, for the most part, did not change in the revised logic model. This is partly due to the fact that these are relatively more predictable factors than others and that existing literature is successful in capturing the majority of them.

Growth rate and the annual water budget of each country were two related issues that were overlooked by the two treaties. Although these elements did not cause any political or practical turmoil during or after implementation for Lesotho and South Africa, these were hot topics that occupied the core of the dispute surrounding the 1994 Treaty between Israel and Jordan. The impact of overlooking these variables in the LHWP negotiation was negligible because the 1986 Treaty was concerned with swapping water, which was abundant in Lesotho, for royalties rather than volumetric water allocation per se. As such, growth rate and water budget seem, for the most part, to be irrelevant in this case. However, these issues created many problems for Israel and Jordan during and after the conclusion of the negotiation. The Treaty did not offer practical directions and guidance on how to cater to the exponentially growing population of the region in terms of water demands. Twenty years from now, the Jordanian population is expected to grow to more than 9 million and the Israeli population is expected to grow by 42 percent, to 8.5 million (Fisher et al., 2005). This significant population growth suggests a different set of water demand figures and allocation approach, which are missing in the Treaty. This increasing demand places a great deal of strain on the available scarce water resources and tends to hinder the implementation of the Treaty in the future. This, as a result, generates a state of resentment and lays the grounds for more complicated disputes over the limited water supply. This is because freshwater is not replaceable, that is to say it has no substitute in most of its uses. It is essential for food production, industrial applications, domestic uses, and human health. Neglecting the demand side of the water resources equation, the Treaty allocated water based only on supply. As these amounts of water allotted by the Treaty based on the supply side are to be delivered to arid and thirsty communities along the Jordan River that are in no position to cope with the current crisis, let alone future demands, the water crisis will remain a looming threat to the security and stability of the entire region. Without concrete estimates of future demands based on precise population forecasts, this increasing hostility will entice each state to provide unrealistic and overestimated figures for their water demand. This has a direct and detrimental impact on the implementation of the Treaty. Regional peace will suffer because of that.

In addition, implementation is often threatened by the lack of adequate resources in the face of growing demand. The confined water resources, along with the lack of accurate and reliable data that reflects precise needs, tends to hinder the implementation of any settlement over water regardless of how equitable it is to all involved parties. The lack of accurate estimates of current and foreseen future water needs puts future implementation of this Treaty in jeopardy and makes it subject to violation. This is because the expected population growth is likely to exceed the amounts of water proposed by the Treaty, which will create more pressure on the current water resources of the region and make the Treaty fragile, especially during times of drought.

The per capita consumption rate, consumption by sector, and the potential use of a water demand model were not considered in either of the treaties. These technical elements were not of particularly great significance in impacting the outcomes, as they are not linked to any of the discussion surrounding issues of water-related outcomes in either Treaty. It is often best to set such issues aside during negotiation in order to avoid opportunities for disagreement and political impasse, as they may tend to complicate the negotiation and cause more distraction. Further, such high-level negotiations are conducted for the most part by politicians who know little about (and might be disinterested in) these technical details. As such, simplicity and a narrowly focused negotiation agenda is identified as a successful strategy for negotiation. For simplicity, supply-side management strategy can be utilized during negotiation, whereas issues related to demand-side management can be addressed by a joint water committee after reaching an agreement. With this in mind, collecting and sharing accurate data about the river system and shared resources are critical during negotiation to be able to consider the available water supply in its entirety. This requires the cooperation of all parties in providing accurate data representative of water availability throughout the river system to conduct realistic analysis, share results, and provide a common operating picture that derives actionable insight and intelligence for the formation of sound strategic and tactical actions.

In the case of the LHWP, negotiators were very prudent in using a needs-based negotiation style, which is superior to the conventional rights-based allocation approach. South Africa needed water to support its industrial activities, whereas Lesotho needed economic and financial aid, and the Treaty was sensitive to and delivered on both countries' needs. This focus on need had a significant impact on the outcome being equitable and acceptable by both parties involved in the dispute. However, the water allocation scheme identified in the 1994 Treaty cannot be said to have advocated for a needs-based allocation plan. This difference cannot be understood without being fully cognizant of the current and historic uses of water in the region.

As shown in Figure 6.1, the first step in water allocation and planning typically encapsulates establishing water needs and attempting to find ways to satisfy them, which is easier said than done. In reality, there are a number of obstacles in the way of satisfying these water needs. Furthermore, there are many steps, concerns, and questions that need to be addressed even before establishing these water needs, such as by whom and on what basis these needs are determined. In addition, satisfying current needs is not enough; it is imperative to cater to both current and future needs. Having said this, it is crucial to verify the accuracy of the data in order to establish accurate water needs. For any water agreement to sustain implementation, it is also critical to go beyond future needs to achieving satiety[1] (level of satisfaction) to both parties, and, in turn, regional stability. Conventional approaches proposed to address the water crises in Jordan, Israel, and other countries in the Middle East have been problematic on these crucial issues. They often tend to establish needs that are either exaggerated or underestimated. The result is a contentious and vicious pendulum-like cycle: unrealistic needs are asserted

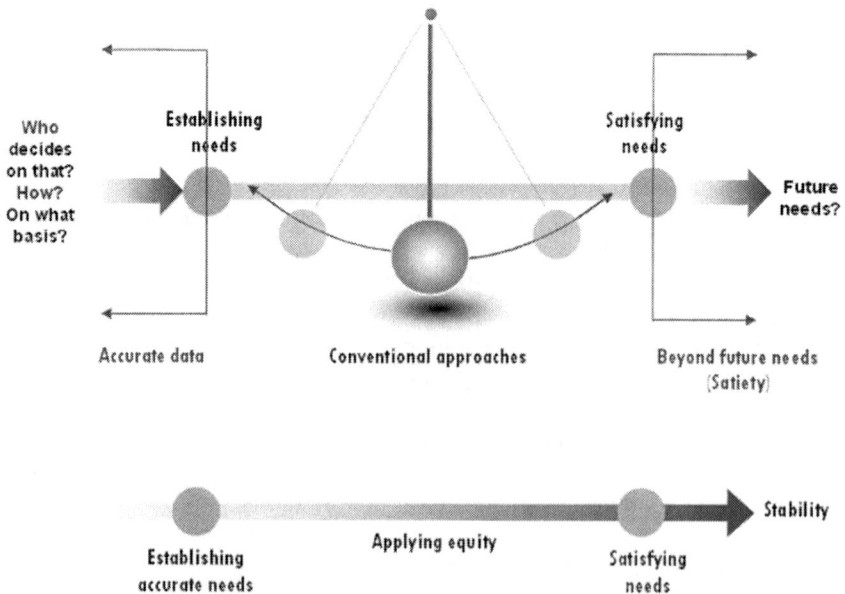

Figure 6.1 Conventional approaches and the role of equity in water allocation.

and attempts to satisfy them are to no avail. Ideally, applying principles of equity to the entire process of water allocation means establishing accurate needs, satisfying current and future needs, and achieving stability and sustainable implementation.

As discussed in Chapter 4, the Treaty between Jordan and Israel was not even at the level of these conventional resolution attempts that tended to establish needs in order to satisfy them. Instead, the Treaty did not seem to have considered water needs as a factor at all in determining the quantities of water distribution rates. Making no reference to actual needs of either state, this Treaty allocated specific amounts of water without providing appropriate grounds to justify the allocation. Although detailed demand-side figures can be addressed jointly after an agreement is reached, it is important to establish some agreement on basic demand figures to justify volumetric water allocation. This shortcoming resulted in an outcome that can only be characterized as not being able to reliably measure equity and perhaps not being equitable, in the end.

In both of these cases evaluated in the book, adequate attention was given to data mediation through the work of the joint fact finding (JFF) teams, although the results of these mediation efforts varied greatly in both cases. Before, during, and even after negotiation, the teams were engaged in joint tasks to collect and validate acceptable hydrologic data related to the available water resources that could be shared by the parties. Joint technical teams from Israel and Jordan were established to figure out how much water was available in the Jordan River and how much of that water could potentially be extracted and shared between the two countries. These efforts started in the early 1980s and continued into the 1990s as the Treaty was being negotiated. These JFF groups, regarding issues of "low politics," offered substantial ideas that helped in initiating and facilitating negotiation over issues of "high politics." For example, Israel and Jordan participated in a study tour of the Colorado River basin management and regional training program and jointly explored technical and legislative ways to resolve water-related conflicts. They also participated in a joint feasibility study on the construction of a canal between the Red Sea and the Dead Sea that would potentially alleviate water stress and benefit both countries (Fuwa, 2003). The technical talks held between Israel and Jordan in the 1980s produced a factual basis regarding available water compared with current and future water demand and deficit. With this in mind, there appeared to be a clear consensus on this factual basis from both sides and neither the Israelis nor the Jordanians questioned the validity of the water-related data. Although this agreement on shared data in turn facilitated smooth negotiation, these data sets were not used as the basis for determining water needs and allocation schemes put forth by the Treaty.

Upon the request of the two governments of Lesotho and South Africa, joint technical teams (known as the Joint Technical Committee) were formed as early as 1978 to study water availability and determine the amount of water to be transferred to South Africa. This resulted in a series of feasibility studies, as discussed in Chapter 5. Although the two countries formed JFF teams to collect and share water-related data, they were, for the most part, in disagreement with each other regarding how much water was available for potential transfer.

This rift was exacerbated by Lesotho's tendency to overestimate available water in the hope that more water could be transferred and more revenue generated. According to Mckenzie (2009), this disagreement continued for 10 years, which called for the need for continuing data mediation and verification on both sides. Exact water availability data was not jointly developed and agreed upon until after the Treaty was signed, which in a way hindered the two countries' ability to exactly determine the amount of royalties. Surprisingly, this disagreement regarding water-related data, although it caused tension whenever it surfaced during negotiation, was not a major obstacle for the negotiation process, nor did it cause significant delays in negotiation. This was because both sets of officials had greater interest in reaching an agreement that would solve their respective problems and, although data was important, both realized that data accuracy was of secondary importance compared with the significant issues at stake and the shared common ground.

It is crucial, however, to note that establishing consensus on accurate and reliable data regarding water availability and consumption prior to negotiation is key in promoting sound water management. It also assists in making informed decisions regarding future water planning and management, building trust among the parties, and enhancing chances of reaching common ground.

Structure of the agreement

Treaties predating 1980 lacked any mention of equitable practice. This was also reflected in the outcome of the treaties, which can be characterized, for the most part, as being less than equitable, favoring one state over another. Owing to the fact that the Israel–Jordan and the Lesotho–South Africa treaties were signed after 1980, they both included language of equity in their texts. The Israel–Jordan Peace Treaty explicitly utilized notions such as "rightful," "just," "equitable," and "practical" (Elmusa, 1995; Louka, 2006). Similarly, the LHWP Treaty emphasized the "the mutual benefits for the Kingdom of Lesotho and the Republic of South Africa to be derived from the enhancement, conservation and equitable sharing of the water resources of the Senqu/Orange River and its affluents" (LHWP Treaty, 1986: Preamble). As discussed earlier in Chapter 4, using language of equity helps to boost equitable outcomes and provides the basis and motive for the parties to work toward a defined mutual goal of attaining water-sharing equity (Lautze and Giordano, 2006).

Although both quality and quantity issues were considered by the negotiators of both treaties, neither Treaty was successful in implementing a potent strategy for preserving water quality. The 1994 Treaty dedicated Article 3 to "Water Quality and Protection," stating that "Israel and Jordan each undertake to protect, within their own jurisdiction, the shared waters of the Jordan and Yarmouk Rivers, and Arava/Araba groundwater, against any pollution, contamination, harm or unauthorized withdrawals of each other's allocations." It further asserts that "the quality of water supplied from one country to the other at any given location shall be equivalent to the quality of the water used from the same location by the supplying country." Similarly, the 1986 LHWP Treaty in Article 15 makes mention of the "protection of the existing quality of the environment" and the issuance of compensation to ensure "the welfare of persons and communities immediately affected by the Project." However, there were no specific guidelines in either treaty for the protection of the environment or the overall water quality. Because the LHWP Treaty focused on technical aspects, whereas the Israel–Jordan Treaty focused on water volumes, neither treaty focused on water quality or the protection of the river systems. As such, both river systems suffered serious environmental ramifications. The over-extraction of water in the Jordan River basin caused the river water to drop below critical levels and the contaminated run-off caused the river to become polluted. This situation resulted in the severe salinity of the Jordan River and the declining of water levels in the Dead Sea. This contributed to the general perception of the Treaty as being unresponsive to the issues of water quality and the deterioration of the ecosystem, which seemed to have been overlooked by the Treaty.

Several environmental and social impact studies were conducted by the LHDA and TCTA to mitigate the adverse impacts of the LHWP project (Fullalove, 1997). These studies concluded that the impact of transferring substantial amounts of water in the wake of the project implementation caused severe impacts in several locations downstream of the project. Another problem is related to the fact that Environmental Action Plans were not completed until 1991, after the main parts of the project were constructed. As such, the outcomes of both treaties were detrimental to the health of the Jordan and Orange rivers' ecosystems. This impacted the general perception of both treaties as being insensitive to the environment and hindered their acceptability and perception accordingly.

Neither treaty considered the needs of different uses and users. There is no indication in the text of either treaty that water was allocated based on priority of use (such as domestic, agricultural, or industrial uses) or users. Prioritization of uses and users is important when water is scarce, as is the case in the Jordan River basin. Interestingly, as one would expect, this element was influential in determining and impacting outcomes and perceptions only in the 1994 Peace Treaty. This issue was irrelevant, however, to the LHWP negotiation, as water is scarce only in South Africa but available in Lesotho for transfer. In this case, water is freely transferred from Lesotho as per South Africa's need and the amount of royalties it has been willing to pay.

Water allocation can be determined on a variety of bases, as illustrated in Figure 6.5 later in this chapter.[2] Reflecting Level 4 of water allocation, the water allotment scheme proposed by both treaties can be best characterized as a combination of time-based and volumetric-based allocation. Both treaties allocated water amounts that were tied to a particular location and during specific times. This was closer to sharing than to dividing, which is a positive quality of both treaties. In some cases, however, the 1994 Treaty of Peace seemed to allocate water based merely on volumetric amounts, namely the additional 50 MCM to be delivered to Jordan without determining when, where, and by whom this water would be delivered. Because of the volumetric nature of such allotment, the agreement does not seem to offer guidance in cases of drought. On the contrary, by relying on the average carrying capacity, reflecting the average annual discharge or the steady state of the river system, the Treaty seems to disregard such extreme hydro-events, although they are bound to occur in the future, as has been the case in the past, namely in 1999. Nonetheless, in both treaties, water was generally allocated based on volumetric and time measures producing an allocation scheme that could be described as sharing rather than dividing, which had positive impacts on the way the treaties were implemented and perceived.

On the whole, cooperation over water resources entails certain benefits as well as costs. Generally, benefits harnessed from a certain treaty or agreement can be categorized as water-related benefits and non-water-related benefits. These can include volumetric water transfers, hydropower, technological and economic incentives, strategic gains in the form of achieving and sustaining cooperation, peace, regional stability, national security, technological knowledge, developing capacity, and other benefits that vary from solidifying international border lines to avoiding potential future invasions and possible military aggression. Costs involved in signing a treaty include compromising existing or historic favorable positions in terms of water access, extraction and consumption, economic cost related to water transfer and the infrastructure of development projects, opportunity cost of unilateral investments in water development projects, deterioration of international relations, and economic sanctions. Unlike the conventional water management approach that views water as a scare resource, this benefit-focused approach views water as a flexible and extendable resource that can be utilized to minimize conflicts by focusing on issues such as virtual (or embedded) water, knowledge sharing, and the purpose of water use for a particular user (Islam and Susskind, 2013).

The LHWP Treaty employed a strategy known as "benefit from water," which was not used in the 1994 Peace Treaty. Benefit from water use was a successful strategy that was deployed in the negotiation and operation of the LHWP. By negotiating the benefit from water use, namely that South Africa needed water and Lesotho needed financial assistance, the Treaty was successful in promoting a win–win situation for the mutual benefit of both countries. This enhanced the perception of the Treaty as being equitable to both parties and therefore helped in its acceptability and implementation. However, the negotiators of the 1994 Treaty were not concerned with benefit from water and did not discuss their ultimate use of water as such. With this in mind, the two countries missed a unique opportunity to discover the other parties' needs and their purposes for using the water from the river. This would have been beneficial in helping the parties prioritize their uses based on urgency and the availability of water. This aspect of the negotiation (benefit from water), however, does not apply to the Middle East, as water shortage per se is the main problem. Since the nature of water crisis in this region is primarily a matter of water supply, water is at the heart of the conflict and cannot be ignored under the premise of "water benefit." Water benefits can most appropriately be considered in situations where basic water needs of the population are fully met. Negotiators recognized the need to resolve the water dispute in a way that ensured the delivery of basic levels of water supply to communities that needed it the most. Water benefit sharing was helpful in the LHWP because the parties were not in need of water for their consumptive purposes. Conversely, both Israel and Jordan needed water for their own consumption and survival. In this case, volumetric water allocation was negotiated to meet both countries' respective consumptive purposes. Therefore, the lack of consideration of water benefit was not of great importance to the outcome of the Treaty and seemed neutral in impacting the overall perception, as it was proven irrelevant.

As shown in previous chapters, the Israel–Jordan Treaty of 1994 was not resilient in the face of droughts and political changes. As discussed in Chapter 4, the drought of 1998/1999 was a vivid indication of the failure of the Treaty in coping with extreme hydro-events. The Treaty's lack of adaptability to drought was detrimental to both countries' internal and external political spheres. The political turmoil and tension that were triggered as a result of this drought signified an important implication for water negotiators: the importance of incorporating flexibility into the negotiation and implementation of a treaty. The fact that the Treaty remains unchanged, may be less a sign of satisfaction in the terms of the Treaty than a result of rigidity and lack of compromise. Various detrimental implications can ensue in the implementation of the Treaty as a result of lack of flexibility (Fischhendler, 2008a). On the contrary, the LHWP demonstrated a great deal of flexibility to political changes and adaptability to droughts. This was made evident by the adoption of several new protocols as part of the Treaty and the creation of new bilateral as well as multilateral water regimes, as discussed in detail in Chapter 5. This embedded resilience in treaty negotiation and implementation had positive implications on the way the Treaty operated. As such, these two elements (flexibility/resilience and adaptability to drought) stand out as crucially important components in enhancing equitable processes and outcomes.

Although non-environmental factors were the prime concerns in these treaties, environmental factors were not adequately addressed. As discussed earlier in this chapter, the Israel–Jordan Treaty provided explicit reference to environmental factors and stressed the need to jointly protect the environment. However, the agreement failed to consider or respond to the adverse consequences of Israel's previous and current water development, namely the construction of its National Water Carrier in the 1950s. The Treaty therefore failed in providing adequate environmental regulations to mitigate environmental harm to the Jordan River ecosystem. Additionally, the Treaty disregarded the health of the ecosystem by considering any unused water remaining in the river system (which is essential to the overall health and integrity of the ecosystem) to be "wasted water." It further promoted the extraction of this residual water stating that, "in order that waste of water will be minimized, Israel and Jordan may use, downstream of point 121/ Adassiya Diversion, excess flood water that is not usable and will evidently go to waste unused" (Treaty of Peace, 1994: Annex II). In contrast, the LHWP Treaty was explicit about "protection of the existing quality of the environment" and the issuance of compensation to ensure "the welfare of persons and communities immediately affected by the Project" (LHWP Treaty, 1986: Article 15). It also states that "the Parties agree to take all reasonable measures to ensure that the implementation, operation, and maintenance of the Project are compatible with the protection of the existing quality of the environment and, in particular, shall pay due regard to the maintenance of the welfare of persons and communities immediately affected by the Project" (LHWP Treaty, 1986: Article 15). However, protection of the environmental and social aspects called for adequate mitigation processes, which were missing. As discussed before, although environmental studies were conducted, these studies were completed after the project was in its advanced implementation phases. At this point, it was too late for them to make a significant impact on the serious environmental and social consequences that the implementation and operation of the LHWP produced. The serious environmental ramifications associated with the implementation of both treaties poisoned public perceptions of the treaties and provided opportunities for opposition.

Explicit contingency planning clauses were missing from the text of both treaties. In this regard, neither agreement offers any contingency planning clauses for how to jointly manage water shortages during times of drought, especially during the summer periods when the river flow drops below a specific threshold. Therefore, the makers of the treaties failed to put in place a strategy for dealing with these impending droughts. For example, Dr. Shamir explains that drought management was not addressed adequately by the 1994 Treaty because "we [negotiators] could not include everything. We had to stop somewhere" (Zawahri, 2004). By so doing, it remains unknown how water will be allocated between the parties in cases of drought and who will bear the brunt of such events. In the case of the LHWP, frequent drought events are likely to impact the amount of water delivered to South Africa, threatening its industrial foundations, and the amount of royalties received by Lesotho, hindering its economic development goals. In short, although both parties are prone to be impacted by frequent drought events, neither treaty included contingency planning clauses to address these potential circumstances.

Adequate enforcement and implementation mechanisms were missing from the Israel–Jordan Treaty. Although the parties have been adhering to the Treaty, in 1999, Israel was unable to abide by its water transfer obligation identified by the Treaty as a result of drought. Political pressure was the only factor that forced Israel to eventually deliver to Jordan the amounts of water that had been agreed on. This indicates that, although identifying enforcement mechanisms is important to ensure implementation, political will has a greater role to play in keeping the parties invested in advancing implementation. Similarly, the LHWP Treaty did not include any enforcement mechanism for the implementation of its terms but it stipulated that persisting conflict, which cannot be resolved by way of direct negotiation, would be deferred to arbitration. Arbitration was never used in this case and the Treaty maintained steady and successful implementation. Interestingly, in both cases, the enforcement mechanism was not influential in determining outcomes or perceptions.

In both cases, deciding on appropriate measures for monitoring of implementation was left to the work of joint water committees. The Joint Water Committee

(JWC) was established to ensure future cooperation and monitor implementation of the Israel–Jordan Treaty. However, the intermittent meetings of the JWC and the uncertainty of the source of the additional water precipitated a "mini-crisis" in the relationship between the two countries, as discussed in Chapter 4. In the case of the LHWP Treaty, regular meetings of the Joint Permanent Technical Commission (JPTC) helped in maintaining adequate monitoring of project implementation. The persistent work of the JPTC was pivotal in facilitating successful implementation.

In light of this cross-case comparison, we can tease out important, unique process equity parameters for the advancement of international water negotiation and hydro-diplomacy. These include the involvement of all stakeholders, trust building, interest-based negotiation, modern and indigenous conflict resolution mechanisms, incremental approach, facilitation/mediation (third-party involvement), public involvement, depoliticizing water (focusing on low politics issues), socioeconomic considerations, reliance on international law, needs-based negotiation, data mediation/joint fact finding (JFF), language of equity, water quantity/quality, prioritization of uses and users, sharing not dividing, flexibility/resilience to droughts and political changes, environmental and non-environmental factors, contingency planning clauses, and monitoring of implementation.

These parameters constitute the cornerstone for a transformative approach to conflict resolution over international water resources, as will be discussed in the following section of this chapter. Certain other factors that were important in each or both of these cases were not considered as generally important because they were accidental factors associated with each of these cases and may not exist in, or be relevant to, other international cases in general. For example, considering average, wet, and drought conditions (which seems to be very important for the Israel–Jordan Treaty) was neutral in the LHWP Treaty because water is abundant in Lesotho. By the same token, although considering growth rate and annual water budget are important elements in the Israel–Jordan Treaty, these were for the most part irrelevant and neutral in the LHWP Treaty, as allocation was not meant for consumptive purposes (although water demand modeling would have been helpful in outlining exact water amounts needed for industrial development in South Africa, and therefore the amount of royalties expected to be generated by Lesotho). The following section expounds these essential factors and provides guidance for developing a transformative approach to conflict resolution that incorporates various guidelines for negotiation and implementation.

An agenda for a transformative approach to conflict resolution

The cross-case analysis outlined above is used as a vehicle to develop a pragmatic transformative approach to conflict resolution. The last section highlighted important process equity parameters; this section will place those parameters in the context of international water negotiation in general. Because of the inadequacy in the current and previous planning practice and conflict resolution processes, many problems related to social justice, inequitable distribution of resources and services, resources depletion, and the unsustainable nature of resources manage-ment have occurred. With the lack of an overarching multiscalar framework, many of these severe problems will continue to occur, grow, and fester (Abukhater, 2009c). A curtailed diminution of a monolithic and homogeneous approach does not suffice to address the heterodox nature of water issues. In light of the afore-mentioned discussion, there is ample justification for the need to develop a new agenda for a transformative, interdisciplinary approach to conflict resolution that recognizes the uniqueness and challenges of each water-related dispute and, at the same time, provides guidance for practical negotiation in terms of enhanc-ing equitable processes and decision making. This entails discussing major policy implications and developing practical boundaries for a transformative approach that moves water conflicts from a non-cooperative to a cooperative state. The usefulness of defining these new boundaries to shape a distinctive path for both hydro-diplomacy in general and negotiation practice in particular lies in their abil-ity, first, to provide a deeper understanding of the processes that water negotiators habitually engage in, and, second, to address the question of how to consciously achieve recognizable and measurable improvement in the quality and outcome of water agreements. Both concerns should be addressed in the light of a contextual understanding of each individual case in arid regions within which water negotia-tors and policymakers operate. Thoughtful formulation, coherent evolution, and adaptive application – these are the guiding principles that tend to continuously direct policymakers' attention and channel their efforts toward important matters. They set the stage for the development of contingent strategies and appropriate responses that ultimately achieve equitable processes and outcomes.

Policymakers and community leaders aspire to achieve desired future condi-tions, but the rapidly changing and multidimensional nature of water poses a great challenge. Two competing factors are contributing to the creation of many continuous environmental and social crises. The first is the natural world with its sensitive areas and limited and disputed resources, including water, and the second is the human footprint with an expanding population that threatens many of these resources. The combination of these factors results in many complex urban and environmental problems that are expanding across multiple regions and scales (Abukhater, 2011b). Because of the boundary crossing nature of water and related complex issues at play, such as social, natural, economic, and political aspects (Islam and Susskind, 2013), decisions regarding water allocation personify a profound challenge of how to adequately respond to a rapid population growth pattern and keep up with growth of water demand, while at the same time safe-guarding social justice for current and future generations and promoting environ-mental protection.

Because water issues are rapidly changing, our future is contingent on what we do today and the nature of our intervention. We need the future not simply to happen to us but rather to be one that we shape and define today with increased knowledge and informed decisions and actions. We need greater knowledge of our environment and our relationship with it as well as awareness of how to protect its valuable natural resources. These problems and challenges suggest the need to rely on new strategies of negotiation that are cross-disciplinary. As a result of the multidimensional aspects of the complex, uncertain, and uncontrollable political, natural, economic and social realities of water, the new approach should reflect the merits and imperatives of a multifaceted approach as well (Islam and Susskind, 2013).

In essence, this approach aims to equally increase the net benefit in terms of water allocation, in all its forms, for all stakeholders involved in a river basin dispute. This objective is not confined to alleviating water stress, but it also includes a variety of strategic, political, economic, and technological gains. To understand how much each country benefits from a certain water treaty, one must determine current water needs met by that treaty (water stress alleviated in the wake of an agreement) on both sides in relation to demand. Ideally, the reduction in water stress should be proportional to the water needs of each country. The goal here is to maximize water stress reduction for each individual country.

Another goal relevant to the scope of this approach is to equally decrease harm, in all its forms, for all stakeholders. With this in mind, a mature adaptive application of equitable water negotiation parameters should incorporate an examination of international water negotiation trends in the past, the relationship between political and economic forces and cultural and social structures, and ways in which power relationships shape political realities and decision making regarding water allocation and distribution. Although discourse can provide part of the explanation, a further step is to engender a deep understanding of the structures of power that not only guide discourse, but in many cases generate it. This suggests the need to be critical and visionary, attentive to both process and discourse, and understanding of the hydro-political structures and dynamics.

Under the auspices of equitable water-sharing processes, this proposed transformative approach in this book encompasses three key components, as shown in Figure 6.2. These include the jurisprudential framework of water law and water institutions amalgamated to adaptive learning and consensus building, equitable water-sharing criteria, and third-party mediation, which makes it exceptionally functional and enhances its applicability to a wide variety of arid regions. Whereas the international water law provides process guidance, or "rules of engagement," adaptive learning and application of equitable water-sharing criteria provides institutional resilience and governance, or "mechanisms of engagement." Additionally, employing effective ADR and the involvement of a neutral third-party mediator are essential to the attainment of win–win cooperative processes and the enhancement of acceptable resolutions for all involved parties.

This approach encompasses important procedural equity aspects identified above, the institutional ability to create a conducive atmosphere for negotiation with equal footing, and the power to influence outcomes with the flexibility to accommodate different political, social, economic, natural, and spatial contexts, and finally the aspiration to promote legitimate public input in pursuit of the common good. Given its holistic merits, this transformative approach to conflict resolution is equally concerned with both short- and long-term consequences

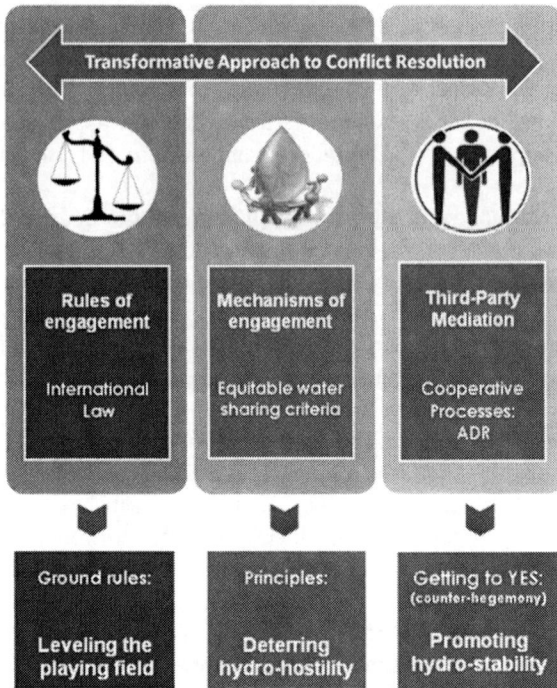

Figure 6.2 Components of a transformative approach to conflict resolution.

of water allocation schemes. Employing and extending available resources and seeking to obtain new resources to satiate current and future water needs, this approach effectively aims to establish a workable, robust, and adaptable framework for equitable water sharing and conflict resolution mechanisms.

Rules of engagement

It is important for disputing parties to develop common ground for negotiations to eliminate any chances for conflict of interests during and after negotiation. This aspect was missing from the Israeli–Jordanian negotiation and caused difficulties at the beginning of the process, resulting in delay and further disagreement. An effective transformative approach to conflict resolution should be based on the adaptive reliance on the guiding principles of the international water law as an overarching and systematic jurisprudential framework of water negotiation. This research suggests that the reliance on principles of international water law helps in providing the necessary rules of engagement to level the playing field and neutralize the impact of power structure imbalance. Woodhouse (2008) suggests that to relinquish international law altogether without exploring another mechanism to govern international negotiation over transboundary water can undermine the overall regional water security as well as political relations among riparian states.

International law is a useful instrument to increase the predictability of states' behavior and decisions regarding cooperation or altercation. In this regard, the research findings indicate that international water law can provide a framework for counter-hegemony processes to mitigate the impact of hydro-hegemony, by promoting transparency, accountability, and predictability of behavioral tendencies of powerful states. It is no surprise therefore that the hegemon often rejects the applicability of international law, as in the Israel–Jordan negotiation. Although the applicability of international law is inconclusive, it at least provides general rules to allow room for flexibility and adaptability so that states can more easily find solutions in various cases. This is not a panacea; in the case of the Arab–Israeli conflict, disagreements arose over the meaning and interpretation of international law, as it was used to justify and legitimize opposing positions. Although international law has been a great area of controversy, as it does not offer a template for negotiation (as was suggested by the Israeli negotiators), it does provide the legitimate framework for negotiation and necessary protection for the weaker party. This conclusion is also echoed by many water experts who believe that relying on international law is a prerequisite to safeguard the process of negotiation and its success. As was discussed in the Israel–Jordan Water Treaty case, the Israeli negotiator believed in applying a more pragmatic approach rather than relying on principles of international law. However, this apparent pragmatism failed to achieve its goals because the outcome of the Treaty was heavily influenced by the most powerful party to serve its own interests. This was a direct byproduct of the fact that the process did not rely on principles of international law, which is a necessary tool to protect weaker states and eliminate general disagreement on basic issues.

International water law, namely the Helsinki Rules of 1966 and the 1997 United Nations Convention on the Non-Navigational Uses of International Watercourses, encapsulates a number of economic, hydrologic, and legal concerns (identified as part of the process equity parameters earlier in this chapter and summarized in Tables 6.1 and 6.2). As indicated by a significant body of literature and the underlying rationale of international water law, principles of international law delineate a set of criteria for equitable allocation of transboundary waters (such as geography, hydrology, and demography). As such, our research suggests that one of the practical benefits of applying principles of international law is that they can be used as a template against which to objectively judge certain water allocation plans and outcomes, based on the extent to which they adhere to these basic principles.

In 1991, the International Law Commission (ILC), an organization associated with the United Nations, further developed the Helsinki Rules and drafted 32 articles on the Non-navigational Uses of International Watercourses (Aliewi *et al.*, 2001; 2003), focusing on enhancing "reasonable and equitable share in the beneficial uses of the waters of an international drainage basin" (International Law Association, 2003). Since both of these laws are considered to be the basis and the guiding spirit of the modern international water law, they carry a significant amount of clout and have a decisive influence on negotiating parties (Shuval,

2007). Nine factors related to "reasonable and equitable share" of water are mentioned in Article 4 as key parameters for equitable water allocation and utilization, including population, hydrology, climate, historical and existing uses, socioeconomic needs, dependent population, comparative costs of alternative means, availability of other resources, and the degree of appreciable harm or damage caused by denial of water rights (International Law Association, 1967). The term "significant harm" (or "appreciable harm") refers to any cost encountered by basin states due to denial of water rights (Goldberg, 1992). Additionally, according to the International Law Association (2003), the term "damage" is generally defined as including loss of life or personal injury; loss of or injury to property or other economic losses; environmental harm; and the costs of reasonable measures to prevent or minimize such loss, injury, or harm. "Environmental harm" includes injury to the environment and any other loss or damage caused by such harm, and the costs of reasonable measures to restore the environment actually undertaken or to be undertaken (International Law Association, 2003).

Providing a robust legal foundation for water negotiation means considering, applying, and further operationalizing these nine factors of equitable water allocation. Using these factors in a meaningful fashion to enhance water allocation equity also means converting these nine factors, which are the key determinants of outcome equity, into operational hydrogeomorphology-based outcome equity drivers, illustrated in Figure 6.3.

By relying on the defining principles of international water law, cooperating riparian states can effectively capture these positive elements (some of which are listed as essential process equity parameters in the logic diagram) as part of the negotiation process as well. There is ample evidence from research findings suggesting the importance of incorporating flexibility, along with consistency, with the principles of the international water law in equal measure. From a pragmatic

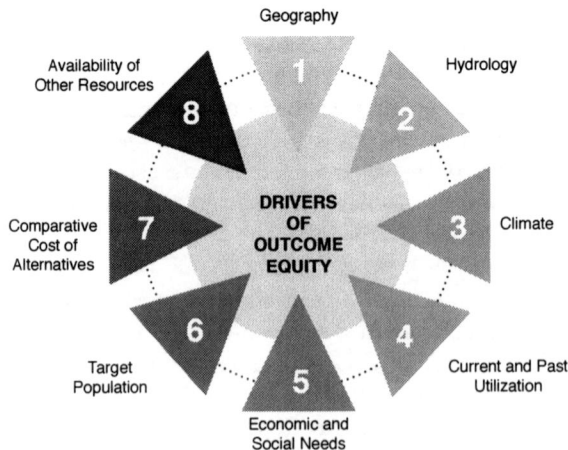

Figure 6.3 Eight operational factors (hydrogeomorphology-based drivers of outcome equity).

point of view, incorporating a certain level of flexibility ensures the practicality, adaptability, and, in turn, adoptability of the terms of the agreement, whereas the international water law ensures that the treaty is developed in accordance with its principles. It is important to note that the international law is necessary yet not sufficient by itself. Goodwill on both sides, the intervention of a third party, and an ongoing negotiation and cooperation process are warranted for the success of effective and environmentally sound water allocation practice. This aspect of negotiation was evident in the LHWP Treaty, as the two parties struggled to reach common ground until the World Bank intervened as a neutral third party. For this purpose, we suggest that the needed reliance on international water law must be coupled with "mechanisms of engagement," discussed below, as well as the involvement of third-party mediation in order to produce an effective, transformative approach to conflict resolution (which will be discussed in the following section).

It is prudent to note, however, that international water law, beneficial as it may be in many cases, is problematic in certain areas. Although it introduces elements of water needs, this international legal framework is more concerned with issues of sovereignty over water resources. Although it provides the ground rules to manage the sharing of disputed water resources, international water law seems ambiguous, especially lacking in a pragmatic definition of the particular terms that it introduces, such as "equitable and reasonable utilization" and the "no harm" principles. It also suffers from the lack of local context-oriented guidelines to ensure its applicability, and proper enforcement mechanisms to ensure its implementation. Notwithstanding these shortcomings, the reliance on principles of international water law is pivotal in advancing negotiation to more plausible and implementable resolutions.

Mechanisms of engagement: a comprehensive approach to conflict resolution

Mechanisms of engagement incorporate a set of elements that reflect important equity parameters identified in previous sections. These key elements are suggested on the basis of the analysis of the two cases and are guided by the logic model. Based on the cross-case analysis provided in the previous section, we believe that, for the conflict resolution process to be effective in producing viable, desired, just, and sustainable solutions, it must rely on a comprehensive conflict resolution approach that encompasses context-sensitive equity and adaptive learning and management, modern and indigenous approaches to conflict resolution, and goodwill/faith-based confidence building measures (CBMs). The reliance on this comprehensive conflict resolution approach also means incorporating an incremental negotiation process, sectoral prioritization of various uses and users, linkage to environmental and non-environmental issues, sharing not dividing, needs-based water allocation, and satiety-based hydro-diplomacy, as illustrated in Figure 6.4.

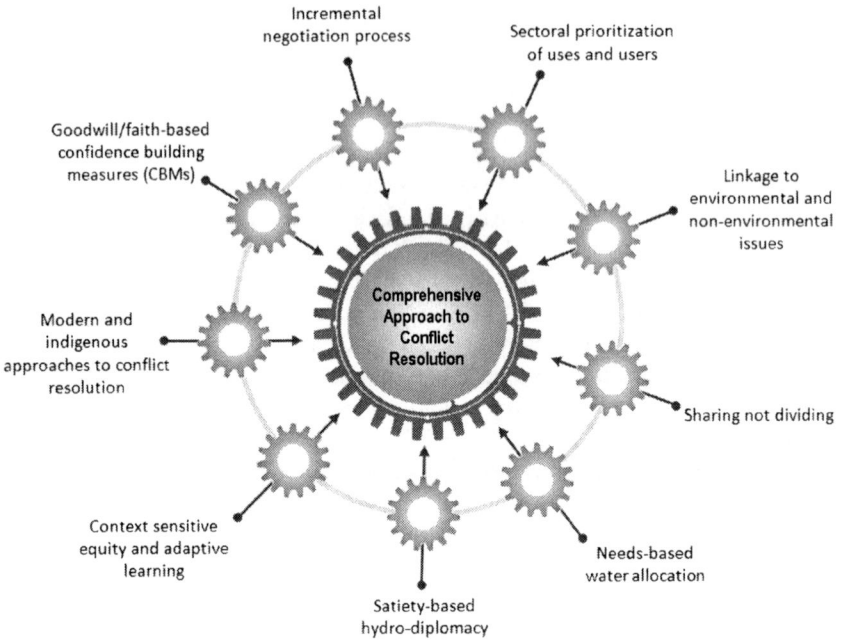

Figure 6.4 Mechanisms of engagement – comprehensive approach to conflict resolution.

Context-sensitive equity and adaptive learning and management

Achieving equity for all disputants in any agreement, a major concern in achieving sound management of shared transboundary watercourses, requires significant dedication to fairness and the ability to consider various interests in the process. This research deduces that sharing an ambiguous and unforeseen risk means that *equity*, not equality, is the underlying principle in water allocation. Thus, all parties should be treated based on measures of equity, rather than equality, which is illusive and illogical considering its implausibility and irrelevance in water resources allocation. Operationally, since ultimate equality is by no means attainable, the issue of equitable water sharing seems, practically, to make much more sense. To help unravel the ambiguity surrounding the notion of water-sharing equity, one might ponder the meaning of the concept of equity.

As a general term, equity means fairness and impartiality. Relying on equitable water allocation that incorporates sustainability and fairness in sharing not only benefits, but also costs and risks, is crucially important in fostering acceptability and implementation. However, new issues arise and new circumstances become apparent on an ongoing basis, demanding innovative and flexible responses based on a situational assessment of the types of issues to be addressed and a prioritization of the most important factors or interests. For example, water demand and supply dramatically fluctuate from one year to the next and the political reality is in flux as well. By relying on the notion of context-sensitive equity (the flexible application of principles of equity by incorporating an ongoing situational assessment and prioritization of related issues and concerns according to their urgency and importance), resolutions can be crafted to guide the decision-making process in the face of the ever-changing realities. Consequently, in keeping with the dynamic and developmental nature of water management, a sound water management approach need not rely on static processes and rigid approaches. As each case is unique, likewise water-related decisions must be unique. Thus, equity entails a system of justice, allowing fair judgment of a certain case based on dynamic and contextual differences that each case represents. For example, although the Israel–Jordan Water Treaty and the LHWP Treaty share common characteristics, each case exhibits contextual differences that are exclusively unique to it. As discussed in Chapter 5, the effective benefit sharing of water allocation achieved in the LHWP Treaty can be attributed to these contextual differences, namely the fact that Lesotho has a variety of water resources relative to water use and water is needed by South Africa. Because the international relationship is based on asymmetry of interests, they were able to reach an agreement whereby South Africa secured the water needed for its industrial development and Lesotho secured economic remuneration for its economic development. These aspects in terms of differential water needs, political relationships, and non-water-related issues that provided a common platform of differential value to trade across do not exist in the Israel–Jordan Treaty. In this case, both countries needed water for their population and lacked sufficient storage capacity and, with such an uneasy political atmosphere, the construction of such capacity for each party was opposed by the other. Thus, a solution that allocated volumetric water rather than differential interests was the answer to resolve the issue. In both cases, equity appears to be context laden and can be attained and perceived differently.

Because water issues are complex in nature, as they exist in a multitude of social, natural, and political domains, experiences and knowledge learned from one spatial or temporal context is not easily transferrable to other contexts or situations, as they may fundamentally vary along these and other aspects (Islam and Susskind, 2013). The boundary crossing nature of water necessitates a fundamental understanding and careful examination of the critical situational and unique aspects of each individual case. This is particularly true as water complexity represents a formidable interaction between numerous issues related to processes, people, and institutions at various levels and scales (Islam and Susskind, 2013). In addition, over time, these protracted water conflicts and their resolutions change along four aspects, including conflict formation (or emergence of new conflicts), conflict exacerbation, conflict mitigation, and conflict resolution or transformation (Mitchell, 2006). Conflicts and resolutions change over time and so should agreements. As such, agreements should not be viewed as static or inelastic mechanisms to resolving conflicts but rather as ever-changing, dynamic, "resolutionary" tools by which conflicts can be transformed to more stable and plausible situations.

The principle of "situational fairness" appears crucial and relevant here. The allocation process, based on this principle, involves an ongoing assessment of each situation, making situationally specific decisions. There is a tradeoff in so far as decision makers have to develop criteria that reflect contextual details unique to each situation, weigh certain important factors based on their degree of immediacy and relevance, and, finally, make decisions that embody this systematic process of prioritization. Contextually identifying elements of equity and how these elements effectively work in certain contexts and in different places helps in identifying the best alternative for particular challenges. This way, compromises are made and benefit can be gained as well.

Having said this, to be realistic and effective and to enhance a treaty's adaptability to, and compatibility with, a wide range of contexts, adopting this transformative approach should incorporate a certain level of flexibility. To this end, it is crucial to incorporate flexibility in water allocation by relying on a dynamic and proactive process of situational assessment in which allocations of water change as a result of new circumstances to achieve regional satiety for all. This can be accomplished through the work of joint river committees, which might meet on a regular basis during the post-negotiation stage and work toward a sound management of the shared water resources. As demonstrated in the LHWP Treaty, the work of the joint river committee was fruitful and resulted in adopting new protocols in response to new political changes. In addition, flexibility and resilience may be incorporated into the text of the treaty to ensure ongoing situational assessment and adequate, fluid response mechanism. In this sense, incorporating language of equity in the text of the treaty, such as the term "rightful allocation" which was inserted into the 1994 Treaty, is very helpful in illustrating the impact of such flexibility in the usage of terminologies to enhance acceptance and promote adoption. However, it is pivotal to be able to strike a balance between achieving flexibility and maintaining compliance.

The success of any treaty in balancing flexibility with compliance is primarily contingent upon the work of the negotiators during all stages of negotiation. Typically, during the pre-negotiation and negotiation stages, both parties work together to utilize appropriate strategies and acceptable ground rules to achieve flexible resolutions. Clear ground rules for water allocation and negotiation must be established during the pre-negotiation stage and confirmed during the negotiation stage. These ground rules should include flexible criteria and fluid mechanisms for equitable water sharing not only during times of normal water supply, but also during times of abundance as well as drought. Lack of joint drought management strategies in the Israel–Jordan Water Treaty caused problems in implementation and political tension in the wake of the 1998/99 drought. Encouraging the parties to mutually formulate and utilize objective and equitable water-sharing standards and flexible procedures is important in enticing coherent and consistent approaches to conflict resolution and producing high-quality agreements. Contrary to popular belief, failure of traditional hydro-diplomatic efforts not only inhibits the involved parties from reaching agreements, but also, and probably most importantly, it produces low-quality agreements that offer little improvement to the status quo (Innes and Booher, 1999). I would argue that leveraging high-quality agreements is a paramount task that can only be attained when the negotiation processes are designed and managed based on principles of context-sensitive equity that recognize the ever-changing hydro-political, socioeconomic, and environmental realities of transboundary water disputes.

Incorporating flexibility also calls for the reliance on an adaptive learning and management approach that offers resilience and joint action. Adaptive management incorporates an assessment of the past and current water management practice and a formulation of a response to change the status quo. The key for the disputing parties is to find ways to be able not only to recognize the problem and its equally deplorable consequents to both of them, but also to identify opportunities for learning and adapting to break away from their historic focus on their differences and age-old conflict to a learning and problem solving opportunity (Islam and Susskind, 2013). This calls for the development of general rules and guidelines for negotiation, free access of information acquisition and dissemination, and a developmentally incremental social learning process that encompasses technological, educational, and informational systems. This, in turn, necessitates the promotion of a fluent institutional structure that provides certain resilience and adaptability to new changes and the ambiguity, complexity, and long-term consequences of water allocation decisions. Here we see the need for integrating new systems of interaction to enhance basin-wide cooperation, coalition, and partnership. Basin-wide cooperation means more than just creating the usual communication channels among multiple riparians, but rather it indicates proactive, multiscalar interactions among all affected and interested stakeholders in the preparation and structuring of activities, which should be geared toward enhancing the process of continuous learning and participation. Although this basin-wide approach was achieved in the LHWP Treaty with the establishment of ORACOM in 2000 as a regional water regime, this element is still missing in the Israel–Jordan Water Treaty to this day. We learned that in the LHWP Treaty of 1986 flexibility was incorporated as part of the post-negotiation effort to enhance continuous cooperation between the parties, a key principle that was not fully adhered to in the Israel–Jordan Peace Treaty of 1994.

In summary, what is called for is a revised process of post-negotiation collaboration to allow for real cooperation, generate bidirectional feedback mechanisms, and meet the need for flexibility and adaptability to cope with the complexity and fluid and changing nature of water disputes.

Incorporating modern and indigenous approaches to conflict resolution

A major concern to the one-sided modern conflict resolution approaches is how they will be received by the other parties that may or may not share the same history and background. This situation was clear in the Israel–Jordan Water Treaty. As shown in previous chapters, many of the previous diplomatic attempts failed because they were perceived as a Western attempt to extend control over the region, its population and resources (Abukhater, 2009b). A common perception was also missing from the 1994 Treaty and was detrimental to the acceptability of the Treaty and its outcome. Thus, an either–or conflict resolution approach that may not have been appropriately attuned to the other party's culture and degree of "Westernization" provided a recipe for negotiation failure. In addition, literature supports this conclusion by suggesting that there is no particular approach that could single-handedly transform a water conflict. Relying on processes that stem only from modern practices will simply deepen the gap between the parties and make it even harder to accept negotiation on this basis.

Having said this, the process of negotiation should apply both modern and indigenous approaches to conflict reconciliation.[3] Combining both of these approaches tends to enhance the acceptability and level of satisfaction in the outcome. Acknowledging the shortcomings of the current exclusively modern approach to dispute resolution employed in both cases suggests the necessity to respect the views and values of others by combining modern and indigenous resolution approaches to facilitate a more mutually acceptable negotiation environment. Otherwise, any resolution attempts will continue to be viewed as a panacea by the other side, which tends to hinder the acceptability and adoptability of any diplomatic effort.

Goodwill/faith-based confidence building measures (CBMs)

Goodwill on both sides is important in leading to an equitable agreement. To gain acceptance and engage all parties in a fruitful negotiation process, goodwill must be demonstrated prior to engaging in any negotiation. This was evident in the Israel–Jordan Peace Treaty of 1994, namely the ability of the negotiating parties to capitalize on mutual goodwill to ameliorate contentious issues that otherwise would not normally have been resolved. It is unfortunate that previous conventional conflict resolution attempts by both countries focused on ethnocentric and entitlement issues, resulting in a situation of talk-fight and ultimately a stalemate, where both parties are put on a collision course. Advancing negotiation toward a more cooperative state calls for facilitating a more inviting, mature, and effective negotiation environment, which first requires bridging the cultural and communication gap prior to entering negotiation. This means more than just bringing the two parties to the negotiation table. It, rather, means respecting the views and values of others and developing relations bonded through mutual trust, even prior to negotiation. Separating the problem from the people helps in resolving the meat of the disputed matter and ameliorating the prejudiced ideologies and negative psychological impacts that tend to perpetuate the violence and conflicts. Engaging different adversaries in acknowledging the "other" party as a partner in peace tends to redirect the focus of negotiation from self-interested positions to broader and more identifiable common goals and interest-based hydro-diplomacy.

It can be asserted that abandoning this ethnocentric approach necessitates finding another plausible, acceptable, and equidistant alternative. To achieve this alternative approach, the parties should work together to enhance confidence building measures and solidify goodwill during the pre-negotiation stage. This book advocates solidifying trust through faith-based negotiation that recognizes both the modern and the indigenous conflict resolution strategies mentioned above. Good faith can be found on both sides of any conflict, but the stalemate in negotiation resides in the inability to operationalize and capitalize on this good faith. Capitalizing on this means fostering channels of negotiation through good faith, whether emanating from religious or cultural heritage, which is inherited in community practice. Investigating common faith-based practices and processes of conflict resolution emerges as a significant factor in bringing disputing parties together to communicate, gain a better understanding of (and enhance trust in) each other, and ultimately explore mutually beneficial solutions. The Israel–Jordan Treaty provides a good example for parties' cooperation prior to the formal negotiation process to build trust and confidence and move beyond ethnocentric divisions. During the "Picnic Table Talks" of the 1980s, the parties cooperated on tangible issues, such as data sharing and the monitoring of the river hydrology. These talks were instrumental in establishing a certain level of trust and acceptance of the other party, which in turn set the stage for them to lead serious negotiation.

In brief, goodwill must be demonstrated by both parties prior to engaging in any negotiation processes. For this to happen, negotiating water allocation plans should be conducted in a coercion-free environment to produce meaningful and robust agreements. Based on the analysis provided in this book, it can be deduced that incorporating and promoting ethically driven guidelines for water sharing in the negotiation process tends to provide effective mechanisms to mitigate pressure on the region's natural resources imposed by the "tragedy of the commons" situation, on the one hand, and sets the stage for a meaningful, transparent, and trust-based relationship, on the other. This cannot be achieved other than by upholding ethical and moral obligations toward each other and a mutual understanding of the necessity to maintain the minimum amounts of water necessary for both populations to survive and flourish and for the environment to preserve its integrity.

Incremental negotiation process

In time, water disputes evolve along two spectra: the number of the parties involved and the number and nature of the issues (Shmueli, 1999). This, and the fact that water disputes are complex in nature, suggests that they evolve in a way that creates new circumstances and different realities, which calls for different strategies for intervention over the course of time. Both the principle and practice of initial agreements and the incremental follow-up process of negotiation reflects this dynamic and fluid nature of water dispute evolution. In the initial stages of water disputes, the parties, having different interpretations of the issues at hand, may not consent to negotiate or even agree upon what issues need to be addressed, let alone reach a final agreement (Shmueli, 1999). In these situations, parties may even tend to shy away from entertaining negotiation as a viable option altogether and instead appear unwilling to cooperate with each other. This was evident in the LHWP Treaty where Lesotho and South Africa were unable to reach an agreement regarding common issues of concern at the beginning of the negotiation process. As such, this key conclusion suggests that negotiation over water should take place through a developmental or incremental process to ensure parties are aligned and issues are collaboratively identified and agreed upon.

Incrementalism is a fundamental principle in water negotiations. It is a recipe for conducting piecemeal negotiation over the course of many years and many rounds rather than an *en masse* approach that attempts to address many issues at one time. This is because creating and adopting new agreements to govern the allocation of disputed water resources cannot all be done at once or in a vacuum. Before reaching their final stage of completion, agreements follow a systematic process of incremental evolution that reflects the very nature of water disputes. Based on this incremental strategy, it is important to utilize both the initial and final agreements in ameliorating conflicts of significant magnitude. The adoption of an incremental negotiation process offers several benefits. First, carving out smaller topics helps parties to better focus their attention and efforts on more simplified topics that are easier to negotiate and resolve, on one hand, and gain common ground before attempting to tackle more complex and controversial topics, on the other. In that sense, reaching an initial agreement over matters of lesser magnitude sets the stage for reaching agreements over larger and more controversial and complicated issues. This is because agreeing on basic issues results in building trust and credibility with the other party and helps in viewing the latter as a partner to cooperate with rather than an opponent to fight with. Although the initial agreements are used to address more dire and immediate needs, comprehensive agreements are concerned with matters of lower priority in order to craft long-term resolutions. Once a specific issue of concern is resolved, it can be used as the basis for resolving more significant matters. It is easier to gradually build on clear and established agreements in order to reach more comprehensive ones, rather than starting from scratch, trying to attain a complete and well-rounded resolution for a certain water conflict, which will often result in a deadlock.

For example, the absence of such an incremental approach in the negotiation and implementation of the 1994 Israel–Jordan Treaty was detrimental to its outcome, the sustainability of diplomatic relations, and achieving a basin-wide multilateral agreement. Conversely, the 1986 LHWP Treaty helped in establishing an acceptable agreement by both sides that not only provided a solid point of reference to resolving the most pressing issues, but also served as a springboard for a more detailed and sophisticated set of protocols that followed its ratification in 1986. Although these six protocols were based on what the parties achieved in the original Treaty, they were added in order to address new changes (Woodhouse, 2008). As discussed in Chapter 5, this incremental approach helped in setting the stage for a more comprehensive multilateral agreement involving all stakeholders within the basin in a subsequent negotiation effort. This steady incremental process slowly but surely moves toward a more stable and permanent resolution.

Second, following an incremental and consensual negotiation process helps in depoliticizing water by isolating the issue of water sharing (or issues of low politics) from the overall political conflict (or issues of high politics). Otherwise, the overall political conflict will deleteriously impact on the negotiation process and cause inevitable political deadlock. In this regard, the Israel–Jordan Water Treaty failed in separating the two aspects of negotiation, resulting in a technical water annex that did not reflect the overall political vision of the two countries. In contrast, the parties involved in the LHWP Treaty did not lose sight of this important issue and were able to separate these aspects of negotiation and focus instead on water to promote common political and socioeconomic concerns.

Third, as negotiation continues, parties can gain a better understanding of the adversarial party and will be more comfortable negotiating their needs. With this in mind, an incremental process of negotiation can take multiple phases before reaching its final stage, at which point negotiating parties become accustomed and sympathetic to each other's needs. This conclusion suggests that it is with such a systematic process of incremental negotiations that trust and confidence can be built and developed.

This conclusion is also supported by literature. For instance, Wolf (2000) asserts that no big settlements can be accomplished in one round and that time is an important virtue. Similarly, Shmueli (1999) suggests that an incremental negotiation process incorporates prioritization of the number of parties involved and the number and types of issues to be addressed in the different phases of negotiation. Cognizant of these important aspects of negotiation, our findings suggest that this prioritization of water-related issues according to urgency through an ongoing situational assessment to guide negotiation is crucial. This could mean addressing and resolving the current and most pressing domestic water needs before any future negotiation over irrigation or industrial uses can take place. Responding to the high level of urgency of water disputes in arid regions, key parties must be identified in the earliest stages of negotiation and some sort of initial water agreement should be sought first between the major disputing parties over these matters of higher urgency. Involving others and considering the region as a whole, although helpful, is not a necessity at this point. In fact, prematurely involving other parties with their own needs in the early stages of negotiation will add more complexity to the issue and tend to distract attention away from the primary water scarcity problem, which adversely impacts the key parties in arid regions the most.

In this regard, we conclude that bilateral agreements are particularly useful as they tend to simplify issues and help parties address their very unique circumstances and resolve them in separation from other parties that might not necessarily be involved or interested (Phillips, 2009). However, bilateral negotiations might appear to be in favor of the strong riparian in some notable cases. As discussed in Chapter 4, Israel succeeded in isolating different countries and exerted pressure on each individual riparian for its own interest. For bilateral agreements to be plausible, they should be coupled with a multilateral approach, on the one hand, and be consistent with any future multilateral agreements, on the other. In bilateral negotiations, parties should not resolve issues related to other parties without taking into account the other parties' needs and recognizing the other parties' share. Otherwise, bilateral treaties will appear to be problematic to the other riparian states that they excluded, as they would not allocate equitable shares for these co-riparians irrespective of how the two involved riparians may define the term (Zeitoun, 2008). For example, there are bilateral treaties over parts of the Jordan

River system, including one between Israel and Jordan and another between Jordan and Syria over the Yarmouk. The segregation of these particular bilateral agreements is likely to cause a conflict of interest, as the Yarmouk is an integral part of the system. Thus, bilateral agreements will not be sustainable on their own in the long run without an incremental multilateral approach that acknowledges and involves all co-riparians as equal partners. Eventually, negotiation should not be limited to only the two key parties, but rather it should involve all impacted stakeholders in the basin as well. Parties sharing the river basin need to come together at some point to jointly address their needs and solidify a multinational agreement. As discussed in previous chapters, after addressing water issues of high priority, the focus can be shifted to include these other co-riparians as part of a multinational approach to achieve a long-term, basin-wide agreement. Involving all regional parties and players should be a priority in the search for a long-term resolution.

Sectoral prioritization of uses and users

As we discussed earlier in this book, there are many intricate realities and competing concerns that need to be adequately addressed and balanced in any negotiation regarding water resources allotment. Economic, social, ethical, and environmental objectives are critical components of water allocation, which greatly influence the extent to which water is distributed in an equitable, efficient, and sustainable way. Two major elements are central to approaching this issue. These include prioritizing water allocation among different uses and striking a balance between productive human needs and environmental water priorities.

First, fostering water-sharing equity through prioritization of water resources allocation between different sectors is a warranted strategy that is critical for achieving sound water allocation and management. The allocation of water between domestic and agricultural uses is a key concern in arid regions. These uses must be prioritized in such a way that ensures that highest priority is given to domestic purposes, followed by irrigation and industrial purposes. Whether agriculture represents the most or the least profitable use of water is an irrelevant question, irrespective of the context, when the issue is related to basic water needs. The key issue remains inexplicably interconnected to the question of necessity and priority. Despite its importance in food production and security, agricultural interests should not be seen as a priority in the allocation of freshwater based on its productivity and economic value. Instead, securing ongoing and uninterrupted access of adequate water (in terms of quantity and quality) for domestic uses at comparatively low prices must be given higher priority.

The second issue concerned with prioritization is the relative distribution of water between productive and environmental uses. As shown earlier in this book, due to the fundamental nature of water as the sustainer of both human life and the health of ecosystems, putting emphasis on securing, managing, and balancing human uses and environmental rights is crucial. Improving water quality and access, while fostering economic viability, is of far reaching importance.

Attempting to engineer what the environment has created often poses a conflict between economical interests and ecological requirements. It is not unusual that the political decision-making process neglects the fact that the environment still has to compete for its share of the water to maintain healthy and functional ecological systems.

For example, in both the 1994 Israel–Jordan Treaty and the 1986 LHWP Treaty, environmental factors were either not adequately addressed or were overlooked altogether. This resulted in the deterioration of the water quality of the aquatic ecosystem of both rivers and further contributed to the negative perception of both treaties as being non-environmentally friendly. It is also clear that, in the LHWP Treaty, the water allocation systems indicate a severe imbalance between environmental and economic concerns, in the sense that economic interests outweighed environmental values (although the makers of the Treaty verbally championed the environment, it mostly amounted to lip service). This problem persisted partially because of the failure of the Treaty to recognize and protect practically those essential environmental rights and the fact that it focused on technical and economic rather than on environmental aspects. In this particular agreement, environmental concerns were an afterthought. Similarly, the 1994 Israel–Jordan Treaty was not explicit about prioritizing various uses and users. Instead, water was allocated in volumetric amounts without specifying details about what the water should be used for, for whom, and in what order of priority.

Based on these important findings, we conclude that it is imperative to maintain the integrity of the environment through sustainable management of water resources. In pursuit of this goal, Sandra Postel (1997) in her book, *Last Oasis*, astutely wrote:

> We have been quick to assume rights to use water but slow to recognize obligations to preserve and protect it. . . . In short, we need a water ethic – a guide to right conduct in the face of complex decisions about natural systems we do not and cannot fully understand.
>
> (Postel, 1997: 184)

Linkage to environmental and non-environmental issues

Negotiation processes over transboundary water resources must be linked to a number of environmental and non-environmental issues. As discussed in the previous section (sectoral prioritization of uses and users), environmental issues are concerned with water conservation and the rights of the environment, whereas non-environmental issues are concerned with attaining political, strategic, and economic gains. In this sense, the environment must be viewed as an entity that vies for its fair and sustainable share of water. Therefore, advocating for the rights of the environment necessitates that allocation plans be conducted in a way that acknowledges and accounts for these environmental precedents. Leaving a great enough volume of water in the aquatic ecosystem must be regarded as a necessity in order to maintain a healthy environment.

In the current and future water allocation and negotiation processes, it is imperative to perceive natural resources as shared properties between humans and the environment. As such, the environment must no longer be viewed as a commodity, but rather as a precious entity that needs to be protected. Finding ways to live peacefully with nature, as opposed to exploiting its resources, is an indispensable request and a crucial parameter of equitable water negotiation and allocation. Under this thought paradigm, equitable sharing of water must be viewed as fostering a fair mechanism of balancing human needs with environmental concerns. As shown in Chapters 4 and 5, both treaties failed to adequately address critical environmental factors. Although the Israel–Jordan Treaty stressed the need to jointly protect the environment, it failed in implanting a functional environmental protection strategy to mitigate environmental degradation to the Jordan River ecosystem. Similarly, the LHWP Treaty included environmental protection clauses but was unsuccessful in implementing them in practice owing to the large size of the project, which resulted in serious environmental consequences.

Non-environmental issues include political, strategic, and economic gains, which are strongly linked to the process of transboundary water sharing. Stability, a political and strategic goal often sought by all parties involved in a water conflict, and economic prosperity should be magnified, incorporated, and embraced in the negotiation process as the best alternative for all involved parties. The makers of the Israel–Jordan Treaty and the LHWP Treaty were very successful in recognizing and capitalizing on the strategic and economic benefits that signing these agreements could potentially offer. Both Jordan and Israel realized the value of cooperation and regional and political stability as part of the overall benefit of reaching an agreement. Similarly, the LHWP Treaty acknowledged and utilized economic gains as an incentive for cooperation. Because of these mutually recognized political, strategic, and economic factors, parties were able to find an acceptable ZOPA (Zone of Possible Agreement), where they succeeded in leading an integrative negotiation style to promote a win–win situation and "increase the pie." Thus, a crucial component of the negotiation process is to convey to both sides that their BATNA (Best Alternative to a Negotiated Agreement) is equally disadvantageous and offer these non-environmental gains as adequate incentive for hydro-diplomacy.

Sharing not dividing

By and large, water can be allocated based on different mechanisms. Depending on the level and method of allocation, transboundary water resources can either be *shared* or *divided*. Based on our research, sharing water resources is the most desirable option, as it tends to promote cooperation and evenly distributes the risk of drought among all parties, whereas dividing is less desirable as an option, as it tends to promote competition and hostility and often overburdens one party with the risk of drought. Sharing also incorporates supply augmentation mechanisms to "increase the pie," rather than the consumptive division of available resources, which can lead to dissatisfaction. This book concludes that there are five levels of

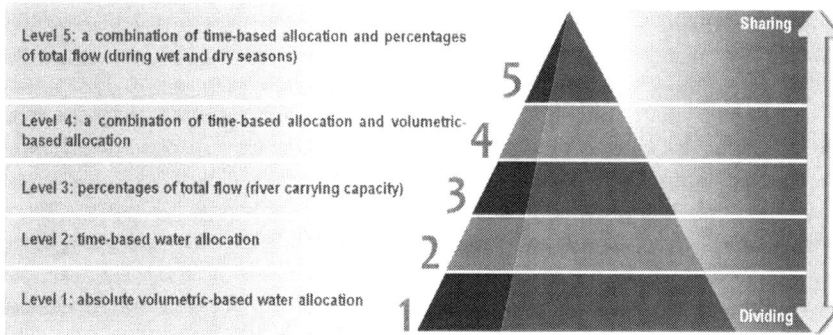

Level 5: a combination of time-based allocation and percentages of total flow (during wet and dry seasons)

Level 4: a combination of time-based allocation and volumetric-based allocation

Level 3: percentages of total flow (river carrying capacity)

Level 2: time-based water allocation

Level 1: absolute volumetric-based water allocation

Figure 6.5 The transboundary water allocation mechanism hierarchy.

water allocation mechanisms, which can be summarized as illustrated by the hierarchical diagram shown in Figure 6.5. Moving from Level 1 to Level 5 represents the spectrum of water allocation schemes from *dividing* to *sharing* respectively. Level 1 represents absolute volumetric-based water allocation; Level 2 represents time-based water allocation; whereas Level 3 represents a percentage of the total flow. Levels 4 and 5 represent a combination of these various water allocation mechanisms.

Based on the analysis of the two cases, coupled with background information, time-based water allocation rather than volumetric-based allocation is beneficial in many ways. It segregates and distributes evenly among all co-riparians the risk of drought caused by the fluctuation of water availability in nature. Conversely, allocating water quantities alone tends to concentrate the risk of drought on the users of one country. It is often the case that the country that carries the burden of delivering specific amounts of water to other countries according to volumetric-based international water treaties is confronted by this predicament of high risk. For example, Israel agreed to deliver additional water from Lake Tiberias to Jordan, but was unable to do so because of the 1999 drought, which posed a great risk to Israel and the overall implementation success. Further, time-based allocation schemes provide co-riparians with an opportunity to store water for times of shortage, although this might seem superfluous given the comparatively low relative risk factors imposed on each user. This offers a high level of water security and provides incentives for conservatory policies. The impacts of such policies are evident in many situations where water usage changes over time due, in part, to the innovative technological methods that provide more efficient measures for conserving water. However, volumetric water allocation stays the same over time regardless of the change in water consumption or natural flow fluctuation.

It is worthwhile noting, however, that both approaches are useful and each is appropriate only in certain contexts and for certain uses. For example, time-based allocation is suitable in long-term water planning, where parties are interested in a more stable and sustainable agreement. This water allocation strategy is conducive when all parties have met their basic domestic water demands and are ready to take risks over agricultural, industrial, and other uses that are more manageable and controllable. This method, involving risk-taking elements, is less conducive when the matter of negotiation is centered on domestic water needs. This is also useful in large-scale water transfer plans from a natural resource such as a river or an aquifer, as demonstrated in the Israel–Jordan Treaty. This way, the sense of winning or losing will be dissipated by evenly dividing the odds of both benefit and risk among all co-riparians. Thus, time-based allocation tends to foster a sense of ownership and responsibility toward the environment and boosts cooperative relations among parties on other matters.

In contrast, volumetric-based allocation is most appropriate in responding to urgent and quantifiable needs, such as critical domestic water demands. The research suggests that this water allocation policy can be used in smaller projects where no party is willing, or can afford, to take the risk of drought. Clearly, this method is well suited to initiate and expedite preliminary agreements responding to the immediate domestic demands, although it is not sophisticated enough to be deployed in larger and more comprehensive agreements. Level 3 represents water allocation that is based on percentages of the total flow, which accounts for and mitigates the impact of the natural fluctuation in the river carrying capacity from one season or year to the next. The outcome of this method is proportional to the overall amount of water available in the system. This, in turn, distributes both water and the risk of drought evenly among all riparians.[4]

It is crucial to note that, in all of the three levels of water allocation discussed above, shared water resources are divided among riparians. We assert, based on the analysis conducted in this book, that a far more advanced water allocation strategy can be realized when we combine these approaches, which results in water *sharing* rather than *dividing*. If done correctly, combining two of these approaches can bring about the advantages of both and in turn produce a very effective and rigorous allocation strategy. To that end, Level 4 joins time-based with volumetric-based water allocation, whereas Level 5 combines time-based water allocation and the percentages of total flow during wet and dry seasons. As it encapsulates the benefits of the two more preferred approaches and further incorporates contingency planning clauses apropos of the joint management of water shortages, especially during times when the river flow drops below a specific threshold,[5] Level 5 is deemed to be the most desirable scheme of water sharing. An example to illustrate Level 5 is when a certain agreement indentifies that country "A" should receive a percentage of the available water resources during a particular season.

In both the 1994 Israel–Jordan Treaty and the 1986 LHWP Treaty, water was allocated based on a combination of time-based and volumetric-based strategies representing Level 4, which is, for the most part, reflective of sharing rather than dividing of disputed water resources. However, a volumetric-based allocation method was deployed in the Israel–Jordan Treaty to address the most pressing water needs of both countries. Although this strategy was successful in alleviating water demand, it was not sophisticated enough to provide a comprehensive settlement regarding the shared regional water resources. Of particular significance was the additional 50 MCM to be delivered to Jordan, which was merely a volumetric-based allocation scheme without identifying the timing, location, or responsible party for the extraction and delivery of this amount. As discussed earlier in the book, this volumetric allocation portion of the Treaty failed to offer sufficient guidance in cases of drought and caused tension and damage to the countries' diplomatic relations.

Needs-based water allocation

Acquiring and transferring water from where it is plentiful to where it is needed is the focus of strategic water management. Given the fact that water is a valuable resource, more so in arid regions, owning the rights to that water is even more precious. In November 2002, the United Nations Committee on Economic, Cultural and Social Rights issued a statement declaring access to safe drinking water to be, like food, a fundamental human right. The declaration reads: "Water is fundamental to life and health. The human right to water [legal water rights] is indispensable for leading a healthy life in human dignity. It is a prerequisite to the realization of other human rights" (http://www.un.org/waterforlifedecade/human_right_to_water.shtml). According to Hodgeson (2004), legal water rights are defined as the right to:

- Abstract or divert and use a specified amount of water from a natural source;
- Impound or store a specified quantity of water in a natural source behind a dam or other hydraulic structure; or
- Use water in a natural source.

According to Chileshe and colleagues (2005), actual ownership right is defined as the totality of the following:

Access rights: the rights to enter a defined physical entity.
Withdrawal: the rights to obtain the benefits from that entity by taking out some of the flow.
Exclusion: the rights to determine who will (or will not) have access to the resource.
Management: the rights to regulate use patterns, thus transforming the resource and potentially altering the stream of benefits from that resource. Management rights also provide the ability to define access or withdrawal rights.
Alienation: the rights to sell, lease, or bequest rights to the resource.

Although the issue of human water rights has normative, pragmatic, and ethical implications, the distinction between water as a human right (legal water rights) and water rights per se is essential to the understanding of water needs. Human water rights (or the right to water) are concerned with acquiring access to sufficient water quality and quantity by securing a basic level of service for all, whereas the concept of water rights represents a much more complicated issue related to the question of sovereignty and riparian ownership of water resources. Water rights are not only concerned with the amount of water that each nation receives, but they are also related to the ease and reliability of access to water and national security concerns, as discussed earlier in the context of Israel's national water policy. This distinction caused the creation and development of a multitude of doctrines to allocate water rights, such as the riparian doctrine[6] and the prior appropriation doctrine.[7]

Research findings based on case analysis suggest that a more effective approach to water negotiation is to consider *needs* as the basis for water allocation in substitution for the argument of rights and entitlement. This conclusion is also supported by many scholars, such as Wolf (1999) and Mohamed-Katerere and van der Zaag (2003). The question then becomes one of how to define needs, and which is the right entity to decide who deserves water, given the ambiguity of the international water law in defining these parameters. This research, historical conditions, and recent trends have shown that water needs is a subjective concept, the meaning of which has always been dependent upon individual interpretation. For example, according to Wolf (1999), water needs can be defined and quantified on the basis of catering to current and future populations' water consumption, irrigable land, and requirements of specific projects.

With this in mind, water allocation can be defined in terms of demographic or economic grounds. For example, certain riparians with a relatively advanced economic base, such as South Africa and Israel, often make the argument of economic-based water needs to justify their higher water consumption for industrial and economic development. In retrospect, economic development itself determines access to water, which creates a cycle of water access monopolization and justification. Cognizant of this dilemma, this book presents the concept of water needs in a different light, suggesting that, generally, humans have the same water needs on a personal level regardless of how robust their society is in terms of economic advancement. Therefore, this book advocates for a per capita needs-based water allocation. This conclusion is also echoed by other scholars such as Philips (2008) and Gleick (1994), who assert that water should be allocated based on demographic grounds regardless of economic and other industrial needs. This approach is also in sync with the research findings regarding the prioritization of uses and users, where satisfying the urgent domestic and potable water needs is given a higher priority than other industrial and agricultural uses. Regardless of the economic conditions of a state, this book suggests the need for water allocation in arid regions to be based on demographic grounds, considering that these are drought prone regions, with domestic water use being the greatest concern.

Unlike the notion of rights-based water allocation, which is hard to quantify and therefore negotiate in a meaningful fashion, the appeal of the notion of needs-based water allocation as an indicator for outcome equity stems primarily from the fact that needs are commensurable; they are quantifiable and negotiable. Relying on needs-based negotiation means that states are able to present their water needs in terms of specific figures and in light of their current and foreseeable population's basic water consumption and negotiate and reach agreement. This approach is also consistent with the notion of interest-based negotiation strategy (also advocated by this book) as opposed to position-based negotiation that focuses on entitlement and the question of water rights, often leading to negotiation deadlocks and sometimes more conflict. This was evident in the analysis of both the Israel–Jordan and the Lesotho–South Africa water negotiations, as the parties demonstrated understanding and consideration of each other's respective interests. By focusing on identifiable interests and quantifiable needs, the parties were successful in mitigating the adverse impacts of entitlement that tend to dominate position-based negotiations. Shifting the focus of the negotiation process from the question of rights and entitlement to the question of water needs means *sharing* water rather than *dividing* it, which is discussed earlier in this chapter and suggested by this book as a preferred water allocation strategy.

Satiety-based hydro-diplomacy

As shown in Figure 6.6, which reflects findings from the case study analysis provided in this book, equitable agreements incorporate elements of both equitable process and equitable outcome.[8] The process and outcome of any treaty cannot be neatly separated from one another. Although one may expect that adhering to adequate process naturally results in a good and acceptable outcome, the circumstances under which water treaties are often negotiated and officiated suggest otherwise. Reaching an agreement is important, yet it does not mean much if the agreement is superficial or inequitable. Similarly, employing an appropriate process is not enough if it does not yield equitable and reasonable

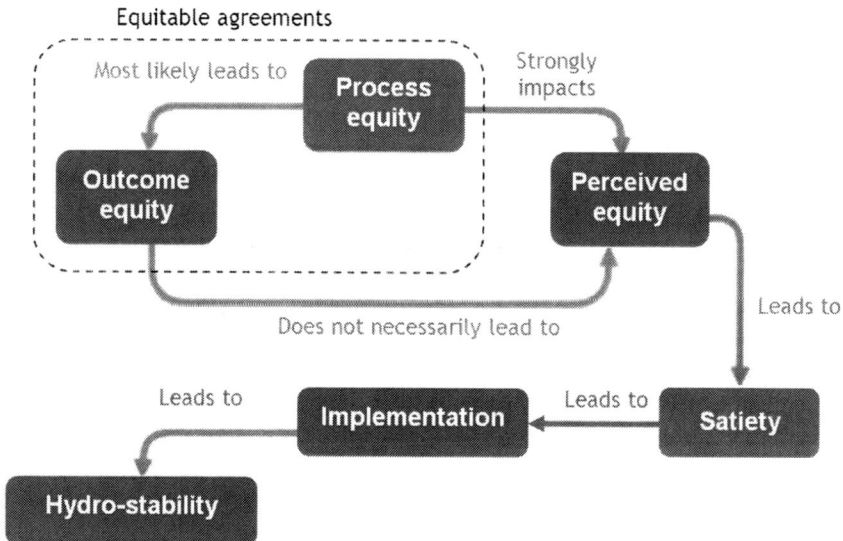

Figure 6.6 The relationship between achieving equitable agreements and hydro-stability.

outcomes. More importantly, having an appropriate process in place is useful in promoting cooperative arrangement, even if it may not produce an agreement. This is because the success of a negotiation process without an agreement stems from the lessons that the process itself can offer, such as information about the real problems and issues, each party's interests, and possible future directions. Without this coordinated process in place the likelihood of reaching an agreement is slim to nonexistent.

By the same token, equitable outcomes do not necessarily lead to high perceived equity of a certain treaty. This is because satisfying basic water needs does not automatically ensure the satisfaction of all disputants involved in water-related conflicts. Rather, certain elements (parameters of process equity)[9] strongly impact perception, which in turn drives the level of satisfaction. Findings from our research lead to the conclusion that the extent to which an agreement is considered equitable is primarily contingent upon the extent to which its development process is perceived as equitable. As such, perceived equity determines whether the agreement is deemed equitable and therefore acceptable. This is particularly true in cases where history of hostility and other negative emotions are attached to the crux of the issues in conflict. We have seen cases where the actual outcome might have been equitable in reality to both parties, but the perception of at least one riparian community did not concur with the actual outcome. We also have seen that because of this negative perception sustaining successful implementation and adherence to the terms of the treaty was greatly challenged and regional stability suffered as well.

Although it is a desirable venue in water management, needs-based water allocation is insufficient alone in promoting cooperation and robust implementation. This is because high levels of hostility result in a zero-sum game situation, where one riparian state's gain is perceived as another's loss (Postel and Wolf, 2001). In this situation, irrespective of their own share of water, relationships corrode and constitute an obstacle to cooperation and implementation. This issue became more relevant in the case of the Israel–Jordan water negotiation, where the conflict was not confined only to water resources but also involved political and cultural dimensions. In such situations, ensuring satiety of all parties involved in water-related conflicts is as crucial as satisfying their very basic needs. As such, it is important to boost a win–win situation, where both parties can satisfy their needs for water, while perceiving the other party as a partner in the process rather than an opponent. It is imperative, therefore, that each disputing party comes (or is brought) to appreciate the perceived needs of others, reducing the most direct reasons for resentment and hostility.

The research introduced in this book infers that the relationship between perceived equity and water *satiety* is a relationship of coexistence. In this sense, satiety is primarily contingent upon perception. Generally, when people perceive their share as equitable, they tend to be comfortable and satisfied. However, the question of what constitutes satiety and whose satiety water treaties should cater to are of gargantuan importance. Case study analysis indicates that perceived equity is synonymous with satiety and can refer to different types of perception, including perception of the general public, perception of the impacted local community, and perception of the negotiators of the treaty.

First, perception of the general public is impacted by knowledge of the treaty, early involvement, media coverage, history of the conflict, and the inherited perception of the "other," which can be a result of various geopolitical, historical, and cultural factors. Outcome equity has little influence on the general public's perception of the treaty as being equitable. The Israel–Jordan Water Treaty emerged as a good example that illustrated the impact of secrecy on the general perception of the Treaty, as there was a lack of media coverage of information critical to the negotiation process. Early and ongoing involvement, transparency of process, and information sharing are helpful in remediating problematic public perceptions. Second, perceptions of the impacted local public (or the end users residing in the target river basin) are directly impacted by the physical outcome of the treaty and the consequent amounts of additional water delivered. For example, the LHWP was seen by the impacted local residents as being destructive because of the consequences of its construction, particularly the displacement of many local communities and inadequate compensation. As such, there is a tie between outcome equity and perceived equity, but only on the local level, where water users can feel the impact of additional water (or lack thereof) on their livelihood. Third, the perception of the negotiators of a treaty is influenced by their political status and international relations, political agenda, and non-water-related gains (including strategic, political, and economic gains). For example, as shown earlier in the book, the Israeli and Jordanian negotiators' perception of the other side and the overall peace process was influenced by their political leaders (such as King Hussein of Jordan and Prime Minister Yitzhak Rabin of Israel), whose involvement in directing the negotiation process resulted in facilitating, expediting, and reaching the final agreement. Akin to the perception of the general public, equitable outcomes are not strongly linked to negotiators' perceived equity. In this case, considering non-water-related items during the negotiation process boosts negotiators' perception of the process and acceptability of the outcome. This suggests that tackling water alone is not enough, but rather addressing water in conjunction with other agenda items (such as economic and strategic elements, international relations, land, and borders) is extremely crucial in providing non-water-related incentives to boost negotiators' acceptability. In all of these types of perception, the actual outcome of a particular water agreement (although influential on the local level) does not seem to steer acceptability and implementation, as demonstrated in various cases.

To better understand these nuances associated with the role of perception in water allocation, the James C. Davies' theory of the "J Curve" seems relevant and appropriate in illustrating the interplay between expectations, perceptions, and acceptance. According to Davies (1962), there is a gap between what people expect and what they actually receive. The size of this gap determines whether the resolution is acceptable (tolerable) or unacceptable (intolerable). Building on this theory, perception is influenced by expectations and actual treaty performance (actual outcome of the treaty). When the gap between expectations and

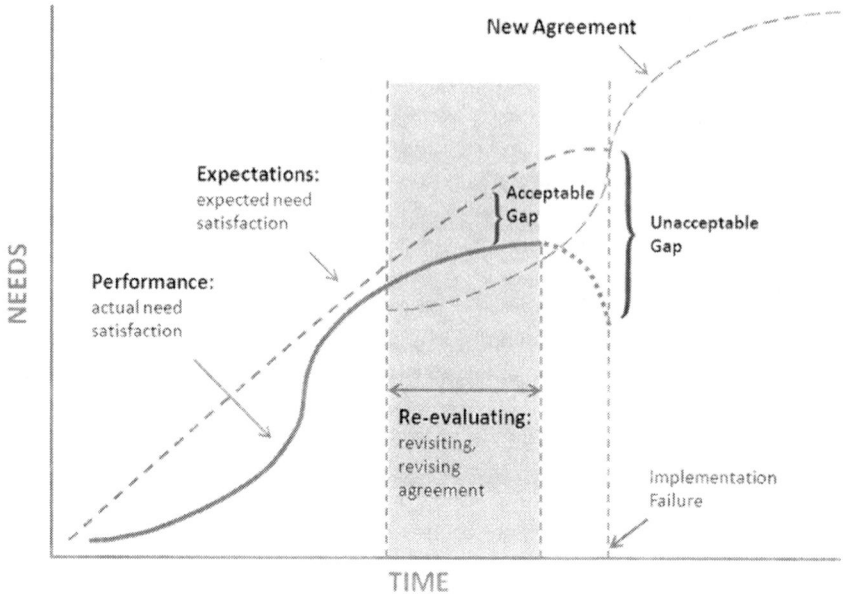

Figure 6.7 Need satisfaction and perception. Source: adapted from Davies (1962).

performance increases, the acceptance of the treaty decreases and vice versa. As shown in Figure 6.7, when expectations exceed the actual outcome or perfor-mance of the treaty, perception and acceptance tend to be low, and therefore treaty implementation is compromised.

When expectations are realistic, the outcome of the treaty will more likely be acceptable to the parties as the gap between the two is minimized. Reflecting conflict dynamics as well as change in the geopolitical landscape, expectations and performance are both moving targets and can change over time. Performance of the treaty can diminish over time and so can expectations. The key is to keep both expectations and performance tightly aligned to maintain acceptance and ensure successful implementation over time. Both expectations and performance can be altered to foster better perception and acceptance. First, expectations need to be realistic and can be altered to align with reality. This can only be accomplished through constant collaboration and communication with the public. Transparency about treaty formation and outcome tends to encourage realistic expectations and keep these expectations realistic throughout the lifetime of the treaty. Proactive public involvement of all stakeholders is key here and will be discussed in the following section. Second, performance can also be altered through sound basin-wide management and collaborative assessment of treaty outcome. Treaties do not and cannot exist forever. They have a start and end point. The lifespan of any treaty is determined by the extent to which the treaty continues to deliver incremental value and yield meaningful outcome to all parties involved in the dis-pute. This incremental value is diminished as time goes by. When this incremental value starts to plateau or fall, treaties need to be revisited and re-evaluated and, if necessary, replaced with new agreements. This will ensure the continuation of cooperation and foster positive perception of the resolution sought.

In short, case analysis of the different treaties provided in this book shows that there is a strong relationship between perceived equity (level of satisfaction) and treaty implementation, and therefore regional hydro-stability. In these cases, perception of the treaty, rather than its outcome, was a fundamental factor in determining whether or not the treaty was acceptable by either party involved in the negotiation. Achieving satiety, which means recognizing one's needs, accepting one's share of water, and acknowledging the needs of others, is crucial in eliminating resentment and hostility and advancing negotiation to implementable and durable agreements. As such, satiety emerges as a key component in successful hydro-diplomacy. Based on these findings, promoting water satiety requires proactive public involvement and the adoption of collaborative agreement seeking processes, as discussed below.

PROACTIVE PUBLIC INVOLVEMENT

Public involvement is key to boosting perceived equity, level of satisfaction, and treaty implementation. Secrecy leads to doubt, suspicion, and distrust (as was evident in the Israel–Jordan Treaty discussed in Chapter 4). The first step to building trust is informing and involving the public from the onset of the negotiation. This includes collecting and sharing information and results of negotiation sessions and casting positive light on the prospect of the negotiation and the potential of reaching an agreement. Highlighting the value of the potential agreement to the community and tying it to substantive, demonstrable, and measurable benefits must be part of this communication plan.

Traditional negotiation and conflict resolution processes can be characterized to promote the "planning for people" approach, where experts and politicians are parachuted in to do the work and make decisions on behalf of their communities or countries. This approach cannot be said to provide an appropriate negotiation strategy when the issues are as contentious, complex, and critical as those often associated with water. It is essential to promote a participatory and engaging negotiation process, collaboratively determining where we are now, where we want to be in the future, and how to achieve the desired future (Abukhater, 2011b). Treaty formation must be viewed as an ongoing process, rather than a product. For this to happen, water negotiators must seek bidirectional community input to make sure that the process is collaborative and transparent. This requires a paradigm shift from "planning *for* people" to "planning *with* people" in a more engaging and collaborative fashion.

Facilitating smooth flow and dissemination of information and data to and from the public in a bidirectional way is crucial in winning the hearts and minds of the people on both sides. For example, in the 1994 Israel–Jordan Treaty, the negotiators of the agreement failed to include the public and keep them informed about the negotiation process. Most of the information related to the development of the Treaty was kept secretive and confidential. This aspect of the Treaty negation influenced the perception of the public and caused opposition on both sides. This calls for a new configuration of communication and decision-making platforms. Effective communication and exchange of information call for expanding the communication footprint, beyond technical or philosophical language boundaries, across the whole community.

Public involvement of diverse stakeholders can be enhanced by utilizing both traditional media and social media and other online tools that are already being widely used and deployed to share data, exchange information and voice their opinions. Social media are already being utilized as effective communication platforms to foster societal infrastructure for human interaction, where the government can obtain feedback from the public with a high level of transparency and accountability (Abukhater, 2009a). This was evident in the most recent Arab Spring, where social media were utilized to their fullest extent as effective means of communication for organizing and uniting the masses. This is not to suggest a replacement of the currently utilized traditional civic engagement processes, but rather an augmentation to these processes. Going beyond traditional public involvement means deploying more sophisticated techniques that promote a hierarchy of engagement opportunities extending across the board. This entails retooling communities with effective means of communication that go beyond the traditional civic engagement venues to a more open dialog facilitating proactive participation (Abukhater, 2011a).

Three levels of implementations are considered elements in proactive public involvement. These include informing, involving, and empowering the public. *Informing* all impacted and interested stakeholders by disseminating critical information about the shared water resources, nature of the conflict, and issues being negotiated through a variety of platforms is key in getting them interested and involved in the process. *Involving* members of the public by capturing their feedback in a bidirectional fashion ensures that their concerns are heard and incorporated in the final resolution. *Empowering* the public means more than just getting their feedback on decisions that are made on their behalf by politicians and water experts. It, rather, means that they can take part in the decision-making process by being able to shape the nature and outcome of the final settlement. This proactive public involvement approach can change how negotiations are being conducted and by whom. Instead of relying on the traditional approach of technocratic, expert-led negotiation and decision making that depends heavily on the opinions of "water experts," this participatory approach utilizes community leadership, joint fact finding, and collaborative decision making. This will help in neutralizing the "duelling expert" syndrome[10] (Jarvis and Wolf, 2010) and transform traditional negotiations that tend to typically focus on merely scientific discussions regarding allocations and flows into more lively and engaging experiences.

International negotiations and agreements over water resources need to incorporate awareness building elements, such as public announcements and outreach programs. The issue of water negotiation is usually perceived as unworthy material to broadcast and, therefore, is conducted in secret meetings where agreements are reached and signed behind closed doors. This practice alienates the public and other important stakeholders from the decision-making arena, where their feedback and consensus are needed the most. An effort to create ceremonies can, in turn, attract media attention and build public awareness.

Informing and involving members of the public, who are most affected by these decisions, not only helps in reaching an agreement and making the right decisions, but also provides a better social climate for implementation, as members of the public feel that they have become an integral part of the process. The absence of public involvement has proven detrimental to both treaties examined above, although less so in the LHWP case. The perception of the treaties on the part of the public in both cases was primarily shaped by the extent to which they were involved in the progression of the treaty negotiation. As such, their general views of both treaties were negatively impacted by the fact that they were not included in, or informed about, the negotiation process or its outcome. In contrast, proactively involving the public in all stages of negotiation fosters public acceptance of the process and overall sense of satisfaction in the outcome.

COLLABORATIVE AGREEMENT SEEKING PROCESS

The processes of collaboration includes identifying all affected and interested stakeholders, informing and involving several sectors of the community, and including representatives from both sides. Bringing appropriately represented parties together entices dialog where all parties are encouraged to freely and clearly express their interests and major concerns. This sets the stage for reaching durable agreements through a collaborative negotiation and consensus building process. The public must be able to attend or at least watch a recording of the negotiation process in order to offer feedback and address the society's needs. It is worthwhile to note, however, an adequate collaborative process does not represent an end in itself, but rather a means to an ever-changing end, where new inputs and outputs coexist and continuously change in fluid and evolving economic, political, cultural, and natural environments.

The international practice of negotiation over water must understand and embrace the importance of collaboration and deliberation. International water dispute resolution practice will benefit from empowering and involving all parties in the making of solutions as they are being formulated by utilizing collaborative agreement seeking processes to reduce post settlement disputes. Collaborative agreement seeking processes provide interested parties with a stake in the decision and the overall negotiation process. There is ample evidence that it is a far more superior approach to the classic distributive bargaining/zero-sum approach, which can lead to political deadlock. Unlike the traditional, adversary methods for resolving conflicts through litigation, multiparty decision making often yields desirable and implementable outcomes. When all stakeholders are involved in the negotiation and decision-making process, they feel responsible to ensure implementation of the agreement. Collaboration means more than just the gathering of people and ideas, rather it can be seen as the symbolic ceremonial commemoration of shared values and the celebration of differences in an interwoven deliberative process that incorporates and caters to all needs.

Given the evolving nature of water-related disputes, collaboration must take into account both bottom-up and top-down approaches to generate concrete ideas that appropriately respond to the complex and ever-changing reality of the contemporary issues surrounding water allocation. By encouraging proactive participation from all entities in a democratic and collaborative fashion in the decision making, law making, and governance, the possibility of everyone getting involved and taking ownership of any decisions made, and, therefore, supporting and pushing for implementation is greatly enhanced. This is a fundamental element and prerequisite for successful cooperation and stability.

Interest-based negotiation is key to the collaboration and consensus building processes. Through collaborative agreement seeking processes and consensus building, voices of all parties can be heard and various interests can be explored. By utilizing this process of collaboration, parties can identify major issues, collaboratively analyze them, generate alternatives, and select options that meet their interests. The Israel–Jordan water negotiation process, exhibiting general characteristics of water negotiation in many cases around the world, showed that riparian states tended to propose solutions that only catered to their own needs, excluding others from the scope of the resolution proposal – essentially offering national solutions to a regional problem. Mindful of this general trend, we suggest the need to incorporate an "all-party resolution" that considers the wellbeing of the region as a whole. Excluding any stakeholder can only lead to adding more complexity to an already complex problem.

Neutral third-party mediation process

Mediation, known also as assisted communications for agreement, is a conflict resolution process where an impartial third party, the mediator, amicably facilitates communication among disputing parties to promote reconciliation, settlement, or agreement. A neutral third party can bring all stakeholders to the negotiation table and mediate the negotiation. Mediation by a neutral third party can have immeasurable influence on treaty formation and outcome. Third-party involvement is key to rebalancing negotiation processes by exposing and modifying the behavior of riparian states and promoting an equidistant approach that is hard to achieve without the proactive involvement of an even-handed, non-allied, and neutral third party. Without the intervention of this third party to initiate and facilitate ongoing negotiation aimed toward finding robust resolutions, both disputing parties will continue to view each other as an adversary and will be less receptive to attempts at reconciliation so long as they are initiated by the "other" party. As such, neutral third-party involvement is warranted to bring representatives from both sides of the conflict together and bridge the communication gap in order to reach plausible, acceptable, and easy-to-implement agreements.

The current international practice of mediation is commonly missing this key component. Even in the few cases where mediation is utilized, the mediator has lacked neutrality and objectivity. The characteristics of the mediator are clearly and precisely outlined in theory, but these essential characteristics are often not fully adhered to in practice. In many cases, the intervening third party appears to have a stake in the outcome of the mediation process and therefore will often tend to push for certain outcomes that serve its own interests. This calls for a change in the current practical utilization of mediation. For the mediation process to be effective and successful, it should maintain neutrality, collaboration, and the option of confidentiality whenever required by the parties. It should also ensure informed collaboration from the beginning to the end of the process and provide commitment to implementation.

It would help immensely for parties to work together in a collaborative fashion to identify this third party and the scope of its involvement. On the whole, the process of mediation can take a number of forms, including cultural dialog mediation, facilitative mediation, and directive mediation.[11] Within such processes, both general and contingent tactics can be combined. General tactics include analyzing the conflict, indentifying key parties, clarifying parties' interests, and developing suitable mediation strategies. According to Moore (1996), these also entail developing a framework to facilitate negotiation by helping the parties generate options, draft and evaluate resolutions, and develop implementation plans. Along with the use of general tactics, contingent tactics can be also used to address special issues arising during negotiation, namely value differences, power structure imbalance, communication problems, conflicting analyses, misrepresentation, and misinformation (SPIDR, 1997). To address these important issues, sound and effective mediation process takes place in all stages of the negotiation, including the pre-negotiation stage, the negotiation stage, and the post-negotiation stage. As was discussed earlier, this continuous mediation process was missing in the Israel–Jordan negotiation. The involvement of this neutral third-party mediation was only confined to the beginning of the formal negotiation process, and mediation was not utilized to its fullest extent and was not used in the pre- and post-negotiation stages.

Once identified, a number of critical tasks are assigned to this mediator. First, all involved, impacted, and interested stakeholders must be identified and included in the process to avoid the adverse impacts of excluding any important players, known as *ex parte*. The exclusion of other key regional players, such as Syria and Palestine in the Israel–Jordan negotiation, appeared detrimental to the outcome and perception of the process and the agreement, as discussed earlier. Second, during this stage, the third party should encourage dialog on issues broader than the issue in question (water) in order to build trust, increase understanding of the other party, and lay the groundwork for meaningful negotiation. Consensus building negotiations between the parties should take place to facilitate and hasten a smooth process of peace and normalization. Mediation and consensus building go hand-in-hand to improve the negotiation atmosphere and engage the parties. To that end, meeting the "other" group, understanding how history is politically manipulated, and examining mistakes made by all parties involved helps to overcome fears and eradicate ungrounded rationales for the annihilation of the "other," while utilizing and embracing the diversity of the parties involved in the enrichment of the cultural heritage of the society as a whole. During these dialog sessions, the parties could share narratives and discuss values important to both of them. With the help and support of a neutral and powerful third party, the parties during this stage also have a chance to mutually establish process ground rules that might reflect and incorporate equitable water-sharing principles identified by international law, as they see fit. Moreover, building public trust in the negotiation process during the pre-negotiation stage is important to its success. This has profound impacts on alleviating the root cause of the conflict by providing opportunities to eliminate misunderstanding.

Third, a major step in cementing relationships based on trust should take place prior to negotiation by offering good-faith economic incentives to all disputing parties. This helps in creating opportunities for multilateral dialog and discussion. Stabilizing the economy not only tends to mitigate poverty, a major reason for clashes, but it also boosts political, economic, and social reforms. Offering economic incentives and addressing economic issues through reconstruction and development alongside the water negotiations are key components of the pre-negotiation process and the overall quality of negotiation. Economic assistance and industrial development were helpful elements in incentivizing both the governments of Lesotho and South Africa respectively to negotiate and enter into an agreement over a water transfer scheme, whereby Lesotho received compensation of transferred water and South Africa received water for its industrial base.

Fourth, credibility is essential in bringing parties to the negotiation table and inspiring an interest in finding a resolution. The mediating third party must build credibility with all disputants in order to enhance confidence in the mediator as well as the mediation process itself. This can be attained by creating a relatively close, comfortable relationship with all parties and educating them about the mediation process, and enthusing them about its potential outcome. It is the mediator's responsibility to acquire informed consent from all parties apropos of process ground rules prior to engaging in any negotiation, and to impartially enforce them once these ground rules have been officiated and mutually agreed upon. Fifth, the mediating third party may initiate contact with the parties to precisely identify major issues, interests, needs, and concerns, which must be consensually outlined and made clear to everyone. Finally, commitment of all parties to mediation must be secured prior to entering into any negotiation process.

"Process neutrality" extends beyond the mediation process and the identity and role of the mediating third party to include also the location where all disputants are willing to meet, negotiate, and seek resolutions. Incorporating neutrality in the process means that this location, which must be collaboratively identified and approved by all parties, should be neutral to ensure a comfortable and intimidation-free atmosphere. Finding a neutral place is one of the major issues that need to be addressed and reconciled with all parties with the help of the third party prior to negotiation. This important aspect was also overlooked in the Israel–Jordan water negotiation, creating tension and intimidation in some instances. Although during the earlier stages of negotiation, the parties met in the

United States as a neutral place for their negotiation, later talks were conducted in the Middle East, namely in Jordan, which could not be regarded as a neutral location considering that Jordan was part of the negotiation.

During the negotiation process, both the mediator and the parties work together to utilize appropriate mediation strategies to achieve resolution. Several tasks are essential to the role of the mediator at this stage in order to assist the parties in choosing the most suitable mediation techniques and collaboratively generate legitimate alternatives for resolution. First, the mediator may work closely with the parties to clarify their interests and re-emphasize their respective goals (SPIDR, 1997). This is a critical step where the parties are reminded of their interests and goals, why they are involved in negotiation, and the importance of resolving them. This is a chance for the parties to reiterate their goals and keep them focused on what matters. Second, the mediator can help the parties explore and generate possible, creative, acceptable, and workable options. The mediating third party has certain obligations to facilitate the invention and assessment of viable options and alternatives. The idea here is to get the parties themselves to be innovative and resourceful in generating these options. Third, by relying on an interest-based negotiation process, the mediator can help the parties in searching for mutual gain and identifying common interests and shared needs and values. This is particularly important in generating the utmost mutual benefits and producing agreement amicably. All parties must be made aware of the ramifications of choosing to walk away from the negotiation process. Although it might seem that the BATNA for the most powerful state is in many cases very advantageous, walking away from the negotiation table means one thing: instability, a price that is too expensive for any state to afford irrespective of its power. Water in this regard should be viewed by all parties as an essential element and thrust for cooperation and regional development and stability. This common interest must be magnified, incorporated, and embraced in the negotiation process as the best alternative for all. Thus, conveying to the disputing parties that the BATNA is equally deplorable for all sides of the table is a crucial component of the process. In both the Israel–Jordan and LHWP treaties, both riparians were aware of the adverse consequences of relinquishing cooperation and viewed water as a catalyst for peace. This helped in providing incentive for both sides to search for workable and acceptable resolutions rather than walking away from negotiation.

Fourth, it is the responsibility of the mediator to implement equitable sharing processes and mechanisms. By encouraging the parties to mutually formulate and utilize objective, fair standards and flexible procedures, the mediator can help the parties to craft the criteria that will guide their choice of appropriate strategies. For example, utilizing criteria for equitable water sharing can be discussed, evaluated, mutually developed, and agreed upon. In this regard, the process equity criteria indentified in this book can generally be adhered to, but also slightly modified or contextualized to fit each unique case. Fifth, the mediator can assist the parties in prioritizing and weighing up their options to craft a mutual strategy and reach decisions. Prioritizing the issues at hand and separately and incrementally focusing on their resolution are central to mediation. Sixth, the mediator should work with the parties to coordinate their strategies and entice coherent and consistent approaches to their conflict.

As discussed earlier in this chapter, pronounced mutual interest in resolving disputes should be coupled with good faith for negotiations to succeed. Therefore, the mediator should also be prepared to address situations where any party may not appear to be acting in compliance with the agreed upon ground rules. Finally, once an agreement is reached and formalized, the mediating third party may assist the parties in developing mechanisms for implementation and strategies for facilitating and putting them to use thereafter. These tasks were not completely and adequately addressed in the Israel–Jordan or LHWP Treaty. For example, although influential third parties (the United States and Soviet Union) were involved at the very beginning of the negotiation, they were not directly involved in facilitating the discussion regarding identifying the parties' needs and interests, generating and evaluating acceptable options, or the prioritizing and weighing up of these options. The early involvement of these powerful parties was helpful in bringing the two sides to the negotiation table. However, this involvement discontinued, causing disagreement and political deadlock as the two sides engaged in a more serious negotiation process. Conversely, the involvement of the World Bank in the LHWP Treaty (both as a mediator and in providing financial support) was very beneficial in moving the negotiation and implementation of the project forward. The bank was successful in working with the parties to coordinate and entice cooperative resolutions to the conflict and encourage other donors to provide financial support for the implementation and operation of the project. In both cases, the parties were made aware of the ramification of relinquishing negotiations and the political, economic, and social benefits of pursuing a final cooperative settlement.

The mediator also has a critical role to play in the post-negotiation stage. The post-negotiation stage includes assessing, facilitating, and overseeing implementation and enticing all parties to maintain continuous relations with each other. Developing a monitoring plan and key performance measures to ensure successful implementation of the agreed upon plans, revisiting the issue, and continuing cooperation must all be part of an ongoing assessment process. Ensuring the implementation of the agreement requires the involvement of a powerful party to influence and persuade the disputing parties to maintain implementation. The process of mediation provides an opportunity for the mediator to utilize external or internal forces to exert pressure on the parties to consistently adhere to successful treaty implementation. Therefore, the involvement of countries such as the United States, being a key influential player with a vast wealth of resources and a leading political role with significant clout, is important to break political stalemates and ensure implementation compliance and long-term relationships, even if it is not the ideal mediator, considering that the ability of the United States to assume the role of a neutral third party, in the Arab–Israeli conflict, for example, has been questioned repeatedly.

As discussed earlier in both cases, monitoring of implementation was addressed by the joint water committees that were formed in the wake of the signing of the treaties. Although this aspect of the post-negotiation process was not addressed properly in the Israel–Jordan Treaty, causing a "mini-crisis" in the two countries' relationship, Lesotho and South Africa were successful in maintaining regular meetings and continuous cooperation to monitor implementation and operation of the project.

Conclusion

There are signs of both water stress and inequitable sharing of ground and surface water resources in many arid regions of the world. There is also ample evidence that the looming threat of water shortage and droughts overburdens the geopolitical and hydro-political landscapes of both the Middle Eastern and South African regions and can easily escalate conflicts. Although critical in this regard, matters related to water shortage alone cannot be viewed as the crux of issues surrounding water conflicts in these regions. Rather, lack of water-sharing equity poses a tremendous challenge to achieving a robust and sustainable peace process.

Comprehensive and collaborative negotiation and mediation are often unrealized, hence the promise of achieving peace is commonly unfulfilled. For any conflict to transform from its current intractable state to a contained, cooperative one, the way the parties view and approach water conflicts must be changed. This requires a change in the parties' mindset and negotiation approaches. Conversely, severe conflicts and outright violence threatening regional stability could ensue if the parties failed to employ new methods of reconciliation. Without relying on new strategies for both negotiation and water allocation, sentiments of resentment and hostility may turn the hope of change to a pure fantasy. Bearing this in mind, any proposed resolution plan must be proportional to such an intricate problem. Otherwise, it will be dwarfed by the magnitude, complexity, and level of urgency of the impending water crisis. These resolution plans must cater and appeal to all parties to foster their implementation.

The interplay of differing degrees of geopolitical power and different cultures and values has profound water policy implications. Bridging the cultural and communication gaps requires more than simply bringing the parties to the negotiation table. Cognizant of these important factors, this chapter has cautioned against continuing to relinquish the option of equitable hydro-diplomacy to alleviate conflicts and attain regional peace. It has provided a cross-case analysis of the Israel–Jordan Peace Treaty of 1994 and the LHWP Treaty of 1986 to identify important parameters of process equity. A multidisciplinary, transformative conflict resolution approach is identified as a useful strategy to advance negotiation toward more plausible agreements. This approach encapsulates rules of engagement, mechanisms of engagement, and neutral third-party mediation. Rules of engagement refer to principles of the international water law, whereas mechanisms of engagement refer to a set of key practical measures (identified as important procedural equity parameters) that collectively comprise a comprehensive approach to conflict resolution. Acknowledging the shortcomings of the current international conflict resolution practice, this chapter has further provided a sequence of procedural steps for third-party intervention and mediation to reflect and reconcile all parties' interests and goals. Under the auspices of this transformative conflict resolution approach, engaging different adversaries in effective hydro-diplomacy can shift the focus of water conflicts from narrow self-interest to identifiable common goals that promote cooperative relationships.

7 Conclusion and future research

Water, after all, is used to extinguish fires, not to ignite them.

(Haddadin, 2002b: 337)

This chapter provides brief concluding remarks and a summation of findings of the research conducted in this book. It also identifies gaps in the current research agenda and provides directions for future research and suggestions for related topics that could be addressed to satisfy these gaps.

Concluding remarks

Contrary to the "water war" paradigm that views water as the source of past conflicts and future wars, this book provides an alternative approach to conflict resolution that, through equitable allocation of disputed natural resources, water can be viewed as a venue for future cooperation (Abukhater, 2010a). The nature and outcome of riparian disputes over international river water resources is largely influenced by the riparian states' political relations, the degree to which states are dependent on and in need of the resource, distribution of power, history of aggression, and third-party involvement. On the whole, conflicts over water resources can be linked to environmental, hydro-hegemonic, psychological and ideological, and political and hydro-diplomatic dimensions. Reflecting one or more of these dimensions, the origin of many international water-related conflicts can be related to a variety of factors, including history of aggression and hostility, power structure and socioeconomic imbalance, limited access to adequate water supply exacerbated by population growth, and value and cultural differences, which collectively can lead to inequitable water utilization and distribution. Similarly, inter-state relations and perceptions of the "other" and of the environment determine the riparian states' behavior and tendencies toward achieving cooperative agreements.

Lack of water-sharing equity is at the heart of many water disputes and emerges as a key factor in sustaining volatile conflicts over the allocation of transboundary water resources, which are not simply a matter of water shortage. To that end, any meaningful resolutions must employ parameters of equity to enhance acceptability and maintain durable implementation and robust relationships among the parties. Mindful of the importance of water-sharing equity, this book aimed at developing key criteria and parameters that constitute water allocation equity (*process equity*) and, further, discussing lessons and parallels that can be learned for the theory and practice of equitable concepts in water negotiation in general.

In keeping with the argument that inequitable water allocation schemes undermine chances for peace, evidence was provided that water has not been the source of conflict, per se, but, rather, in many cases it acted as an obstacle to peace. If water is in fact an obstacle to peace, resolving the water issue means removing a major obstacle to attaining peace. Given this, this book suggests that water can be a catalyst for cooperation, peace, and regional stability, and the reason for the disputing nations to come together at the negotiation table. The demise of the environment, as well as the peace process being at stake, emerged as the impetus to promoting cooperation. According to Haddadin (2002b: 337), "water, after all, is used to extinguish fires, not to ignite them."

Nine cases were analyzed and two selected for detailed scrutiny and analysis. The two detailed cases were the Israel–Jordan Peace Treaty of 1994, which exhibited low outcome and perceived equity, and the LHWP Treaty of 1986 between Lesotho and South Africa, which was successful in delivering equitable outcomes and perception. Based on the analysis of lessons learned through these cases, parallels were outlined to inform theories of negotiation and equitable water allocation.

In both cases, the parties seemed cognizant of the unforeseen, yet impending, consequences of reaching no settlement. The fact that the parties agreed to enter into negotiation and sign a water treaty was indicative of the amount of appreciable harm encountered by both parties in the absence of a water treaty. This suggested that the net benefits for the parties were considered to outweigh the costs of not signing a treaty. In that sense, the parties realized that they would encounter a certain degree of harm, whether water-related or non-water-related, if they relinquished the diplomatic option. In other words, the BATNA was identified as an undesirable course of action for all parties. This was because the appreciable harm for the parties was not confined only to loss of access to water resources, but also included strategic harm in terms of impending conflict that was likely to occur in the absence of a treaty in place.

Policymakers and natural resource managers need to understand the role of process equity in attaining acceptable settlements over disputed water resources. Much can be learned by examining the international aspects of transboundary water allocation in arid regions where water shortages, and most importantly inequitable utilization of water, have sparked international tension in the past and are likely to poison inter-state relationships in the future, putting regional peace and stability at great risk. By developing process equity parameters, the goal is to study how best to negotiate water allocation for sustained peace and prosperity based on a multifaceted, multiscalar approach that recognizes the complexity of water disputes.

Although process equity is an influential factor in hydro-diplomacy, it is critical also to recognize that process equity is not the only important factor in this regard. Process equity constitutes a key factor of influence on equitable outcomes and perceptions, but certainly not the only one. There are many other contextual differences and important factors that are determinative of the outcome and the ways in which a certain treaty is perceived. An example of these important circumstantial differences is the impact of severe power structure imbalance in the way a negotiation is conducted and the nature of the outcome. In situations where the two parties are severely mismatched in power, wealth, and economic development, the possibility of attaining process equity is admittedly quite remote. Even the appearance of equity may be deceiving in such cases. The impact of this factor is found in the Israel–Jordan Treaty, where the more powerful state was able to harness the greater benefit from particular agreements. To that end, the impact of the asymmetry of power is profound in leading to a certain outcome.

Another example of a contributing factor, other than process equity, is the geographic context and spatial configuration of the region in relation to the respective needs of each riparian state. The relationship between Lesotho and South Africa is built on asymmetry of interests, represented in the geographic location of the water supply being situated in Lesotho and the greater proportion of water needs being situated in South Africa. Such hydropolitical and geopolitical settings are not replicated in other cases and cannot be said to be neutral in impacting the tendency of riparian states to consider cooperation as a platform for resolving their dispute in a way that is beneficial to both parties. Collectively, the combination of these and other important aspects along with process equity are shaping factors of outcomes and perceptions.

By and large, the success of international water treaties can be judged not only by their respective outcomes, but also, and most importantly, by the manner in which they were negotiated (referred to in this book as treaty formation). First, two types of treaty outcomes have been identified: tangible and intangible products (Innes and Booher, 1999). Tangible products are easy to identify, since they are the direct benefit of any negotiated agreements. These are concerned with the physical outcome of the negotiation, such as formal agreements, treaties, plans, and policies pertinent to the volumes of water allocated for each co-riparian, the time period, and the location of the withdrawal. They are also related to strategies put forth to maintain successful implementation processes. Intangible products, which in many cases are regarded as more important than the tangible products, are intellectual capital and non-water-related matters, including mutual understanding of each other's interests and needs, a shared definition of the problem and an agreement on accurate data and quantitative models, and diplomatic and political capital[1] (Innes and Booher, 1999). In evaluating and analyzing the aforementioned key treaties, value was placed on both tangible and intangible outcomes.

However, a fundamental distinction must be made between attaining adequate processes and desirable outcomes. The tension between means and ends reflects the narrow focus of many theorists and practitioners on either procedural or substantive frameworks (Abukhater, 2009c). This also prompted many to discuss what we do, or ought to do, to yield desirable future conditions with little reference to contextual differences. This procedural tendency disregards the extent to which successful results are attained and maintained as long as the process utilized follows rational justification. In this sense, successful outcomes are regarded as trivial as long as a "good process" is followed. On the contrary, many others tend to justify their process (although it may appear to be devious or unethical) on the basis of achieving desirable outcomes.

It is critical to note that following equitable processes does not necessarily yield equitable outcomes. Similarly, equitable outcomes do not necessarily always coincide with equitable perception. Many agreements fail to sustain implementation notwithstanding their outcome.[2] This is largely because of the intricate nature of water conflicts, which poses tremendous challenges not only in terms of reaching plausible agreements, but also in terms of accepting and, in turn, implementing them. In other words, process does not guarantee outcome. However, it is important to point out that following appropriate process, akin to democracy, for example, provides the necessary platform for enhancing consistency and acceptability of the outcome. It also enhances the likelihood of achieving desired, acceptable, and implementable outcomes by all parties. To that end, it is helpful to think of the democracy metaphor to understand the importance of the notion of process and its different and defining characteristics and implications in contrast with the notion of outcome. Similar to democracy, which is concerned with the process and procedures delineating and distinguishing democratic from non-democratic discourses, equity in the same fashion is, for the most part, concerned with processes rather than outcomes. In light of this rationale, equitable events and agreements take place in a developing, fluid context, notwithstanding their outcomes. This does not suggest that treaty outcomes are not important, but it rather implies that outcomes are trivial when compared with procedures, as processes are influential factors in the way a given treaty is perceived as equitable or not. Focusing on outcomes alone as the key determinant to what constitutes equitable agreements is an unfortunate misunderstanding of the dynamics of negotiation and would certainly tend to eviscerate these agreements. This is because equity entails more than just outcomes: it is a procedure, perception, and continuity.

Whether equitable outcomes have been developed or not, process for the most part plays a key role in determining the extent to which a positive-sum (win–win) situation has been attained, as opposed to a zero-sum situation (win–lose), where to be a winner there have to be others who lose, or, worse yet, a negative-sum (lose–lose) situation. It is therefore in the interest of all parties to focus their negotiating efforts on developing, following, and attaining important process equity components outlined in this book to foster a positive-sum situation, where everyone is a winner in such somber geopolitical discourses. This way, agreements can be perceived as equitable and satisfactory, which tends to enhance their ratification and implementation, regardless of their outcome. Thus, it is reasonable to deduce that adopting and applying principles and parameters of process equity is more likely to lead to a more equitable and robust settlement and produce feasible and implementable agreements that are viewed as equitable, and in turn are satisfactory to both sides of the water conflict. This is because utilizing these parameters of process equity (identified in Chapter 6) leads to enhancing the level of satisfaction, durable implementation, and regional hydro-stability. It is critical, however, to point out that satiety may mask some disagreements that are quite thorny and some entrenched political positions that fall back on blaming the other side. So, crucial questions remain: Who defines satiety? How do they define it? How can their definition be brought into line with a more neutral, third-party perspective?

Considering the larger peace process as a whole, the relationship between equity, or justice, and peace is a relationship of coexistence; you cannot have one without the other. To that end, Morton Deutsch (2000), in the first chapter of *The Handbook of Conflict Resolution: Theory and Practice*, astutely observed that:

> The relationship between conflict and justice is bidirectional. Injustice breeds conflict and destructive conflict gives rise to injustice . . . [P]reventing conflict requires more than training in conflict resolution . . . It requires reducing gross injustice . . . Such reduction require changes in how various institutions of society [political, economic, educational, familial, and religious] function so that they honor and value human equality, shared community, non-violence, fallibility and reciprocity.
>
> (Deutsch, 2000: 67)

Cognizant of this pivotal relationship, findings from this book on process equity suggested employing a multidisciplinary, transformative approach to conflict resolution. This proposed transformative approach incorporated three key components: rules of engagement, mechanisms of engagement, and neutral third-party mediation. Incorporating rules of engagement calls for a full utilization and reliance on principles of international water law to mitigate the impact of power structure imbalance. Utilizing mechanisms of engagement advocates for a comprehensive approach to conflict resolution and calls for adaptive reliance on important parameters and principles of procedural equity identified in Chapter 6. Neutral, trusted, and influential third-party involvement is also warranted to bring representatives from all parties together in an attempt to bridge ethnic and cultural gaps, and therefore the consequent communication gap and mediate a long-term peace negotiation. Understanding and meeting the "other" is central to any conflict resolution attempt. Both political and social reforms are needed in order to address ethnic divides, public opinion, and preconceptions. Through educational campaigns and public awareness programs, along with cross-ethnic dialogue, the parties can understand each other's different needs and viewpoints. Offering good-faith economic incentives to all disputing parties is a major step in cementing relationships based on trust. Riparian states' pronounced mutual interest in resolving their water disputes is a prerequisite for fruitful negotiation and sustained cooperation (Abukhater, 2006).

Evidently, context does matter. Solutions that are equitable in specific cases might not work somewhere else or, worse yet, may produce catastrophic ramifications. Similarly, what is deemed to be equitable in one context might not be viewed as such in other contexts. Therefore, contextual differences are profound and influential and cannot be ignored or evaded by claims of universality (Abukhater, 2009c). Neglecting context and assuming that resolutions and decisions can be made in a vacuum strikes a utopian chord and reflects an unequivocal misunderstanding of the intricacy and the heavily context-dependent nature of water disputes. To be equitable, practical and relevant, any resolution must first be flexible enough to recognize the continuous, evolving, and interlinked networks of the deeply contextualized water resources problems that could be systematic or *ad hoc* depending on the environment within which they originate and operate. Any resolution approach should define and diagnose these problems in light of a contextual understanding of their environment in order to promote equitable remedies accordingly.

In a nutshell, achieving water-sharing equity is a major concern and employing an equitable sharing process is essential in any water agreement to enhance acceptable resolutions for all involved riparians. To put this pursuit into perspective, innovative measures and proactive steps must be taken toward collaborative, transparent, and satiety-oriented negotiation processes over water allocation. Deterring the impacts of the "tragedy of the commons" by promoting cooperative settlement necessitates a process of consensually developing and adopting criteria for equitable water-sharing and acceptable outcomes and incorporating an adequate level of resilience in their applicability that accounts for contextual differences. Only when parties acknowledge the right of the other to access and utilize water, promote equitable mechanisms for water-sharing, and foster a vision for collaboration will they be able to move from rhetoric to execution.

Directions for future research

The completion of this book has unearthed a number of possibilities and ideas for future research. First, the footprint of similar research can be extended to include other cases in various geopolitical contexts. This can be done by combining qualitative and quantitative methods and expanding the number of cases used to make significant statistical inferences about equity parameters. Relying on a mixed-method research design allows researchers to examine this phenomenon from different angles and perspectives. Combining both approaches, if done correctly and successfully, can produce a very effective and rigorous research method with a high level of validity, reliability, and generalizability. Researchers should be cautioned, however, that modeling equity across time and space can be a challenging and difficult task, particularly because of data privacy, confidentiality, and sovereignty limitations in many international contexts. By so doing, future researchers can expand the scope of the study, range of inquiry, and the applicability of the results to a larger population.

Second, further research is needed and can be conducted to examine and measure the spatial and temporal diffusion of hydro-equity. Researchers can utilize geographic information systems (GIS) and Location Intelligence procedures and technologies to model and quantify commensurable equity by relying on multiple criteria evaluation (MCE) methods and the triangulation of various typologies and measurements developed by the literature. Various typologies of commensurable equity should include equity across space and equity across time (intergenerational equity). These studies will be able to answer the question of how to mitigate the spatial and temporal non-movement of environmental equity by fostering hydro-equity across space and time. The goal is to develop an equitable structure of future water distribution by creating *assets, needs,* and *conflict maps.* Spatial information provides context, enables relevance, and enhances understanding of critical issues, such as the current and future needs based on population growth forecast (*needs maps*), current and potential sources of water available for communities in different localities to meet this demand (*assets maps*), and the difference between water availability and deficit and where resources are disputed (*conflict maps*). This difference in supply and demand will constitute the amounts of water to be negoti-ated with other co-riparians to collaboratively figure out feasible ways to acquire these amounts. In addition, there should be demographic, socioeconomic, spatial, technological, hydrographic, and other such factors of influence in allocation decisions that can be quantified and spatially and temporally analyzed to shine light on important and interesting findings. This will help in deriving insights from data, developing actionable intelligence for negotiation, and providing a strategy and platform for execution (Abukhater, 2012a). This growing technology provides a platform for a more efficient and effective decision-making process, not only mapping and visualization but also data management and analysis, collaborative Web 2.0 services, insight, strategy, and communication (Abukhater, 2012b).

Third, future research can greatly benefit from tackling the question of quan-tifying equitable principles introduced by international law applicable in water-related negotiation to formulate systemic commensurable equity standards. In order to provide agreed upon common ground for negotiation, there is a pressing need to develop agreed upon criteria for outcome equity principles as defined by international law, which can potentially play a leading role in resolving water conflicts and reducing associated hostility among riparian nations (Mimi and Sawalhi, 2003). It is necessary for future researchers, therefore, to develop models for measuring commensurable equity (*outcome equity*) in light of the available literature and previous research conducted on the topic. Reflecting the recent overarching shift in international law from water rights to needs, the development of these models should ensue based on the notion of water needs rather than right or entitlement.

By translating the key descriptive and textual principles for equitable allocation of international water resources outlined by international law, these models can be very useful in providing an objective measurement of outcome equity by estimat-ing the degree to which a certain treaty attained equitable outcomes in terms of its proposed volumetric water allocation. This can be accomplished by testing for the difference in the amount of water allocated based on a particular treaty and the amount of water allocated based on equity criteria introduced by international law between two or more riparians. However, researchers tackling this topic should be mindful of a number of statistical and inferential problems associated with utilizing such statistical models. First, it is hard to develop a single and cumulative model able to provide accurate measurement of outcome equity in all contexts. Another challenge is related to the dynamic nature of the negotiation and decision-making process that leads to certain outcomes in water allocation, in addition to the spatial and geographic patterns of water distribution, which makes even the most sophisticated, dynamic spatial models inefficient and uncertain. The inability to quantify benefits from water use and other non-water linkages poses another challenge. In addition, researchers should be aware of the inherent difficulties of making generalizations about binational water treaties, particularly considering the limited number and scope of cases and the unique nuances and contextual differences that each case represents on both sides. Further, the amount and nature of information available varies from one case to another, which may cause data incompatibility and methodological inconsistency issues. Despite these statistical challenges, these models have the potential to provide systematic guidelines and decision tools for future research by approaching the issue in an objective, legislative stance.

Appendix A: research methodological design (*sequential exploratory design*)

> [Q]ualitative and quantitative methodologies are not antithetic or divergent, rather they focus on the different dimensions of the same phenomenon. Sometimes, these dimensions may appear to be confluent: but even in these instances, where they apparently diverge, the underlying unity may become visible on deeper penetration . . . The situational contingencies and objectives of the researcher would seem to play a decisive role in the design and execution of the study.
>
> (Mathison, 1988)

Given the kinds of questions that this research poses, the complexity of the issues, and the great degree of uncertainty, a single methodological approach may not be effective enough to yield rigorous and fruitful findings. One source of this complexity stems from the multiple environmental, political, technological, and cultural dimensions of this study. These multiple layers, although contributing to the complexity of the issue at hand, provide grounds for this research to rely on a triangulation of a variety of methods representing a multidesign approach. This multidesign approach is intended to provide a more holistic and comprehensive outlook on the issues and dynamics of the subject and to produce a more effective and rigorous study with stronger validity and reliability.

The research methodology is designed in a way that seeks to construe a better understanding of the issue of process equity in different arid contexts, on the one hand, and to clarify the significance of process equity in influencing perception and boosting future cooperation between states on the other. Having established this need for a new understanding and expansion of research, a sequential exploratory design approach serves as a complete and well-rounded research approach. Although the book relies on a mixed method approach, it is mainly concerned with qualitative analysis. There is ample evidence that qualitative research is superior, compared with quantitative research methods, in tackling issues that are hard to quantify in numbers, such as human feelings, perceptions, and behavioral data as well as complex causality. The qualitative approach is intended to unravel the mystery and ambiguity surrounding the theoretical aspects of the research topic and develop a better understanding of complex issues related to such an elusive concept. The descriptive component of qualitative case study analysis and story-telling helps reveal important details providing a comprehensive understanding of the contextual details of each case as the research gradually unfolds. As discussed in detail in the closing chapter, an additional quantitative component is identified as a potential long-term research strategy to be conducted in the future, and would constitute a supplementary phase of the research to highlight and inform a specific area of interest with higher precision (such as mapping out the spatial and temporal diffusion of equity and developing measurements for *commensurable equity* standards).[1]

It is crucial to adequately address and mitigate the potential shortcomings concerned with the reliability and validity of the research, particularly because it is concerned with developing a better understanding of issues related to policy and decision-making processes. Relying on a triangulation of several methods of inquiry brings to this research the required objectivity, on the one hand, and richness of detail and context on the other. The research findings are less vulnerable to charges of subjectivity and bias, which should enhance their acceptability. Based on all factors discussed above, this research employs both a within-method and between-method of triangulation to overcome the reliability and validity problems that would affect a study employing a single methodology.

The research employs a mixed method approach that incorporates various methods of inquiry, illustrated in Figure A.1, including (1) literature review, which is critical in identifying parameters of process equity and triangulating measurements of commensurable equity (outcome equity); (2) collecting expert opinions, querying water treaty databases, and reviewing other secondary sources, which are utilized to conduct a first-cut selection of the nine cases and triangulate and measure perceived equity and the degree of implementation; (3) content analysis of secondary literature and media sources and initial screening of these nine cases of international water treaties, which helps in developing rigorous understanding of the issues and forces impacting international water allocation in different regions, on the one hand, and informing the selection of in-depth cases on the other; (4) comparative study and analysis of these detailed cases and pattern matching, which informs the development of process equity parameters to inform theories of water conflict resolution; and (5) in-depth explanatory one-on-one interviews in each of the detailed cases selected for case study analysis to identify key parameters of process equity, inform the theory of equitable water negotiation, and explicate the findings learned from relevant cases to a future context.

The objective of this mixed method research approach is to identify significant process equity factors that boost the propensity of producing cooperative and sustainable agreements that support the sharing of water. This part of the research, which constitutes the major focus of the book, incorporates first-cut selection and analysis of nine cases, content analysis of two key case studies, comparative cross-case analysis, and in-depth explanatory one-on-one interviews. Together, these components of the research agenda establish a clear delineation of process equity parameters.

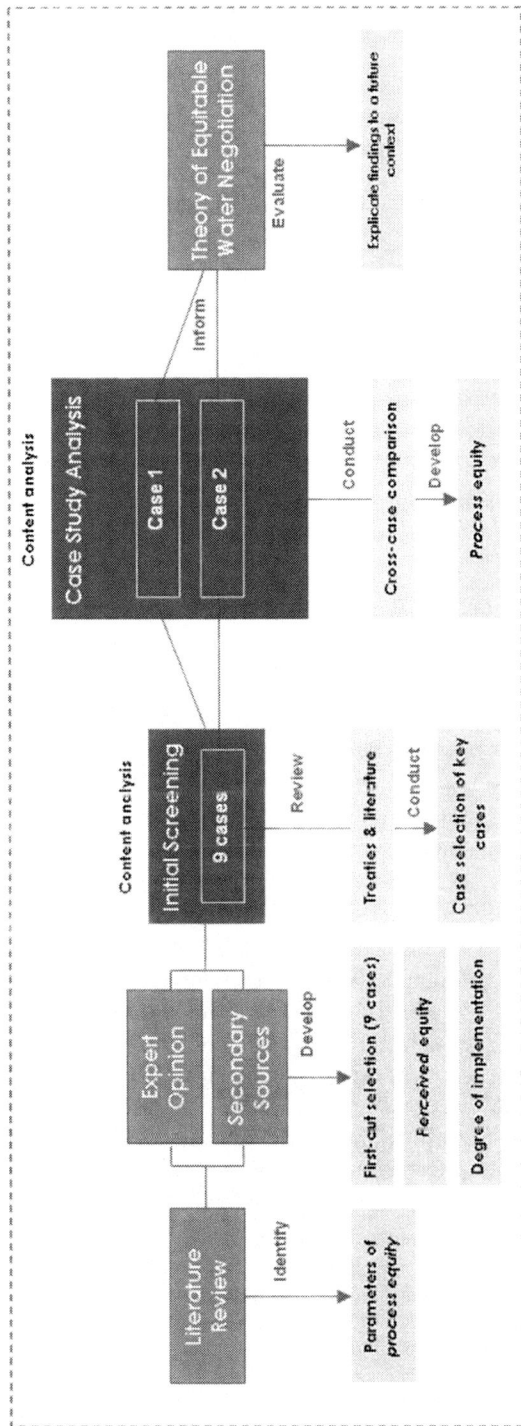

Figure A.1 Research design development.

First-cut case selection and initial screening (snapshot case analysis)

The research analyzes nine ratified international water allocation treaties that are systematically selected based on contextual characteristics, discussed in detail in Chapter 3 (and based on the logic model presented in Chapter 2). In addition, the research engages world acclaimed experts in water resources management and conflict resolution in selecting and rating these nine cases in terms of their level of equity attainability (or lack thereof). Other sources are also utilized for the selection of these cases, such as water treaty databases and other secondary sources.

The purpose of the snapshot case analysis of these nine compatible cases in the preliminary screening process is to provide a brief description of these cases and an adequate understanding of key global issues related to water disputes and hydro-diplomacy, based on which tangible conclusions can be drawn. Relying on a thorough review of the literature, this part of the research aims to (1) provide more breadth and less depth for analyzing the nine cases, and (2) help in selecting the two key cases for in-depth case analysis and cross-case comparison.

In-depth case study analysis

Creswell (2003), author of *Research Design*, listed a number of techniques for qualitative research that include ethnography, grounded theory, case study, phenomenological research, and narrative research. However, it seems that the only method that allows us to effectively answer the primary questions posed in this research, which represent the "why," "what," and "how" categories, is the case study research method (Yin, 2003). According to Yin (2003: 13), "a case study is an empirical inquiry that investigates a contemporary phenomenon within its real-life context, especially when the boundaries between phenomenon and context are not clearly evident." Stake (1995: xi) contends that "a case study is expected to catch the complexity of a single case . . . We look for the detail of interaction with its contexts." He further adds, "We are interested in [case studies] for both their uniqueness and commonality. We seek to understand them. We would like to hear their stories." As such, case study analysis is an effective tool for tackling complex issues by relying on an exploration of causal paths, influences, and interaction effects that may eventually reveal patterns or correlations with other cases. Case study provides a rich opportunity to examine parameters associated with equity attainability, or lack thereof, through a comparative cross-case analysis. Beyond this, the rich details that case study research can provide permit us to explore a variety of topics, issues, and factors, and their interrelations that could not be quantified in a statistical study.

Comparative cross-case analysis

Based on a holistic approach that uses a single unit of analysis (ratified, bilateral, international water allocation treaties) and multiple case studies, many issues become clear even with two distinct cases. Accordingly, we study two key case studies for the purpose of cross-unit comparison.

Comparison of cases has been proven effective in allowing for more control over exogenous factors, which permits us to focus on the target factors of interest (Sartori, 1991). Although employing a comparative cross-case analysis tends to boost the external validity of the research, relying on theory to derive and structure the research is also important in enhancing the internal validity of the research (Yin, 2003). To that end, independent factors (process equity parameters) are selected based on a theory driven method and are informed and validated by expert opinion. This comparative cross-case analysis is employed to identify important lessons for the theory and practice of equitable concepts in water negotiation and potentially contextualize the findings of the case analysis and explicate them to a future context. Based on this analysis, we are able to map out both commensurable and perceived equity (outcome equity) and the dynamics (process equity) across these cases. A systematic examination of equity patterns and processes across the key cases, based on the negotiation process (*treaty formation*), planning analysis, and structure of the agreement, provides a documentable research methodology that illuminates an operational set of process equity parameters. These process equity parameters constitute the cornerstone of a transformative approach to conflict resolution in this book.

In-depth explanatory one-on-one interviews

> If you want to know how people understand their world and their life, why not talk with them?
>
> (Kvale, 1996)

I conducted in-depth explanatory one-on-one interviews with key stakeholders and officials, including (1) stakeholders negotiating on behalf of the state and (2) governmental and/or academic water resources experts. These leading water experts were identified based on a thorough review of the most cited literature and a snowball sampling (they were asked to suggest names of other experts in international water treaties). Some of these interviews were conducted in person, whereas others were conducted over the phone, as some interviewees were located in different parts of the world. As part of the detailed case study analysis, the purpose of these interviews was to develop a contextual understanding of process equity, as well as perceived equity, and of the level of satisfaction. These were semi-structured interviews (utilizing both precise and open-ended questions) to allow for consistent investigation of particular issues and afford flexibility to engage in natural conversation that provided deeper insight. However, as the issue progressed during the interviews, other backup questions were asked when needed, to clarify certain issues or to obtain more relevant information. The interviews were transcribed and analyzed to extract significant lessons and relevant content.

Appendix B: summary of the conclusions of the water meetings

Four meetings were held at the Foreign Ministry to discuss the basis of the Jordan position in negotiations over water. There was agreement on these bases. These are as follows:

1 Adoption of the Johnston Project for water sharing.
2 Amendment of the 1987 agreement with Syria to utilize the Yarmouk waters.
3 Adoption of the Mukheiba site in lieu of Maqarin site for the construction of a dam with suitable storage capacity.
4 Doing away with whatever results came out of the Armitage intermediary efforts in connection with the Wehda dam.
5 Discussing the rights of the West Bank and the Palestinians in the Jordan Basin.

Wednesday, 7/4/1993.

(Haddadin, 2002a)

Notes

Introduction

1 From the message of Secretary General Kofi Annan on the World Day for Water, 2001.

1 Water past and present

1 In this context, the right to water refers to the legal water rights of accessing sufficient water quality and quantity, which is considered a fundamental human right by the United Nations (for more information, refer to Chapter 6, page 225, "Needs-based water allocation").

2 Quoted in the *New York Times*, August 10, 1995.

3 In this book, the terms *process equity* and *outcome equity* will be used in reference to *procedural* justice and *distributive* justice respectively.

4 The Arab–Israeli conflict represents ample example of the extent to which these international laws are related to disputed water resources.

5 Also referred to as the doctrine of absolute territorial sovereignty.

6 These are considered extreme principles because of their strict regulations and lack of consideration for equitable water allocation.

7 Rights-based equity in this context refers to rivalries' allegations of territorial sovereignty and the exclusion of other parties sharing the river transboundary water resources. Needs-based water allocation is grounded on the real needs of the population based on per capita water consumption requirement. For more information, please refer to the section in Chapter 6. Needs-based allocation strategy is regarded as a desirable venue in water management and more effective than rights-based allocation in reaching consensus because needs are quantifiable and can be easier to negotiate over.

8 The concept of *needs* refers to per capita water needs, whereas *rights* refer to territorial sovereignty. For more information, please refer to the section in Chapter 6 on needs-based water allocation.

9 The notion of *needs* (as well as *necessity*) is considered conducive to more equitable negotiations and outcomes.

3 Case selection and first-cut analysis

1 Proceeding of the World Bank Seminar, 1998.

2 Product of the Transboundary Freshwater Dispute Database, Department of Geosciences, Oregon State University. Additional information about TFDD can be found at: www.transboundarywaters.orst.edu

3 Proceeding of the World Bank Seminar, 1998.
4 Product of the Transboundary Freshwater Dispute Database, Department of Geosciences, Oregon State University. Additional information about TFDD can be found at: www.transboundarywaters.orst.edu
5 This survey contained three parts. First, they were asked to provide nine or more cases of bilateral international water treaties that complied with the following criteria (previously identified by the logic model as control factors): (1) arid regions context; (2) bilateral treaties; (3) cultural variation between the riparian countries; (4) history of aggression and/or hostility; (5) power structure imbalance between riparians; and (6) full/partial implementation of treaty. Second, based on a Likert scale from one to five, five being the most successful, interviewees were asked to rank the success of these treaties in attaining equitable outcomes for all signatories to the treaty by assigning a numerical value representing their view of how equitable each treaty was. Third, depending on a snowball sampling, they were asked to suggest names of a few other experts in international water treaties.
6 The construction of the Dam meant the creation of a large reservoir that extended into Sudan, causing the displacement of human settlement and degradation of the ecosystem of the whole basin.
7 This situation reflects the use of water as a "weapon" by Turkey to exert pressure on Syria and gain political influence, as was discussed earlier in Chapter 1.
8 In 1980, Iraq and Turkey established the JTC and later in 1983 Syria joined the committee.
9 The Indus basin has one of the world's most extensive irrigation systems.
10 Water was needed in West Bengal and for keeping Calcutta Port operational.
11 Detrimental environmental impacts included high salinity and environmental degradation of the river. In 1976, Bangladesh protested against India at the United Nations (Priscoli and Wolf, 2009).
12 Previous diplomatic efforts resulted in a number of short-term agreements.
13 Although the shortfall in Bangladesh's share in the first 10-day period of April 1997 was for safety concerns, the public in either country was not aware of such details and therefore viewed this as violation of the Treaty terms (Iver, 2003).
14 This treaty outcome is conclusively reflective of the status quo of a power structure imbalance (Beyene and Wadley, 2004). Although the two countries promoted shared visions and cooperation through the work of the Nile Basin Initiative (NBI), there are still many unresolved environmental issues (Willemse, 2007).
15 This database was developed by the Oregon State University Department of Geosciences in collaboration with the Northwest Alliance for Computational Science and Engineering.
16 This database includes data, books, journal and newspaper articles, and documents on political, socioeconomic, demographic, and legal issues of water in the Middle East from a variety of publishers and national and multinational agencies and organizations.
17 This database contains full text documents of over 400 international freshwater-related treaties.
18 This is a searchable online database containing over 6,400 historical international water relations. This includes the basins and countries involved, the dates at which they occurred, the issue areas, BAR Scale (Basins At Risk – Water Event Intensity Scale) ratings, and detailed summaries of the events.
19 TBAR Scale (Basins At Risk – Water Event Intensity Scale) ratings and detailed summaries of the events. This database contains the world's international river basins, delineated by continent, including basin name and total area and a list of the countries associated with the basin and their areas located within the basin.

20 This bibliography provides a compilation of references and publications related to national and international river basins (including books, journal and magazine articles, conference proceedings, and institutional reports). It was developed by the Transboundary Freshwater Dispute Database based on the bibliography compiled for Beach *et al.*, 2000.

21 This database compiles indicator variables for analyzing international river basins and provides spatial and tabular searchable data and biophysical, socioeconomic, and geopolitical data relating to the world's international river basins.

22 This atlas, published in cooperation with the United Nations Environment Programme (UNEP) and the Food and Agriculture Organization of the United Nations (FAO), contains a detailed listing and historical overviews of more than 300 international freshwater agreements and various relevant thematic maps.

23 This database includes electronic and hardcopy versions of data and GIS maps and files.

4 The Israel–Jordan Peace Treaty of 1994

1 Map created and designed by the author based on data from the Transboundary Freshwater Dispute Database, Oregon State University.

2 The current Palestinian population in the West bank and Gaza Strip is estimated at 4.1 million people, of which 57 percent is urban (Hummel, 2006; Dinar, 2009). The Palestinian population density is the highest in the region and one of the highest in the world with 1,584 people per square kilometer. (Hummel, 2006). The population is composed of Arab Palestinians and refugees from pre-1948 Palestine. With an estimated growth rate of 3.5 percent, the highest in the region, the Palestinian population is expected to grow to nearly 7.4 million people in 2025 and 11.9 million people in 2050 (Hummel, 2006).

3 Blue water is the water available in rivers, lakes and groundwater for human use, whereas green water is the water held in soils, supporting rain-fed agriculture and all terrestrial ecosystems (GLOWA, 2008).

4 According to the World Business Council for Sustainable Development (WBCSD), 2005, humans need a minimum of two liters of drinking water per day to survive.

5 *De facto* water use refers to the actual use in practice, irrespective of whether or not it is ordained by law.

6 Israel has the highest per capita GDP of about USD 28,900, followed by Lebanon (USD 6,200), Jordan (USD 5,000), Syria (USD 3,900), and Palestine (USD 920) (McKinney, 2008). With an unemployment rate of only 6.1 percent, Israel's economy rivals other developed countries and can be characterized as a technologically advanced market economy (CIA, 2008). With a soaring unemployment rate of 30 percent, Jordan is a developing country challenged today by reducing dependency on foreign aid, attracting investments, and creating jobs (CIA, 2008).

7 This armed conflict was interrupted by the 1979 Peace Treaty between Israel and Egypt and later by the 1994 Peace Treaty between Israel and Jordan.

8 Product of the Transboundary Freshwater Dispute Database, Department of Geosciences, Oregon State University. Additional information about TFDD can be found at: www.transboundarywaters.orst.edu

9 In that sense, Israel envisioned constructing its National Water Carrier to divert water for out-of-basin uses in the coastal plain and the Negev, a Zionist vision that was originally developed by Theodore Hertzl and adopted by Bourcart in 1899 (Shamir and Haddadin, 2003). On the other hand, the plans proposed by the Arab countries emphasized and adhered to the principles introduced in the McDonald Plan, namely the assertion that "the general principle, which to our mind has an

undoubted moral and natural basis, is that the waters in a catchment area should not be diverted outside this area unless the requirements of all those who use, or genuinely intend to use the waters within the area have been satisfied" (McDonald, 1951 cited in Haddadin, 2002a).

10 Product of the Transboundary Freshwater Dispute Database, Department of Geosciences, Oregon State University. Additional information about TFDD can be found at: www.transboundarywaters.orst.edu. Proceeding of the World Bank Seminar, 1998.

11 The United Nations Relief and Works Agency.

12 The final version of the Plan allocated water as shown in Table 4.11.

13 This was particularly true for Lebanon, which gained little water (only 35 MCM) by signing the Plan.

14 Declaring this "piecemeal approach" adopted by the United States and highlighting the intended equitable nature of the Plan, the US administration contended: "In the period 1953–1955, the United States Government, through the work of Ambassador Eric Johnston, made a strenuous and generous effort to devise a plan for unified development of the Jordan Valley that would be equitable and acceptable to all parties. Regrettably, the comprehensive [Johnston] Plan drawn up by Ambassador Johnston failed to win the immediate approval of the parties. However, in the absence of later studies of an equally comprehensive nature, we have considered that it remained useful as a model for full, orderly, and equitable utilization of an immensely important international water resource. We remain convinced of the benefits of the types of unified development embodied in the [Johnston] Plan. In the absence of agreement thereon, we have supported the right of the riparians to proceed with national water programs provided these did not conflict with the general principles and allocation patterns envisioned in the [Johnston] Plan" (Whiteman, 1964).

15 The Israeli and Jordanian delegations included more members in various groups and stages of negotiations.

16 The Palestinian team was headed by Dr. Haidar Abdul Shafi, a medical doctor from Gaza.

17 Recognizing equal partners necessitates the recognition of economically equal partners with equal access to natural resources of the region, including scarce water.

18 This stipulation of maintaining national security stands out as one of the most prominent intentions of the Treaty and a desired outcome.

19 These diplomatic efforts led by the United States were the foundation for the Madrid Peace Process.

20 Jordan and Palestine were jointly represented under the Jordanian state umbrella (Haddadin, 2008). In addition, other neighboring countries also attended, namely Egypt, Saudi Arabia, and Morocco, representing the Arab Cooperation Council, the Gulf Cooperation Council, and the Maghreb Union respectively. The Conference also invited international presence, namely representatives of the European Union and the United Nations, in addition to the sponsors of the conference: the United States and the Soviet Union (Haddadin, 2002a).

21 The separation of bilateral and multilateral negation tracks was a direct result of Israel's position toward the Palestinian representation (Shamir and Haddadin, 2003).

22 Negotiation started as multilateral and ended with separate bilaterals, an approach that was proposed and favored by Israel (Daoudy, 2009).

23 Negotiations regarding low politics issues started in January 1992, over two years prior to reaching the Treaty.

24 Although both the Jordanians and Palestinians were very critical of Israel's policy and activities on the ground regarding their unlawful utilization of Palestinian water, the Jordanian negotiator felt that the "water front was making good progress" (Haddadin, 2002a).

25 Dr. Munther Haddadin, former president of the Jordan Valley Authority, served as the senior water negotiator on behalf of the Jordanian side, whereas Mr. Noah Kinarti was the chief water negotiator on behalf of Israel.

26 The work of the Multilateral Working Group on Water Resources resulted in the recognition of the Palestinian water rights and the creation of the Palestinian Water Authority and an Executive Action Team (EXACT), a multinational team of Israeli, Jordanian, and Palestinian officials to enhance regional water data acquisition, standardization, and dissemination that would support water decision making and planning.

27 According to Haddadin, 2002b, "Those multilateral talks, as their objective stated, were not meant to resolve disputes, but were meant to enhance the environment of the bilaterals, and were, in fact, ineffective and almost unproductive" (Shamir and Haddadin, 2003).

28 For example, the Israeli delegation clashed with the Jordanian delegation in Moscow, Vienna, and Washington, and with the Palestinian delegation in Geneva (Haddadin, 2002a).

29 This arrangement, which was also proposed and supported by King Hussein, was followed in the multilateral conference and extended into the bilateral negotiations as well (Shamir and Haddadin, 2003).

30 There were two Palestinian representatives along with nine other Jordanian negotiators in the Jordanian bilateral talks with Israel and, similarly, two Jordanian representatives accompanied the other nine Palestinian negotiators in the Palestinian bilateral talks with Israel (Haddadin, 2002a; Shamir and Haddadin, 2003).

31 For example, please see *Breaking the impasse: consensual approaches to resolving public disputes* by Susskind and Cruikshank (1987).

32 For Jordan and Israel to participate in serious talks for a final settlement, the Palestinians and Syrians were left out of the diplomatic arrangement.

33 Negotiators, mainly scientists and engineers, from the Israeli and Jordanian sides met, shared hydrologic data, and discussed options for reconciliation and settlement along the Jordan River Valley.

34 These agreements, although sanctioned by both countries' leaders, were not considered formal and therefore binding agreements.

35 There was enough determination and political will on both sides to enter into formal negotiation with the purpose of developing a comprehensive peace treaty but trust was, and is still, a missing element, as the parties have a long history of conflict, aggression, and hatred that was not adequately resolved.

36 Many critics further accused Israel of being behind the pollution in Amman.

37 Offering good-faith economic incentives to all disputing parties is a major step in cementing relationships based on trust prior to negotiation.

38 This part of the negotiation was more factual in nature and politically neutral for the most part.

39 For example, the data regarding the sustainable yield of the groundwater were provided to the Palestinians by Israel with little verification or modeling by the Palestinian side. This was not the case during the Israeli–Jordanian negotiations, as data regarding Jordan's groundwater use (namely its reliance on the Desi aquifer shared with Saudi Arabia) is known to the Jordanians but remained unknown to and unshared with the Israelis (Phillips, 2008). It is worthwhile mentioning that, in addition to data accuracy issues, Israel managed to avoid discussing the surface water of the Jordan River during the Palestinian–Israeli negotiations, as they claimed that the Syrian and Lebanese uses of the river were unknown and unpredictable.

40 The deal achieved in Madrid in 1991 between the Israelis and their Palestinian counterparts paved the way for an open negotiation process between Israel and Jordan.

41 On September 13, 1993, the Oslo Accords, officially known as the Declaration of Principles on Interim Self-Government Arrangements, were finalized in Oslo,

Norway, and later officially signed by Yasser Arafat and Israeli Prime Minister, Yitzhak Rabin, in Washington, DC, in the presence of US President Bill Clinton (Haddadin, 2002a). The Common Agenda had been reached many months before its ratification date but awaited an opening on the Israeli–Palestinian track, which was achieved on October 13, 1993, with the ceremony on the White House lawn sponsored by President Clinton, in which both Israel and the PLO signed the DOP (Shamir and Haddadin, 2003).

42 These diplomatic efforts were the basis for terminating the state of belligerency between Jordan and Israel and rejuvenating the desire of both states to seek a just, lasting, and comprehensive peace based on UN Resolutions 242 and 338. Communications and bilateral negotiations extended between the two parties as they proceeded to meet on July 18–19, 1994, at Ein Avrona, located in the boundary area north of Aqaba of Jordan and Eilat of Israel.

43 The initiation of the Treaty occurred on October 17 at the Hashimiyya Palace in Amman by King Hussein and other high officials from the two countries, including the prime ministers of Israel and Jordan, Mr. Yitzhak Rabin and Dr. Abdul Salam Majali respectively.

44 During the bilateral negotiation, the Israeli and Jordanian delegations were composed of a number of experts in many fields and were headed by influential high officials, such as Dr. Abdul Salam Majali and Ambassador Fayez Tarawneh (later prime ministers) from Jordan, and Ambassador Elyakim Rubinstein, secretary of the government (later attorney general) from Israel (Shamir and Haddadin, 2003). The members of the Israeli and Jordanian delegations included groups of experts on many related topics, namely security, land/borders, energy, environment, police, economics, and water.

45 Although the structure of the bilateral arena changed over the course of negotiation, ranging from formal face-to-face negotiation to informal corridor meetings and chats, these groups were maintained throughout the negotiation for the most part (Shamir and Haddadin, 2003).

46 Shamir (2008) explains, "We are big boys; we can handle ourselves on both sides. Nobody wanted to have a third party, but when we met in Washington, we had the Americans and Russians hovering in the hallways. Negotiation was usually held in the State Department. And so these were the formal channels and once in a while there was informal consultation or informal briefing to the Russians and to the Americans and maybe words that was provided by them and small role to facilitate, but it was basically bilateral inside the room of negotiation . . . it was only between the two parties."

47 The Jordanians stayed in the Willard International Hotel and the Palestinians stayed in the Grand Hotel. This created an atmosphere that was not conducive to collaboration. The two parties, as such, seemed detached from one another and missed an opportunity to unify their negotiating agenda.

48 As the joint delegation waited patiently, staring at the empty Israeli seats, the State Department did not allow the press into the negotiation room to avoid documenting the absence of the Israeli team, which was referred to by the State Department spokesman as "hitting below the belt" (Haddadin, 2002a).

49 The Israelis wanted to meet on December 9th, but the Arabs preferred not to meet on this day, which memorializes the fourth anniversary of the Palestinian Intifada (uprising) of 1987.

50 The Israelis wanted to give more emphasis to the Jordanian role and authority to handle and negotiate issues related to Palestinian affairs. Both Palestinians and Jordanians rejected the involvement of Jordan in Palestinian manners, as a way to assert Palestinian independence, although it was known to the Israelis that the joint delegation was one entity with which they were negotiating.

51 The approach suggested by the joint Arab delegation was in accordance with the "Letter of Invitation" and reflected the Jordanian strategy of disengagement from the Palestinian negotiations.

52 They also concluded that significant progress on the bilateral front was needed to move the peace process forward.

53 This delay was due to the late arrival of the Arab delegation, who had protested against the expulsion of 12 Palestinians from the Occupied Territories by Israel (Haddadin, 2002a).

54 As mentioned before in this chapter, the negotiation took place along two bilateral tracks: the Israeli–Jordanian and Israeli–Palestinian tracks. In the Israeli–Jordanian track, nine Jordanian negotiators and two Palestinian representatives participated in the negotiation, whereas two Jordanian representatives accompanied the nine Palestinian negotiators participating in the Israeli–Palestinian track (Haddadin, 2002a; Shamir and Haddadin, 2003).

55 Only matters related to the Palestinian Self Governing Authority (PISGA) were discussed.

56 These combined groups, along with the heads of the delegations, who provided general guidance, convened to change the dynamics of the single-domain negotiations.

57 On the issue of environment, namely the demise of the Dead Sea, the delegations discussed the option of linking the Dead Sea to the Red Sea, as proposed by the Jordanians, and to the Mediterranean Sea, as proposed by the Israelis (Haddadin, 2002a).

58 This negotiation strategy is recommended by Raiffa (1982) to keep the negotiation efforts focused, save time and effort, and mitigate misunderstandings and impasses.

59 This strategy is proven effective, when compared with the shuttle diplomacy approach used by Eric Johnston, who met with each party separately and, based on his own notes and interpretation, communicated to the other parties what he discussed with their counterparts. As such, he communicated different versions of the proposals made by each party, in which situations confusion and misunderstanding are bound to occur. Therefore, unlike Johnston's efforts, using STN tends to narrow the communication gap between the parties (Shamir and Haddadin, 2003).

60 This text was a modification of the original text drafted and proposed by Jordan and was mutually identified and informally agreed upon by both parties.

61 The parties also discussed and reached informal consensus on considering environmental protection and quality factors and the issue of the Dead Sea water level was included as part of another item in the Common Agenda (Haddadin, 2002a).

62 The proposed new agenda was approved by the other side after some negotiations and changes on April 26th, 1992.

63 This strategy was justified based on the fact that many Palestinian refugees reside in Jordan and still need water for their own uses. According to Haddadin (2002a), this strategy was used to avoid dividing Jordanian water share between the West Bank and East Bank and was successful in keeping the Palestinian water share in the hands of Jordan, rather than Israel.

64 In addition, the issue of controlling the declining Dead Sea water level was informally discussed and further joint committees and studies were identified by both parties as necessary measures to identify the best strategy for saving the Dead Sea.

65 This party, unlike the previous rightist government led by the Likud Party (led by Yitzhak Shamir), held a much more moderate agenda toward peace.

66 On one hand, Mr. Noah Kinarti (the Israel water negotiator) objected to the amount allocated for Israeli use by the Johnston Plan (25 MCM), arguing that Israel never consented to such an amount and demanding that Israel's share from the Yarmouk should be 25 MCM during the dry season alone (May 15–October 15) (Haddadin, 2002a). He also demanded an additional amount of 20 MCM (more than the 15 MCM that Israel had always been asking for since the Johnston Plan), which would make

Israel's share 45 MCM from the Yarmouk. On the other hand, Dr. Haddadin (the Jordanian water negotiator) argued otherwise and was quick to point out the reasons for his argument for relying on the Unified Plan as the basis for water allocation. Two key reasons were provided to prove the acceptability of the Plan by Israel: first, the declaration of Plan's acceptance made by then Israeli Prime Minister, Mr. Levy Eshkol, in 1964; and, second, the Israeli Prime Minister Mr. Rabin's statement to King Hussein affirming that Israel was not interested in one drop of Jordan's water (Haddadin, 2002a).

67 For Israel, environmental concerns such as the protection of the Gulf of Aqaba and the issue of mosquito and houseflies in the Jordan Valley were at the top of their negotiation agenda (Haddadin, 2002a). For Jordan, the diversion of water at the Adasiyya point into the KAC was insufficient to meet the domestic demand in Amman, which was described as tight (Haddadin, 2002a).

68 During this round, Bill Clinton won the American election, which meant that the Republican administration involved in the negotiation was about to leave office and the new Democrat administration, which was not tested on issues related to the ongoing Middle East peace process, was about to take over.

69 The latter included strategies to achieve sustainable and comprehensive peace based on the Security Council Resolutions 242 and 338, as well as security, water, refugees, borders and occupied lands, fields of potential cooperation, phasing of implementation, and establishment of appropriate mechanisms for negotiations, matters related to the Jordanian and Palestinian tracks, and the conclusion of a "peace treaty" (Haddadin, 2002a).

70 However, the approved Common Agenda was published in Jordan, which created a chain reaction in both Jordan and Israel.

71 Dr. Haddadin, however, remained the head of the Jordanian delegation to the Water Resources Working Group in the multilateral negotiations.

72 The negotiation strategy for Jordan, which validated the adoption of the water-sharing formula provided by the Unified (Johnston) Plan, is outlined in Appendix B.

73 Jordan's strategy focused on matters related to Jordan's rightful share of water (including both quantity and quality), the construction of the diversion weir and storage facilities at Adasiyya, and the development of the river basin.

74 To that end, Jordan was hoping to address water, refugees, and the occupied Jordanian territories. In contradiction, Israel was concerned about, and was pushing the negotiation toward focusing on, security issues and achieving cooperation and normalization.

75 These were: (A) water, energy, and environment; (B) security and borders; and (C) refugees and displaced persons, banks and economics, tourism, health, and drug controls (Haddadin, 2002a).

76 These meetings coincided with the secret negotiations of the Oslo Accord, in which Israel and the PLO affirmed mutual recognition. As the Oslo Accord secret negotiations and the signing of the Declaration of Principles (DOP) by Israel and the PLO caused an impasse in the bilateral track, the United States sponsored a trilateral economic conference in February 1994 to rejuvenate the stalled bilateral negotiation track.

77 The Oslo Accord between Israel and the PLO in August 1993 and the signing of the Declaration of Principles (DOP) on September 13, 1993, provided Jordan with the opportunity to move forward with its negotiation agenda with Israel.

78 According to Haddadin (2002a), the king, urging Dr. Haddadin to disengage, said, "take good consideration of the decision of disengagement with the West Bank. Palestine has its own men to defend Palestinian rights. We limit ourselves to the Jordanian rights as Jordan stands today." When Dr. Haddadin pointed out the water needs of displaced West Bankers living in Jordan, the king further explained, "but there is land on the West Bank that did not cross over to the east, and that land has irrigation needs, please disengage" (Haddadin, 2002a).

79 The deal achieved in the Oslo Accord between Israel and the PLO provided legitimacy and justification for this Jordanian decision to disengage and separately negotiate with Israel without having to worry about representing the Palestinian interests.

80 As the tone of the negotiations grew more serious, the negotiations were moved closer to home on July 18th, 1994, and the atmosphere was conducive for a separate Jordanian–Israeli track.

81 This location was selected carefully as it was located on the uncontested portion of the Armistice Line between Israel and Jordan.

82 With the presence of the United States delegation, headed by Secretary of State Christopher, the Jordanian delegation, headed by Prime Minister Dr. Majali, and the Israeli delegation, headed by Mr. Shimon Peres, the delegations discussed projects that were cooperative in nature and agreed to develop a draft master plan for the development of the Jordan River Valley.

83 This momentous event that took place in Washington, DC, changed the pace of the negotiation and paved the way for more proactive rounds of negotiation. The White House meeting brought King Hussein and Mr. Yitzhak Rabin under the sponsorship of President Bill Clinton, who was able to announce the "Washington Declaration" on July 25th, 1994.

84 In an informal talk with the head of the Israeli water team, Dr. Knarti, Dr. Haddadin explained, "whatever we agree to shall in no way prejudice the rights of the other riparian parties on the Jordan River basin, particularly the Palestinians" (Haddadin, 2002a).

85 The Jordanians advocated for Jordan's growing municipal and irrigation needs in Amman that had changed since the Johnston Plan's conclusion, requesting 200 MCM from the Jordan River (Haddadin, 2002a). On the other hand, Israel wished to maintain its current use from the Yarmouk of about 75 MCM (65 MCM, according to Kinarti), but the Jordanian negotiator insisted that Israel's share should not exceed 25 MCM in accordance with the Unified Plan (Haddadin, 2002a; Haddadin, 2008).

86 The water–land (border) and water–environment nexus in the negotiation made it easy for the parties to consider other benefits.

87 The parties agreed to involve the World Bank to conduct a study of the Jordan Rift Valley and the Red Sea–Dead Sea Canal, which was included in the "Terms of Reference" (Haddadin, 2002a).

88 The next round of bilateral negotiation, which tackled issues related to water and shared borders, was held in September at Beit Gabriel on the shore of Lake Tiberias and in Aqaba.

89 These groundwater resources are of crucial importance to Israel as they are located in the south, beyond the reach of its National Water Carrier (Fischhendlar, 2008a).

90 Despite the fact that no time was given to the preparation of the Water Annex of the Treaty, the use of STN was beneficial and effective in refining the common text. These meetings allowed both teams to mutually inspect the articles of the Water Annex, prepared by the Jordanian team, and insert agreed upon water allocation figures. The two teams were open to changes, as they operated under the direct auspices of the two countries' leaders and the involvement of the World Bank.

91 Mr. Kinarti made an astute observation to Dr. Haddadin that was proven to be correct later, that these water issues "will be resolved by the two leaders" (cited in Haddadin, 2002a).

92 "This is good news," as Dr. Haddadin described the border adjustment proposal, and further explained, "at least they now recognize the line of lowest points in Wadi Araba is its center as is meant by the text of the Mandate definition of borders, not the median line as the Israelis always insisted. They tried to throw that line to the geographic mean, which would have taken quite a bit of Jordanian territory, with fossil water aquifers under it" (cited in Haddadin, 2002a).

93 Haddadin (2002a) explains, "it was not the time, nor was it the forum, to address Syrian actions on the Yarmouk. Rather, the time and the forum were for the conflict with Israel, not Syria, to be resolved through bilateral negotiations. In other words, the Syrian infringement on Jordan's share, as stipulated in the Unified Plan, was not a subject to be settled in the bilateral negotiations with Israel."

94 According to Shamir (2008), the use of arbitration was ruled out as a viable option for the negotiation as no agreement was reached on the issue. Rather, the parties believed that only through direct negotiation could conflict be managed and amply settled (Haddadin, 2008).

95 It was agreed, and documented in the text of the Treaty, that the issue of refugees will not be discarded. Rather, the Treaty identified refugees as an issue to be resolved in the final status negotiations between Israel and the Palestinians and Israel and Jordan, where the issue of the Palestinian refugees and the refugees in Jordan would be simultaneously addressed (Haddadin, 2002a).

96 Although the Treaty was ratified on October 17th in Amman, the ceremonial signing of the Treaty took place in Wadi Araba on October 26, 1994, where US President Clinton attended (Shamir and Haddadin, 2003).

97 This is because many displaced Palestinians are currently residing in Jordan and in need of water.

98 For example, it was primarily developed to address the refugee problem at that time and the need for freshwater for agricultural uses only, and further disregarded groundwater availability (Phillips, 2008; Zeitoun, 2008b).

99 Namely the 1997 UN Resolution on the Non-Navigational Uses of International Watercourses.

100 In particular, Jordan was interested in collecting and storing the winter surplus flow in either the Wehda Dam, which seemed to be handicapped by the reduced river flow due to the Syrian water use, or the Mukheiba site, located in the occupied Golan Heights (Haddadin, 2002a).

101 However, the JWC seemed week in monitoring the co-riparians' water extraction compliance and river development (Zawahri, 2004).

102 Although it convenes weekly during times of drought, it may convene monthly otherwise (Zawahri, 2004).

103 In contrast with the term "constructive ambiguity" introduced by Fischhendlar (2008a).

104 An alternative solution, suggested by Israel, emerged from further negotiation by increasing the efficiency of water use by the Israeli fisheries industry in the Beit Shean Valley.

105 As the two sides failed in addressing this issue, even after forming a task force specifically to handle this additional water supply, the issue was referred to a higher level of negotiations (Fischhendlar, 2008a).

106 Although the cost of the transfer was equally shared between Israel and Jordan, the deal was criticized by the Israeli Ministry of Finance and the Ministry of Defense (Haaretz, 1997).

107 At that point, Jordan, in turn, threatened to end any diplomatic relations with Israel as a strategic response to exert political pressure on Israel to abide by the Treaty.

108 The items of discussion and negotiation that led to the adoption of the term "rightful allocations" are included in the "Common Sub-Agenda" that was agreed on June 7, 1994, between the parties (Haddadin, 2002a).

109 Shamir (2008), who believes that "theory generates nothing" substantial in negotiation, gives an example to illustrate how the parties managed to be pragmatic in resolving some of the most controversial issues: "Because the Jordanians insisted on the very beginning to talk about water rights, it was made abundantly clear that based on the concept of water rights no practical outcome will be achieved. Therefore, the agreement uses a new term as it states that both sides should recognize each others'

'rightful' allocation. 'Rightful' allocation, you know is related to the actual quantity of water, and the reason for sticking this word [rightful] in there is to sound good . . . and to satisfy this wish to have your rights back. But both sides understood that using the term 'water rights' cannot be translated into anything practical and tangible."

110 Israel has been increasingly proposing regional water imports from Egypt, namely the Nile water, to provide additional supply of water to the Negev Desert.

111 "In light of paragraph 3 of this article, with the understanding that cooperation in water-related subjects would be to the benefit of both parties, and will help alleviate their water shortages, and that water issues along their entire border must be dealt with in their totality, including the possibility of transboundary water transfers, the Parties agree to search for ways to alleviate water shortage and to cooperate in the following fields:

112 (a) development of existing and new water resources, increasing the water availability including cooperation on a regional basis, as appropriate, and minimizing wastage of water resources through the chain of their uses;

113 (b) prevention of contamination of water resources;

114 (c) mutual assistance in the alleviation of water shortages;

115 (d) transfer of information and joint research and development in water-related sub-jects, and review of the potentials for enhancement of water resources development and use" (Treaty of Peace, 1994: Article 6(4)).

116 According to Annex II of the Treaty, which outlines water allocation and related matters, Israel is entitled to pump 12 MCM of the Yarmouk water during the summer period (May 15th to October 15th of each year) and 13 MCM during the winter period (October 16th to May 14th of each year), while Jordan takes the remainder of the flow in both periods.

117 Such water shortages are expected to occur frequently in such a semi-arid region.

118 Annex II, Article 1(2) states, "Israel concedes to transfer to Jordan in the summer period (20) MCM from the Jordan River directly upstream from Deganya gates on the river."

119 The Treaty states, "Jordan is entitled to an annual quantity equivalent to that of Israel, provided however, that Jordan's use will not harm the quantity or quality of the above Israeli uses" (Annex II, Article 2(C)).

120 As such, Jordan's use of the Jordan River seems to be conditional and comes second in priority.

121 The senior water negotiator on behalf of Jordan.

122 According to Haddadin, who decided to resign as minister, rumors of Israeli poison-ing spread through the media and caused panic that "was orchestrated from within Jordan to bring down the government" (Otchet, 2001). The time given to establishing gradual and long-term peace negotiation was used by the opponents of normalization on each side to derail the opportunities for peace.

123 The loss of 1 percent of the body's water causes one to feel thirsty; loss of 5 percent of the body's water causes loss of consciousness; and loss of 10 percent of the body's water may lead to death (Water Care, 2004).

124 For instance, the World Health Organization (WHO) specified a minimum baseline water requirement of 100 liters/person/day for domestic water consumption and 150 liters/person/day (l/c/d) as the preferred requirement (WHO, 2003; Phillips *et al.*, 2007b). These values reflect optimal level of service (100–200 l/c/d). Opting to advocate the same needs-based approach, Wolf (1995) also advocated for a total of a 100 m^3/year per person (about 275 l/c/d) as the minimum threshold allocation for the Middle Eastern urban dweller. Gleick (1994) specified 75 m^3/year (about 205 l/c/d) as the recommended per capita basic domestic water requirement in the Middle East. This amount was later reduced, in 1996, to 18.25 m^3/year, which is about 50 liters per day (a little more than one gallon), to ensure good hygiene (Gleick, 1998). This value (50 l/c/d) reflects intermediate level of service. Likewise, Shuval (1992; 2000; 2005)

suggested a minimum level of water consumption for Israel, Jordan, and the Occupied Palestinian Territories of about 125 m³/year per person (equivalent to 342 l/c/d) for domestic, urban, and industrial uses (Wolf, 1999). Isaac (1994), working in this vein, also suggested similar concepts that he termed "water equity" (Phillips *et al.*, 2007b). Similarly, Haddadin (2002a) suggested a compensation strategy for the Palestinians of the West Bank based on an annual per capita requirement of 50 m³ (about 137 l/c/d).

125 This segmented management of the river system is a byproduct of the fragmented Arab position and tends to favor Israel due to its advantageous hydro-strategic position (Elmusa, 1995).

126 Although Israel's share stipulated by the Treaty exceeds what was allocated by the Plan, this water gain was comparatively insufficient and made little difference to Israel, considering its annual water budget (Elmusa, 1995).

127 This amount included water for the Palestinians of the West Bank, which is estimated at 215 MCM/year.

128 According to the agreement, Israel was allocated fixed amounts, whereas the quantities allocated for Jordan were for the most part estimates. Giving a higher priority to Israel's use also seems inconsistent with the Plan, which gave a higher priority to the other co-riparians by allocating fixed amounts for the Arab states and the residuals for Israel (Elmusa, 1995).

129 Conversely, Israel relies on Lake Tiberias for its storage needs, which provides 25–30 percent of its annual water budget (Shamir and Haddadin, 2003).

130 The text of the Treaty seems vague regarding the source of water to be delivered to Jordan. It instead makes a vague reference to the lower end of the lake as the source of water, stating "directly upstream from the Deganya gates on the river" (Annex I(2) (a)).

131 The Treaty allocated 90–315 MCM and 625–665 MCM for Jordan and Israel respectively.

132 Water supply figures offered in this Treaty did not reflect the variation in consequent water demands of the two nations.

5 The Lesotho Highlands Water Project Treaty of 1986

1 Product of the Transboundary Freshwater Dispute Database, Department of Geosciences, Oregon State University. Additional information about TFDD can be found at: www.transboundarywaters.orst.edu

2 Map created and designed by author, based on data from the Transboundary Freshwater Dispute Database, Oregon State University.

3 Product of the Transboundary Freshwater Dispute Database, Department of Geosciences, Oregon State University. Additional information about TFDD can be found at: www.transboundarywaters.orst.edu

4 In reference to the Bantu-speaking peoples that started to gradually grow, forming the Basotho ethnic group (the indigenous inhabitants of Lesotho) in the early seventeenth century.

5 The annual precipitation at the source of the Orange River in the Lesotho Highlands to the east reaches as much as 2,000 mm, topping the annual evaporation rate of 1,200 mm (Heyns, 2004). It is estimated that the annual precipitation at the River's mouth decreases to less than 50 mm, while the annual evaporation increases to 3,000 mm, exceeding the limited rainfall in some areas (DWAF, 2005; Earle *et al.*, 2005). However, as the River flows westward, precipitation decreases below 100 mm, while evaporation increases.

6 Lesotho receives the highest amount of rainfall ranging from 575 to 1,040 mm annually (755 mm per year), totaling twice as much as the MAP of the entire basin, whereas South Africa receives slightly more than the MAP (365 mm). In contradistinction, the groundwater resources and aquifer yields are generally low in Lesotho.

Poor groundwater supply in South Africa, contributing only 15 percent of the country's annual water budget, is also to blame for the water scarcity faced by the country (Scudder, T. [2005]. The Lesotho Highlands Water Project [2003] and Laos' Nam Theun Dam. Unpublished manuscript).

7 Located in the Free State province in South Africa (Turton, 2003), the Gariep Dam (which was named after the original name of the River and considered the main storage structure along the Orange River) is considered the largest storage reservoir in South Africa, standing nearly 88 metres high and containing approximately 1.73 MCM of concrete. With a total storage of approximately 5,500 MCM and a surface area of more than 370 km^2 at full capacity, the dam delivers water at a flow rate of 800 m^3/s. Water is then diverted into two directions. The first direction is westward along the Orange River to the Vanderkloof Dam using hydro-electric power generators, and the second is southward to the Eastern Cape via the Orange-Fish Tunnel.

8 The Vanderkloof Dam, located approximately 130 km downstream from the Gariep Dam, is the second largest dam with the highest dam wall (of about 108 metres) in South Africa and a total capacity of about 3,237 MCM (FAO-Aquastat, 2007). Together, both of these dams are used to manage the downstream flow in the lower Orange River, forming major storage reservoirs along the River path. Water is stored and transferred from these dams for irrigation and hydropower generation in the Eastern Cape Province and the city of Port Elizabeth. Other notable dams are the Katse Dam, Mohale Dam, Vaal Dam, Vanderkloof Dam, Sterkfontein Dam, Pongolapoort Dam, Bloemhof Dam, Welbedacht Dam, Hardup Dam, and Naute Dam. Combined, both the Katse (the highest dam in Africa) and the Mohale dams are the key components of the LHWP (Turton, 2003).

9 For example, 76 percent of households in the central mountains and 57 percent of households in the eastern mountains lacked access to clean water in 1999 (Turner, 2003).

10 These four rivers include the Orange, Limpopo, Incomati, and Maputo and cover a significant part (approximately 60 percent) of South Africa's total land area (Turton, 2003).

11 South Africa transfers water from the Vaal to the Crocodile (part of the Limpopo) to elevate the extreme water demand imposed by these development needs (Basson, 1999). Similarly, groundwater is limited and extensively utilized for domestic and agricultural uses.

12 More than 25 percent of South Africa's GDP and about 85 percent of its electricity is produced in the Vaal River basin and the Gauteng region respectively (Turton, 2003; Earle *et al.*, 2005).

13 Product of the Transboundary Freshwater Dispute Database, Department of Geosciences, Oregon State University. Additional information about TFDD can be found at: www.transboundarywaters.orst.edu

14 Product of the Transboundary Freshwater Dispute Database, Department of Geosciences, Oregon State University. Additional information about TFDD can be found at: www.transboundarywaters.orst.edu

15 This drought, which was particularly long and threatened the Vaal Dam to run dry, forced the authorities to impose strict water use policy to promote water conservation and reduce wasteful practices in domestic and industrial water uses (Mirumachi, 2007; Lang, 2008).

16 Founded in 1910, the Southern African Customs Union (SACU) is an African regional economic organization that includes Botswana, Lesotho, Namibia, South Africa, and Swaziland.

17 The term "significant harm" (or "appreciable harm") refers to any cost encountered by basin states due to denial of water rights (Goldberg, 1992). Additionally, according to the International Law Association (2003), the term "damage" is generally defined as inclusive of the loss of life or personal injury; loss of or injury to property or other

economic losses; environmental harm; and the costs of reasonable measures to pre-vent or minimize such loss, injury, or harm. "Environmental harm" includes injury to the environment and any other loss or damage caused by such harm; and the costs of reasonable measures to restore the environment actually undertaken or to be undertaken (International Law Association, 2003).

18 These bilateral agreements were achieved between South Africa and one of the other weaker co-riparians.

19 Compensation covers (1) the loss of physical assets; (2) the loss of agricultural resources; and (3) the loss of community resources (EIB, 2002).

20 According to Whann (1995), many questioned the validity of the Treaty as King Moshoeshoe, who in the wake of the coup was given all executive and legislative authority by the Military Council, did not sign the treaty and instead Colonel Letsie did.

21 For example, the 465.5 m long and 85 m high bridge over the Malibamatso River, as part of the 120 km long access road, was constructed through the rugged terrain from Pitseng to the Pelaneng and then to the Katse Dam area (Wallis, 1996; Fullalove, 1997).

22 About 90 percent of the country's electricity came from South Africa (Poivey, 1997).

23 The former CEO of the LHDA was accused of receiving bribes from 12 multinational corporations contracted under Phases IA and IB of the project, over the course of 10 years.

24 This alternative scheme, which was more expensive than the LHWP, involves col-lecting water eventually draining into South African territories and pumping it back up to the Vaal River basin where it is most needed.

25 This includes costs related to the construction, operation and maintenance of the Muela hydropower plant, which constitutes 5 percent of the project total cost (Mirumachi, 2007). Additional projects, such as development projects, compensation of affected populating, and environmental conservation programs to be implemented in Lesotho are also the responsibility of Lesotho.

26 *Pari-passu* means equal footing and without preference to any particular party in asset and financial allocation.

27 Refer to Table 5.9 which shows the scheduled water deliveries and associated royalty payments.

28 It is also to the detriment of the environment that no comprehensive environmental impact assessment was conducted for Phase 1A (Hildyard, 2000).

6 Comparative cross-case analysis, hydro-political implications, and lessons learned

1 The notion of water satiety and its implications will be discussed in more detail later in this chapter.

2 This is discussed in the "sharing not dividing" section of this chapter, which discusses five levels of water allocation ranging from level one (representing dividing) to level five (representing sharing).

3 Indigenous approaches are the locally practiced conflict reconciliation methods of water dispute resolution by the local populations who inhabit arid regions throughout the world (Wolf, 2000). These are an unspoken, yet well known, set of principles that govern the societal interaction and community discourse, particularly in resolv-ing water disputes, and can emanate from the local cultural or religious practices (Abukhater, 2009b).

4 This strategy was used in the India–Bangladesh Treaty over the Ganges.

5 By focusing on times of drought and putting in place a contingency plan, the parties can develop a strategy to guide their joint effort to effectively manage available water during these drought periods.

6 The riparian doctrine is based on the principle that riparians bordering a naturally flowing river or stream have a right to use its water. Further, all rights to water use by a riparian state depend upon the equal, correlative rights of other riparians to the use of the common resource (Commonwealth of Pennsylvania, Department of Environmental Protection, 2006).

7 In the prior appropriation doctrine, alternatively known as the historic rights doctrine, water rights are awarded based on seniority or first in time, first in right.

8 Refer to the logic model, shown in Figure 2.1 in the methodology chapter.

9 These were identified as important parameters of process equity and further discussed in detail earlier in this chapter, such as context-sensitive equity and adaptive learning and management, modern and indigenous approaches to conflict resolution, and goodwill/faith-based confidence building measures (CBMs).

10 As described by Wade (2004: 420), the "duelling expert" syndrome involves certain characteristics related to parties' and experts' behaviors, such as each party hires its own expert ("ours is the best in the field"), who often times is not impartial to the party's preferred outcome ("reputational partiality"); communicates different stories to its own expert ("garbage in, garbage out"); makes explicit or implicit reference to particular desired advice it wants from its expert ("remember who is paying you"). Behavioral characteristics of the "duelling experts" include unwillingness to consult with each other ("delusionary isolation"); tendency to tell their clients what they want to hear ("you get what you pay for"); unwillingness to provide a clear list of practical alternatives ("delusionary certainty"); unwillingness to share draft reports ("no early doubts or compromises"); and tendency to produce long and ambiguous reports ("mysterious complexity").

11 Although directive mediation might not be suitable for certain conflicts, such as the Arab–Israeli conflict, cultural dialog and facilitative mediation can be employed as part of a consensus-based mediation process.

7 Conclusion and future research

1 The diplomatic and political capital includes international relations, genuine communication, trust building, joint problem solving, conflict prevention, strategic gain and political stability, and, in turn, regional peace.

2 The problem lies, for the most part, in the disconnect between reaching an agreement and its implementation.

Appendix A

1 To enhance its efficiency and validity, this future research will particularly be concerned with acquiring accurate data regarding water availability and consumption in each of the key cases. In illustrating this point, Amaratunga and colleagues (2002) state that "there is a strong suggestion within the research community that research, both quantitative and qualitative, is best thought of as complementary and should therefore be mixed in research of many kinds." The goal is to quantify outcome equity standards by relying on multiple criteria evaluation (MCE) methods and the triangulation of various typologies and measurements developed by the literature and the researcher. This potential research strategy would constitute a long-term research agenda that could be conducted in future research venues and would use a significant number of cases.

References

Abdalla, A. and Eldaw, A. (2002). Changes and Future Opportunities in the Nile Basin. In From Conflict to Co-operation in International Water Resources Management: Challenges and Opportunities, International Conference, November 20–22, 2002. Delft, the Netherlands: UNESCO-IHE Institute for Water Education.

Abukhater, A. (2006). Water for Peace, *Media Monitors Network*, August 5. Retrieved June 4, 2013, from: http://canada.mediamonitors.net/layout/set/print/content/view/full/33707

Abukhater, A. (2009a). Fostering Citizen Participation, GIS Development, *Neogeography & Participatory GIS: Power to People*, 14 (2), pp. 32–33.

Abukhater, A. (2009b). Through the Eyes of the Indigenous Population: Innovative and Comprehensive Approaches to Mediation and ADR in the Middle East, *Peace Conflict & Development Journal*, 14, pp. 1–28.

Abukhater, A. (2009c). Rethinking Planning Theory and Practice: A Glimmer of Light for Prospects of Integrated Planning to Combat Complex Urban Realities, *Journal of Theoretical and Empirical Researches in Urban Management*, 2 (11), pp. 64–79.

Abukhater, A. (2010a). On the Cusp of Water War: A Diagnostic Account of the Volatile |Geopolitics of the Middle East, *Peace and Conflict Studies Journal*, 17 (2), pp. 378–419.

Abukhater, A. (2011a). Planning 2.0: A Collaborative Platform for Actionable Intelligence, *Next City Magazine*. Retrieved April 10, 2012, from: http://nextcity.org/daily/entry/planning-2.0-a-collaborative-platform-for-actionable-intelligence

Abukhater, A. (2011b). GIS for Planning and Community Development: Solving Global Challenges, *Directions Magazine*, January 2.

Abukhater, A. (2012a). Using Location Intelligence to Create and Enable a Smarter Enterprise, *Engage Today*, August 28.

Abukhater, A. (2012b). Harnessing the Power of Your Data Through Enterprise Location Intelligence, *Engage Today*, August 14

Alatout, S. (2000). "Water Balances in Palestine: Numbers and Political Cultures in the Middle East." In D. Brooks and O. Mehmet (eds.), *Water Balances in the Eastern Mediterranean*. Ottawa: IDRC.

Alatout, S. (2003). Imagining Hydrological Boundaries, Constructing the Nation State: A "Fluid" History of Israel, 1936–1959. PhD Dissertation, Cornell University.

Alatout, S. (2006). Towards a Bio-Territorial Conception of Power: Territory, Population, and Environmental Narratives in Palestine and Israel, *Political Geography*, 25, pp. 60–621.

Aliewi, A., *et al.* (2003). Legal Framework for shared Groundwater Resources Development and Management in Palestine. A country paper presented to workshop of legal framework for shared groundwater development and management in the ESCWA region, Beirut, Lebanon, June 10–13, 2003.

Aliewi, A., Mimi, Z., and Sawalhi, B. (2001). Multi-Criteria Decision Tool for Allocating the Waters of the Jordan Basin between all Riparians. In Proceedings of the Globalization and Water Resources Management: The Changing Value of Water. AWRA/IWLRI: University of Dundee International Specialty Conference, August 6–8.

Allan, T. (2001). *The Middle East Water Question: Hydropolitics and the Global Economy.* London: I.B. Tauris.

Allouche, J. (2005). Water Nationalism: an Explanation of the Past and Present Conflicts in Central Asia, the Middle East, and the Indian Subcontinent? PhD Dissertation, Université de Genève.

Amaratunga, D., Baldry, D., Sarshar, M. and Newton, R. (2002). Quantitative and Qualitative Research in the Built Environment: Application of "Mixed" Research Approach, *Work Study,* 51 (1), pp. 17–31.

Ashton, P. (2000). "Southern African Water Conflicts: Are they Inevitable or Preventable?" In *Green Cross International, Water for Peace in the Middle East and Southern Africa.* Geneva: Green Cross International, pp. 94–98.

Balkind, E. and Meir, B. (1996). *Summary of telephone conversation between Balkind and Ben Meir,* Assistant to the Minister of National Infrastructures, 21 October.

Basson, M. S. (1999). South Africa Country Paper on Shared Watercourse Systems. South African Department of Water Affairs and Forestry, presented at the SADC water week workshop in Pretoria, South Africa, September 16, 1999.

BBC News (1999). World: Middle East – Drought "forces Israel to break treaty." Retrieved April 10, 2009, from: www.news.bbc.co.uk/2/hi/middle_east/296797.stm

Beach, H., Hammer, J., Hewitt, J., Kaufman, E., Kurki, A., Oppenheimer, J. A. and Wolf, A. (2000). *Transboundary Freshwater Dispute Resolution: Theory, Practice and Annotated References.* Tokyo: United Nations University Press.

Beaumont, P. (1997). Dividing the Waters of the River Jordan: An Analysis of the 1994 Israel–Jordan Peace Treaty, *Water Resources Development,* 13 (3), pp. 415–424.

Berman, I. and Wihbey, P. (1999). *The New Water Politics of the Middle East* [electronic version]. Division for Research in Strategy Research and Analysis, Institute for Advanced Strategic and Political Studies. Retrieved July 3, 2009, from: www.iasps.org/strategic/water.htm

Beyene, Z. and Wadley, I. L. (2004). Common Goods and the Common Good: Transboundary Natural Resources, Principled Cooperation and the Nile Basin Initiative. In Proceedings of Breslauer Symposium on Natural Resource Issues in Africa, March 5, 2004. Berkeley, CA: University of California.

Beyth, M. (2006). "Water Crisis in Israel." In M. Leybourne, and A. Gaynor (eds.), *Water: Histories, Cultures, Ecologies.* Crawley, WA: UWA Press.

Bhatti, H. (1999). *Pakistan's Accommodative Moves vis-à-vis India: A Case Study of the Dynamics of Accommodation in the Developing World.* Montreal, QC: Department of Political Science, McGill University.

Biliouri, D. (1997). Environmental Issues as Potential Threats to Security. Paper presented to the 38th Annual Convention of the International Studies Association, Toronto, March 18–22, 1997.

Biswas, A. K. and Bino, M. J. (2001). In Conversation, IDRC Reports. Retrieved January 10, 2009, from: www.idrc.ca/en/ev-43225-201-1-DO_TOPIC.html#Amman

Blaikie, P. and Muldavin, J. (2004). Upstream, Downstream, China, India: The Politics of Environment in the Himalayan Region, *Annals of the Association of American Geographers,* 94 (3), pp. 520–548.

Brams, S. J. and Taylor, A. D. (1996). *Fair Division: From Cake-cutting to Dispute Resolution.* Cambridge: Cambridge University Press.

Brown, C. (2008). Let It Flow: Lessons from Lesotho, *International Rivers*, 23 (2), p. 6.

Bureau of African Affairs (2008). Background Note: Lesotho and South Africa. US Department of State. Retrieved July 29, 2008, from: www.state.gov/r/pa/ei/bgn/2831. htm

Bureau of Statistics, Ministry of Finance and Development Planning (2007). *2006 Lesotho Census of Population and Housing: Preliminary Results Report*. Maseru: Bureau of Statistics.

Burgess, H. and Burgess, G. (2007). Constructive Confrontation: A Transformative Approach to Intractable Conflicts, *Conflict Resolution Quarterly*, 13 (4), pp. 305–322.

Caponera, D. A. (1985). Patterns of Cooperation in International Water Law: Principles and Institutions, *Natural Resources Journal*, 25, pp. 563–588.

Cascao, A. and Zeiton, M. (2010). "Changing Nature of Bargaining Power in the Hydropolitical Relations in the Nile River Basin." In A. Earle, A. Jägerskog, and J. Öjendal (eds.), *Transboundary Water Management: Principles and Practice*. Washington, DC: Earthscan.

Chakela, Q. K. (ed.) (1999). *State of the Environment in Lesotho 1997*. Maseru: National Environment Secretariat, Ministry of Environment, Gender and Youth Affairs.

Chileshe, P., Trottier, J., and Wilson, L. (2005). Translation of Water Rights and Water Management in Zambia. International workshop on "African Water Laws: Plural Legislative Frameworks for Rural Water Management in Africa," Johannesburg, South Africa. Newcastle: University of Newcastle.

Chimni, B. S. (2005). "A Tale of Two Treaties: The Ganga and Mahakali Agreements and the Watercourses Convention." In Suvedī Sūryaprasāda (ed.), *International Watercourses Law for the 21st Century: The Case of the River Ganges Basin*. Burlington, VT: Ashgate Publishing, Ltd.

CIA (Central Intelligence Agency) (2008). *The World Factbook* [electronic version]. Retrieved July 29, 2008, from: www.cia.gov/library/publications/the-world-factbook/countrylisting.html

Ciccolo, A. (1992). Environmental Responsibility under Severe Economic Constraints: Re-Examining the Lesotho Highlands Water Project, *Georgetown International Environmental Law Review*, 4, pp. 447–467.

Cohen, R. L. (1987). Distributive Justice: Theory and Research, *Social Justice Research*, 1 (1), pp. 19–40.

Coman, K. (1911). Some Unsettled Problems of Irrigation, *American Economic Review*, 1 (1), pp. 1–19 [reprinted in *American Economic Review*, 101 (1), pp. 36–48].

Commonwealth of Pennsylvania, Department of Environmental Protection (2006). *Can I Use Water From This Stream? Riparian Doctrine Fact Sheet*. Retrieved December 21, 2005, from: www.dep.state.pa.us/dep/subject/hotopics/drought/facts/RipDoc_factsheet. htm

Conley, A. and van Niekerk, P. (1998). "Sustainable Management of International Waters: The Orange River Case." In H. Savenije and P. van der Zaag (eds.), *The Management of Shared River Basins: Experiences from SADC and EU*. Delft, the Netherlands: Netherlands Ministry of Foreign Affairs.

Conley, A. H. and Niekerk, P. H. (2000). Sustainable Management of International Waters: The Orange River Case, *Water Policy*, 2, pp. 131–149.

Creswell, J. W. (2003). *Research Design: Qualitative, Quantitative, and Mixed Methods Approaches*, 2nd edn. Thousand Oaks, CA: Sage Publications.

Crow, B. (2008) (Associate Professor, Department of Sociology, University of California at Santa Cruz) personal phone interview by author, Austin, TX, October 31, 2008.

CSIS (Center for Strategic and International Studies) (2003). *U.S.–Mexico Transboundary Water Management.* Washington, DC: Center for Stategic and International Studies.

Daoudy, M. (2008). A Missed Chance for Peace: Israel and Syria's Negotiations Over the Golan Heights, *Journal of International Affairs,* 61 (2), pp. 215–234.

Daoudy, M. (2009) (Lecturer, Department of Political Science, the Graduate Institute of International Studies, Geneva) personal phone interview by author, Austin, TX, January 29, 2009.

Datta, A. (2005). "The Bangladesh–India Treaty on Sharing of the Ganges Water: Potentials and Challenges." In Suvedī Sūryaprasāda (ed.), *International Watercourses Law for the 21st Century: The Case of the River Ganges Basin.* Burlington, VT: Ashgate Publishing, Ltd, pp. 63–104.

Davies, J. C. (1962). Toward a Theory of Revolution, *American Sociological Review,* 27 (1), pp. 5–19.

Davis, R. and Hirji, R. (2003). *Water Resources and Environment, Environmental Flows: Case Studies.* Washington, DC: The International Bank for Reconstruction and Development, The World Bank.

De Châtel, F. (2007). "Perceptions of Water in the Middle East: the Role of Religion, Politics and Technology in Concealing the Growing Water Scarcity." In H. Shuval, and H. Dweik (eds.), *Water Resources in the Middle East.* New York, NY: Springer.

Dellapenna, J. W. (2002). Markets-Ethics-Law: What can each Contribute? In Conference Proceedings: From Conflict to Co-operation in International Water Resources Management: Challenges and Opportunities, International Conference, November 20–22, 2002. Delft, Netherlands: UNESCO-IHE Institute for Water Education.

Department of Water Affairs, Republic of South Africa (2012). *Proposed National Water Resources Strategy 2 (NWRS2)* [electronic version]. Retrieved October 12, 2012, from: www.dwaf.gov.za/nwrs/NWRS2012.aspx

Deutsch, M. (2000). "Justice and Conflict." In M. Deutsch, and P. Coleman (eds.), *The Handbook of Conflict Resolution: Theory and Practice.* San Francisco, CA: Jossey-Bass Publishers, Inc.

Diabes-Murad, F. (2004). *Water Resources in Palestine: A Fact Sheet and Basic Analysis of Legal Status.* Dundee: International Water Law Research Institute, Dundee University.

Dinar, S. (2009). Water Use Trends and Infrastructure Needs in the Jordan River Watershed. In Transboundary Water Crises conference, New Mexico State University, January 22, 2009.

Dolatyar, M. and Gray, T. S. (2000). *Water Politics in the Middle East.* London: Macmillan Press Ltd.

DWAF (Department of Water and Forestry), Republic of South Africa (2005). *Orange River Project* [electronic version]. Retrieved June 23, 2012, from: www.dwaf.gov.za/orange/

Earle, A., Malzbender, D., Turton, A., and Manzungu, E. (2005). *A Preliminary Basin Profile of the Orange/Senqu River.* Pretoria: African Water Issues Research Unit, University of Pretoria.

Earthtrends (2008). *Water Resources eAtlas: Orange River.* Retrieved July 29, 2008, from: www.earthtrends.wri.org/maps_spatial/watersheds/africa.php

Eckstein, G. (1995). Application of International Water Law to Transboundary Groundwater Water Resources and the Slovak–Hungarian Dispute over Gabcikovo-Nagymaros, *Suffolk Transnational Law Review,* 19, 67–116.

Eckstein, G. (2002). "Development of International Water Law and the UN Watercourse Convention." In A. Turton and R. Henwood (eds.), *Hydropolitics in the Developing World: A Southern African Perspective.* Pretoria: African Water Issues Research Unit, University of Pretoria.

Eckstein, H. (1975). "Case Study and Theory in Political Science." In F. I. Greenstein and N. W. Poisby (eds.), *Handbook of Political Science, Vol. VII.* Boston, MA: Addison-Wesley.

EBC (Educational Broadcasting Corporation) (2007). *The Demand: Five Controversial Dams – Lesotho and South Africa*, Wide Angle, PBS. Retrieved from: www.pbs.org/wnet/wideangle/episodes/the-damned/five-controversial-dams/lesothosouth-africa/3106/

EIB (European Investment Bank) (2002). *Lesotho Highlands Water Project* [electronic version]. Retrieved July 24, 2008, from: www.eib.org/projects/news/lesotho-highlands-water-project.htm

El-Fadel, M., Quba'a, R., El-Hougeiri, N., Hashisha, Z., and Jamall, D. (2001). The Israeli Palestinian Mountain Aquifer: A Case Study in Ground Water Conflict Resolution, *Journal of Natural Resources and Life Science Education*, 30, pp. 50–61.

El-Fadel, M., El Sayegh, Y., Abou Ibrahim, A., Jamali, D., and El-Fadl, K. (2002). The Euphrates–Tigris Basin: A Case Study in Surface Water Conflict Resolution, *Journal of Natural Resources and Life Science Education*, 31, pp. 99–110.

Elhance, A. (1999). *Hydropolitics in the Third World: Conflict and Cooperation in International River Basins.* Washington, DC: US Institute of Peace Press.

Elmusa, S. (1995). The Jordan–Israel Water Agreement: A Model or an Exception? *Journal of Palestine Studies*, 24 (3), pp. 63–73.

Elmusa, S. (1998). "Toward a Unified Management Regime in the Jordan Basin: The Johnston Plan Revisited." In J. Coppock and J. A. Miller (series eds.), *Transformations of Middle Eastern Natural Environments: Legacies and Lessons.* Bulletin Series, Number 103, Council on Middle East Studies, Yale Center for International and Area Studies, and the Yale School of Forestry and Environmental Studies. New Haven, CT: Yale University Press.

Esterhuysen, P. (1992). *Africa at a Glance.* Pretoria: Africa Institute.

EXACT (Executive Action Team) (2005a). *Surface Water: Upper Jordan River.* Multilateral Working Group on Water Resources. Retrieved January 10, 2009, from: www.exact-me.org/overview/

EXACT (Executive Action Team) (2005b). *Water RainCatcher—A Rain Harvesting Pilot Project.* Multilateral Working Group on Water Resources, Middle East Peace Process. Retrieved June 4, 2013, from: http://exact-me.org

Falkenmark, M. (1989). The Massive Water Shortage in Africa – Why Isn't It Being Addressed? *Ambio*, 18 (2), pp. 112–118.

FAO (Food and Agriculture Organization) (1997). *Irrigation Potential in Africa: A Basin Approach* [electronic version]. Retrieved June 30, 2012, from: www.fao.org/docrep/W4347E/w4347e00.htm#Contents

FAO-Aquastat (1995). *FAO's Information System on Water and Agriculture: Lesotho* [electronic version]. Retrieved June 30, 2012, from: www.fao.org/ag/agl/aglw/aquastat/countries/lesotho/index.stm

FAO-Aquastat (2002). Water Availability per Person per Year. Retrieved June 2, 2013, from: www.fao.org/nr/water/aquastat/data/query/index.html

FAO-Aquastat (2007). *Geo-Referenced Database on African Dams.* Water Development and Management Unit (NRLW), Land and Water Division (NRL), and Food and Agriculture Organization of the United Nations (FAO). Retrieved June 30, 2008, from: www.fao.org/ag/agl/aglw/aquastat/countries/lesotho/index.stm

FAO-Aquastat (2008). *FAO's Information System on Water and Agriculture.* Retrieved January 30, 2009, from: www.fao.org/nr/water/aquastat/countries/israel/index.stm

Faurès, J. and Santin, G. (eds.) (2008). *Water and the Rural Poor: Interventions for Improving Livelihoods in sub-Saharan Africa.* Rome: Food and Agriculture Organization.

Fischhendler, I. (2008a). When Ambiguity in Treaty Design Becomes Destructive: A Study of Transboundary Water, *Global Environmental Politics*, 8 (1), pp. 111–136.

Fischhendler, I. (2008b). Ambiguity in Transboundary Environmental Dispute Resolution: The Israeli–Jordanian Water Agreement, *Journal of Peace Research*, 45 (1), pp. 79–97.

Fischhendler, I. and Feitelson, E. (2004). The Short-Term and Long-Term Ramifications of Linkages Involving Natural Resources: the US–Mexico Transboundary Water Case, *Environment and Planning C: Government and Policy*, 22, pp. 633–650.

Fisher, F. (2006). "Water: Casus Belli or Source of Cooperation?" In K. Hambright, F. Ragep, and J. Ginat (eds.), *Water in the Middle East: Cooperation and Technological Solutions in the Jordan Valley*. Brighton: Sussex Academic Press.

Fisher, F. and Huber-Lee, A. (2005). *Liquid Assets: An Economic Approach for Water Management and Conflict Resolution in the Middle East and Beyond*. Washington, DC: RFF Press.

Frey, F. and Naff, T. (1985). Water: An Emerging Issue in the Middle East? *The Annals of the American Academy of Political and Social Science*, 482, pp. 65–84.

Fullalove, S. K. (1997). Lesotho Highlands Water Project, *Proceedings of the Institution of Civil Engineers, Vol. 120, Special Issue 1*. London: Thomas Telford.

Fuwa, Y. (2003). Natural Resources Management from a Conflict Prevention Perspective, *JBICI Review*, 8, pp. 35–58.

Geddes, B. (1990). How the Cases You Choose Affect the Answers You Get: Selection Bias in Comparative Politics, *Political Analysis*, 2, pp. 131–152.

Gibson-Graham, J. K. (2006). *A Postcapitalist Politics*. Minneapolis, MN: University of Minnesota Press.

Giordano, M. and Wolf, A. (2001). Incorporating Equity into International Water Agreements, *Social Justice Research*, 14 (4), pp. 349–366.

Glassner, M. and Fahrer, C. (2004). *Political Geography*, 3rd edn. New York: John Wiley & Sons, Inc.

Gleditsch, N., *et al.* (2007). "Conflicts in Shared River Basins." In I. V. Grover (ed.), *Water: A Source of Conflict or Cooperation?* Hamilton, ON: Natural Resource Consultant.

Gleick, P. H. (1992). "Effects of Climate Change on Shared Fresh Water Resources." In I. M. Mintzer (ed.), *Confronting Climate Change: Risks, Implications and Responses*. Cambridge: Cambridge University Press.

Gleick, P. H. (ed.) (1993a). *Water in Crisis: A Guide to the World's Fresh Water Resources*. New York: Oxford University Press.

Gleick, P. H. (1993b). Water and Conflict: Fresh Water Resources and International Security, *International Security*, 18 (1), pp. 79–112.

Gleick, P. H. (1994). Water, War, and Peace, in the Middle East, *Environment*, 36 (3), pp. 6–42.

Gleick, P. H. (1998). The Human Right to Water, *Water Policy*, 1, pp. 487–503

Gleick, P. H. (2009). Are Water Wars a Myth or an Imminent Threat to Global Security? Panel discussion, seven experts debate the past and present existence of water wars, consider the difficulty of owning a fluid resource, and examine the hot spots for future conflict, *Seed Magazine*, May 14.

GLOWA Jordan River (2008). *Introduction to GLOWA JR Project*. Retrieved August 30, 2008, from: www.glowa-jordan-river.de

Goeller, S. (1997). *ICE Case Studies, TED Conflict Studies: Water and Conflict in the Gaza Strip*. Retrieved February 20, 2008, from: www1.american.edu/TED/ice/GAZA.HTM

Goldberg, D. (1992). *Projects on International Waterways: Legal Aspects of the Bank's Policy*. Washington, DC: World Bank.

Goldreich, Y. (2003). *The Climate of Israel: Observation, Research, and Application*. New York: Springer.

Grover, V. (2007). *Water, A Source of Conflict or Cooperation*. Enfield, NH: Science Publishers.

Gurr, T. R., Marshall, M. G., and Khosla D. (2001).*Conflict and Peace*. College Park, MD: University of Maryland.

Haaretz (1995). A Message from Amman to Israel: the Disappointment from the Peace Treaty in Jordan Grows, *Haaretz*, June [Hebrew].

Haaretz (1997). Distorted Agreement with Jordan, *Haaretz*, June 13 [Hebrew].

Haddadin, M. J. (2000). Negotiated Resolution of the Jordan–Israel Water Conflict, *International Negotiation*, 5, pp. 263–288.

Haddadin, M. J. (2002a). *Diplomacy on the Jordan: International Conflict and Negotiated Resolution*. Norwell, MA: Kluwer Academic Publishers.

Haddadin, M. J. (2002b). Water in the Middle East Peace Process, *The Geographical Journal*, 168 (4), pp. 324–340.

Haddadin, M. J. (2003). *The Jordan River Basin: Water Conflict and Negotiated Resolution, Potential Conflict to Cooperation Potential*. Paris: UNESCO.

Haddadin, M. J. (2006). *Water Resources in Jordan: Evolving Policies for Development, the Environment, and Conflict Resolution*. Washington, DC: Resources for the Future.

Haddadin, M. J. (2008) (former water minister of Jordan and Head of the Jordanian Negotiation Team in the Israel–Jordan Peace Talks) face-to-face personal interview by author, Dallas, TX, October 26.

Haftendorn, H. (2000). Water and International Conflict, *Third World Quarterly*, 21 (1), pp. 51–68.

Hamner, J. H. and Wolf, A. T. (1998). *Trends in Transboundary Water-Disputes and Dispute Resolution*. Corvalis, OR: Oregon State University.

Hardin, G. (1968). The Tragedy of the Commons, *Science*, 162, pp. 1243–1244.

Hassan, F. M. A. (2002). *Lesotho: Development in a Challenging Environment. A Joint World Bank–African Development Bank Evaluation*. Washington, DC: The World Bank.

Heyns, P. (2004). Achievements of the Orange-Senqu River Commission in Integrated Transboundary Water Resource Management. Paper presented at the General Assembly of the International Network of Basin Organisations, January 24–28, 2004.

Heyns, P., Patrick, M., and Turton, A. (2008). Transboundary Water Resource Management in Southern Africa: Meeting the Challenge of Joint Planning and Management in the Orange River Basin, *Water Resources Development*, 24 (3), pp. 371–383.

Hildyard, N. (2000). *LHWP: What went Wrong?* [electronic version]. Retrieved July 29, 2009, from: www.probeinternational.org/odious-debts/lhwp-what-went-wrong

Hiniker, M. (1999). *Sustainable Solutions to Water Conflicts in the Jordan Valley*. Geneva: Green Cross International.

Hirji, R. and Grey, D. (1998). "Managing International Waters in Africa: Process and Progress." In S. M. Salman and L. B. D. Chazournes (eds.), *International Watercourses Enhancing Cooperation and Managing Conflict*, World Bank Technology Paper no. 414. Washington, DC: World Bank.

Hitchcock, R., Inambao, A., Ledger, J., and Mentis, M. (2007). *The Lesotho Highlands Water Project: Report 46*. Lesotho: LHWP.

Hodgeson, S. (2004). *Land and Water – The Rights Interface*. FAO Legal Papers. Retreived February 2, 2006, from: www.fao.org/legal/prs-ol/lpo36.pdf

Housen-Couriel, D. (1994). *Some Examples of Cooperation in the Management and Use of International Water Resources*. Jerusalem: Hebrew University of Jerusalem, Truman Research Institute for the Advancement of Peace.

Howard, G. and Bartram, J. (2003). *Domestic Water Quantity, Service Level and Health.* Geneva: World Health Organization.

Hughes, S., Johnston, E., Bruce, G., Vogel, H., Hoffmann, P. Q., and Meinier, B. (2010). "Orange-Senqu River Awareness Kit: Supporting Capacity Development." In A. Earle, A. Jägerskog, and J. Öjendal (eds.), *Transboundary Water Management: Principles and Practice.* Washington, DC: Earthscan.

Hummel, D. (2006). Population Dynamics and Conflicts on Water Resources in the Jordan River Basin. In The Third Environmental Symposium of the German-Arab Society for Environmental Studies, Environmental Protection in the Middle East and North Africa. Frankfurt am Main, Germany September 18–19, 2006.

Ilter, K. (2000). Analysts Expect no Drastic Change in Turco–Syrian Relations, *Turkish Daily News,* June 12, 2000.

Innes, J. E. and Booher, D. E. (1999). Consensus Building and Complex Adaptive Systems: A Framework for Evaluating Collaborative Planning, *Journal of the American Planning Association,* 65 (4), pp. 412–423.

International Law Association (1967). The Helsinki Rules on the Uses of the Waters of International Rivers. Adopted by the International Law Association at the fifty-second conference, held at Helsinki in August 1966. London: Report of the Committee on the Uses of the Waters of International Rivers.

International Law Association (2003). *Helsinki Revision: Sources of the International Law Association Rules on Water Resources.* Berlin: International Law Association.

Irani, R. (1991). Water Wars, *New Statesman and Society,* 4 (149), pp. 24–25.

Isaac, J. (1994). *Core Issue of the Palestinian–Israeli Water Dispute.* Jerusalem: Applied Research Institute (ARIJ).

Isaac, J. (1999). *The Palestinian Water Crisis.* Washington, DC: The Jerusalem Fund for Education and Community Development.

Islam, S. and Susskind L. E. (2013). *Water Diplomacy: A Negotiated Approach to Managing Complex Water Networks.* New York: RFF Press/Routledge.

Israel Ministry of Foreign Affairs (1995). *The Bilateral Negotiations, Israel–Jordan Negotiations.* Retrieved February 24, 2009, from: www.mfa.gov.il/mfa/peace%20process/guide%20to%20the%20peace%20process/Israel–Jordan%20negotiations

Israel Ministry of Foreign Affairs (2000). *From Contention to Cooperation: A Case Study of the Middle East Multilateral Working Group on Water Resources.* Retrieved from: www.mfa.gov.il/mfa/foreignpolicy/peace/guide/pages/from%20contention%20to%20cooperation-%20a%20case%20study%20of%20th.aspx

The Israeli–Palestinian Interim Agreement on the West Bank and Gaza Strip (1995). Signed in Washington D.C., on September 28, 1995.

Jarvis, T. and Wolf, A. (2010). "Managing Water Negotiations and Conflicts in Concepts and in Practice." In A. Earle, A. Jägerskog, and J. Öjendal (eds.), *Transboundary Water Management: Principles and Practice.* Washington DC: Earthscan.

Jarvis, T., Giordano, M., Puri, S., Matsumoto, K., and Wolf, A. (2005). International Borders, Ground Water Flow, and Hydroschizophrenia, *Ground Water,* 43 (5), pp. 764–770.

Kally, E. (1993). *Water and Peace: Water Resources and the Arab–Israeli Peace Process.* Tel Aviv: Tel Aviv University.

Kerr, M. H. (1969). *Regional Arab Politics and the Conflict with Israel.* Washington, DC: Rand Corporation.

Kibaroglu, A. (2002). Settling the Dispute over the Water Resources in the Euphrates–Tigris River Basin. In Conference Proceedings: From Conflict to Co-operation in International Water Resources Management: Challenges and Opportunities, International Conference, November 20–22, 2002, Delft, the Netherlands: UNESCO-IHE Institute for Water Education.

Kibaroglu, A. and Unver, O. (2000). An Institutional Framework for Facilitating Cooperation in the Euphrates–Tigris River Basin, *International Negotiation: A Journal of Theory and Practice*, 5 (2), pp. 311–330.

Klare, M. T. (2001). "Water Conflict in the Jordan, Tigris–Euphrates, and Indus River Basins." In M.T. Klare (ed.), *Resource Wars: The New Landscape of Global Conflict*. New York: Owl Books.

Kliot, N. (1994). *Water Resources and Conflict in the Middle East*. London: Routledge.

Korf, B. and Funfgeld, H. (2005). War and the Commons: Assessing the Changing Politics of Violence, Access and Entitlements in Sri Lanka, *Geoforum*, 37, pp. 391–403.

Kuttab, J. and Ishaq, J. (2000). *Approaches to the Legal Aspects of the Conflict on Water Rights in Palestine/Israel*. Bethlehem, West Bank. Jerusalem, Israel: Applied Research Institute.

Kvale, S. (1996). *Interviews – An Introduction to Qualitative Research Interviewing*. London: Sage Publications.

Lang, S. (2008). *Unexpected Benefits of Lesotho Highlands Water Project*, Inter Press Service News Agency (IPS) [electronic version]. Retrieved June 17, 2009, from: www.ipsnews. net/africa/nota.asp?idnews=41005

Lascurain, C. F. (2001). The Performance of the Mexico–U.S Environmental Regime: Managing the Water of the Rio Grande and the Colorado River, Unpublished Doctoral Dissertation, University of Essex, UK.

Lascurain, C. F. (2002). Conflict and Cooperation: Managing the Water of the Rio Grande and the Colorado River at the Mexico–US Border. In Conference Proceedings: From Conflict to Co-operation in International Water Resources Management: Challenges and Opportunities, International Conference, November 20–22, 2002, Delft, the Netherlands: UNESCO-IHE Institute for Water Education.

Lautze, J. and Giordano, M. (2006). Equity in Transboundary Water Law: Valuable Paradigm or Merely Semantics? *Colorado Journal of International Environmental Law and Policy*, 17 (1), pp. 89–122.

Leach, M., Mearns, R., and Scoones, I. (1999). Environmental Entitlements: Dynamics and Institutions in Community-Based Natural Resource Management, *World Development*, 27 (2), pp. 225–247.

Leboela, T. and Turner, S. D. (2003). *The Voice of the People: Report on Community Consultations for the National Vision and the Poverty Reduction Strategy Paper*. Maseru: Ministry of Finance and Development Planning.

LeMarquand, D. (1977). *International Rivers: The Politics of Cooperation*. Vancouver, BC: Westwater Research Centre.

Libiszewski, S. (1995). *Water Disputes in the Jordan Basin Region and their Role in the Resolution of the Arab Israeli Conflict*. Zurich: Center for Security Studies and Conflict Research.

Lillian Goldman Law Library (2008). *The Israel–Jordan Common Agenda*, September 14, 1993, Washington, DC. Retrieved January 24, 2009, from: www.avalon.law.yale. edu/20th_century/pal07.asp

Ling, M. L., Jr. Letter from a Birmingham Jail. Retrieved May 27, 2013, from: http://www. africa.upenn.edu/Articles_Gen/Letter_Birmingham.html

Lipchin, C. D. (2003). *Public Perceptions and Attitudes Toward Water Use in Israel: A Multilevel Analysis*. Ann Arbor, MI: Department of Natural Resources and Environment, University of Michigan.

Lipchin, C. D., Antonius, R., Rishmawi, K., Afanah, A., Orthofer, R., and Trottier, J. (2005). "Public Perceptions and Attitudes toward the Declining Water Level in the Dead Sea Basin: a Multi Cultural-Analysis." In S. Schoenfeld (ed.), *Palestinian and Israeli Environmental Narratives*. Toronto: Centre for International and Security Studies, York University, pp. 263–300.

Lipper, J. (1967). "Equitable Utilization." In Garretson *et al.* (eds.), *The Law of International Drainage Basins*. Polisher: International Water Law Project, IWLP, pp. 104–179.

LHDA (Lesotho Highlands Development Authority) (2002). *Final Report: Summary of Main Findings for Phase 1 Development. Kingdom of Lesotho*. Lesotho: LHDA, Metsi Consultants.

LHDA (2003). *The Ombudsman Listens to Grievances against the LHDA: Press Statement by the Lesotho Highlands Development Authority (LHDA) on Concerns tabled before the Ombudsman and background information on the compensation policy and the Project* [electronic version]. Retrieved June 2, 2009, from: www.lhda.org.ls/news/archive2003/may03/press22052003b.htm

LHWP (Lesotho Highlands Water Project) (1986). *Treaty Between the Government of the Kingdom of Lesotho and the Government of the Republic of South Africa*, signed at Maseru, October 24, 1986.

LHWP (2009). *Overview of the Lesotho Highlands Water Project* (LHWP) [electronic version]. Retrieved February 13, 2009, from: www.lhwp.org.ls

Louka E. (2006). *International Environmental Law: Fairness, Effectiveness, and World Order*. Cambridge: Cambridge University Press.

Lowi, M. R. (1991). West Bank Water Resources and the Resolution of Conflict in the Middle East. Paper presented for the project on Environmental Change and Acute Conflict, June 15–17, 1991.

Lowi, M. R. (1995). *Water and Power: the Politics of a Scarce Resource in the Jordan River Valley*. Cambridge: Cambridge University Press.

McDermott, T. (2009). Water, Economics and the Jordan Valley. In Transboundary Water Crises conference, New Mexico State University, January 22.

Mckenzie, R. (2009) (a technical advisor for South Africa on the implementation of the LHWP, and also South African representative to negotiating royalty payments) personal phone interview by author, Austin, TX, March 5.

McKinney, D. C. (2008). *Transboundary Water Challenges: Case Studies*. Austin, TX: University of Texas at Austin.

Majali, A. S., Anani, J. A., and Haddadin M. J. (2006). *Peacemaking: The Inside Story of the 1994 Jordanian–Israeli Treaty*. Norman, OK: University of Oklahoma Press.

Makovsky, D. (1997). The U.S. Promised to Recruit Funding for Water Desalination, with Obtaining Agreement between Israel and Jordan, *Haaretz*, May 11 [Hebrew].

Malzbender, D. and Kranz, N. (2010). "Stakeholder participation in the Orange-Senqu River Basin, South Africa." In A. Earle, A. Jägerskog, and J. Öjendal (eds.), *Transboundary Water Management: Principles and Practice*. Washington, DC: Earthscan.

Mathison, S. (1988). Why Triangulate? *Educational Researcher*, 17 (2), pp. 13–17.

Mechlem, K. (2002). Water as the Vehicle for Inter-state Cooperation: a Legal Perspective. In Conference Proceedings: From Conflict to Co-operation in International Water Resources Management: Challenges and Opportunities, International Conference, November 20–22, 2002, Delft, the Netherlands: UNESCO-IHE Institute for Water Education.

Meissner, R. and Turton, A. R. (2003). The Hydro Social Contract Theory and the Lesotho Highlands Water Project, *Water Policy*, 5 (2), pp. 115–126.

Mekonnen, K. (1999). *The Defects and Effects of Past Treaties and Agreements on the Nile River Waters: Whose Faults were they?* Claremont, CA: The Blue Nile Resource Center, Pitzer College.

Mimi, Z. and Bassam, S. (2003). A Decision Tool for Allocating the Waters of the Jordan River Basin between all Riparian Parties, *Water Resources Management*, 17, pp. 447–461.

Mirumachi, N. (2007). "River Development and Bilateral Cooperation: The Lesotho Highlands Water Project Case Study." In V. Grover (ed.), *Water: A Source of Conflict or Cooperation?* Enfield, NH: Science Publishers.

Mirumachi, N. and Allan, J. A. (2007). "Revisiting Transboundary Water Governance: Power, Conflict, Cooperation and the Political Economy." In International Conference on Adaptive and Integrated Water Management, November 12–15, Basel, Switzerland.

Mirza, M. *et al.* (2004). The Ganges Water Diversion. In M. M. Q. Mirza (ed.), *The Ganges Water Diversion: Environmental Effects and Implications.* Toronto, ON: Springer.

Mitchell, C. R. (2006). "Conflict, Social Change and Conflict Resolution. An Enquiry." In D. Bloomfield, M. Fischer, and B. Schmelzle (eds.), *Social Change and Conflict Transformation.* Berlin: Berghof Research Center for Constructive Conflict Management.

Mitchell, T. (2002). *Rule of Experts: Egypt, Techno-Politics, Modernity.* Berkeley, CA: University of California Press.

Moeti, L., Mphale, M., Makoae, M. and Tango International Consultants (2003). *Lesotho Emergency Food Security Assessment Report.* Maseru: Lesotho Vulnerability Assessment Committee.

Mohamed, A. E. (2003). "Joint Development and Cooperation in International Water Resources." In M. Nakayama (ed.), *International Waters in Southern Africa.* Tokyo: United Nations University Press.

Mohamed-Katerere, J. and van der Zaag, P. (2003). "Untying the 'Knot of Silence,' Making Water Policy and Law Responsive to Local Normative Systems." In F. A. Hassan, M. Reuss, J. Trottier, C. Bernhardt, A. T. Wolf, J. Mohamed-Katerere, and P. van der Zaag (eds.), *History and Future of Shared Water Resources.* IHP Technical Documents in Hydrology – PCCP series. No. 6. Paris: UNESCO.

Naff, T. and Matson, R. C. (1984). *Water in the Middle East, Conflict or Cooperation?* Boulder, CO: Westview.

NBI (Nile Basin Initiative) (2006). *NBI Background* [electronic version]. Retrieved April 15, 2009, from: www.nilebasin.org

Newton, J. T. and Wolf, A. T. (2008). "Case Studies of Transboundary Dispute Resolution." In J. D. Priscoli and A. T. Wolf (eds.), *Managing and Transforming Water Conflicts.* Cambridge: Cambridge University Press.

Nicol, A. (2003). *The Nile: Moving Beyond Cooperation.* Water Policy Programme, ODI. Retrieved June 4, 2013, from: http://webworld.unesco.org/water/wwap/pccp/cd/pdf/case_studies/nile.pdf

Nishat, A. (2001). "Development and Management of Water Resources in Bangladesh: Post-1996 Treaty Opportunities." In A. K. Biswas and J. I. Uitto (eds.), *Sustainable Development of the Ganges–Brahmaputra–Meghna Basins.* Tokyo/New York: United Nations University Press.

Nishat, A. (2008). (Professor of Water Resources Engineering, IUCN Bangladesh Country Office, and water negotiator on behalf of Bangladesh in the Ganges Water Treaty negotiations) personal phone interview by author, Austin, TX, September 30.

Nishat, A. and Pasha, M. F. K. (2001). A Review of the Ganges Treaty of 1996, Globalization and Water Resources Management: The Changing Value of Water, AWRA/IWLRI-University of Dundee International Specialty Conference, August 6–8.

Nozick, R. (1974). *Anarchy, State, and Utopia.* New York: Basic Books.

Otchet, A. (2001). Striking on Trou, *The UNESCO Courier: Striking Peace on Troubled Waters,* October, p. 16. Paris: UNESCO.

Pallett, J., Heyns, P., Falkenmark, M., Lundqvist, J., Seeley, M., Hydén, L., Bethune, S., Drangert, J. -O., and Kemper, K. (1997). *Sharing Water in Southern Africa.* Windhoek: Desert Research Foundation of Namibia.

PASSIA (Palestinian Academic Society for the Study of International Affairs) (2002). *Water: The Blue Gold of the Middle East.* Palestine: PASSIA.

Peet, R. and Watts, M. (2004). *Liberation Ecologies: Environment, Development, Social Movements.* London: Routledge.

Phillips, D. (2008). (Freelance Consultant, England) personal phone interview by author, Austin, TX, December 26.

Phillips, D., Daoudy, M., McCaffrey, S., Öjendal, J., and Turton, A. (2006). *Trans-boundary Water Cooperation as a Tool for Conflict Prevention and for Broader Benefit-sharing.* Windhoek: Ministry for Foreign Affairs, Sweden Phillips Robinson and Associates.

Phillips, D. J. H., Attili, S., McCaffrey, S., and Murray, J. S. (2007). The Jordan River Basin: 2. Potential Future Allocations to the Co-Riparians, *Water International,* 32 (1), pp. 39–62.

PNIC (Palestinian National Information Center) (1999). *Dangers Threatening Water Resources in Palestine.* PNA, Ramallah: States Information Service.

Poivey, Y. (1997). *TED Case Studies: Lesotho Water Exports to South Africa.* The School of International Service, American University [electronic version]. Retrieved May 5, 2009, from: www.gurukul.ucc.american.edu/ted/lesotho.htm

Population Reference Bureau (2009). *Data by Geography.* Retrieved March 24, 2009, from: www.prb.org/

Postel, S. (1997). *Last Oasis: Facing Water Scarcity.* New York: W.W. Norton & Company, Inc.

Postel, S. and Wolf, A. (2001). Dehydrating Conflict, *Foreign Policy,* September/October, pp. 60–67.

Princeton University, Woodrow Wilson School of Public and International Affairs (2006). *Water Rights in the Jordan Valley.* Retrieved May 18, 2008, from: www.wws.princeton.edu/wws401c/geography.html

Priscoli, J. D. and Wolf, A. T. (2009). *Managing and Transforming Water Conflicts.* Cambridge: Cambridge University Press.

Rahaman, M. (2006). The Ganges Water Conflict: A Comparative Analysis of 1977 Agreement and 1996 Treaty, *Asteriskos,* 1 (2), pp. 195–208.

Raiffa, H. (1982). *The Art and Science of Negotiations.* Cambridge, MA: Belkamp/Harvard University Press.

Ramaswamy, R. I. (2003). *Water: Perspectives, Issues, Concerns.* New Delhi: Sage Publications.

Ramaswamy, R. I. (2008). (Former Secretary Water Resources in the Government of India and Research Professor at the Centre for Policy Research, New Delhi) personal face-to-face interview by author, The University of Texas, Austin, TX, November 1.

Rawls, J. and Kelly, E. (2001). *Justice as Fairness: A Restatement.* Cambridge, MA: Harvard University Press.

Rawls, J. M. (1971). *A Theory of Justice.* Cambridge, MA: Harvard University Press.

Ribot, J. and Peluso, N. (2003). "A Theory of Access," *Rural Sociology,* 68 (2), pp. 153–182.

Robbins, D. (1983). Drought Plagues South Africa, *Africa Report,* 28 (4), pp. 56–58.

Robbins, P. (2004). *Political Ecology.* Oxford: Blackwell.

Rogers, P., Burden, R., and Lott, C. (1978). Systems Analysis and Modeling Techniques Applied to Water Management, *Natural Resources Forum*, 2, pp. 349–358.

Rothert, S. (1999). *When Big Dams Spell Disaster: Assessing the Lesotho Highlands Water Project*, Institute for Global Dialogue [electronic version]. Retrieved May 12, 2009, from: www.probeinternational.org/odious-debts/when-big-dams-spell-disaster-assessing-lesotho-highlands-water-project

Rouyer, A. R. (2000). *Turning Water into Politics: The Water Issue in the Palestinian–Israeli Conflict*. Moscow, ID: Department of Political Science, University of Idaho.

Roy, R. (1997). *WOW Case Studies, India–Bangladesh Water Dispute* [electronic version]. Retrieved July 5, 2009, from: www1.american.edu/ted/ice/indobang.htm#rl.

Salame, L. and van der Zaag, P. (2010). "Enhanced Knowledge and Education Systems for Strengthening the Capacity of Transboundary Water Management." In A. Earle, A. Jägerskog, and J. Öjendal (eds.), *Transboundary Water Management: Principles and Practice*. Washington, DC: Earthscan.

Salameh, E. (1997). *The Climate of Jordan*. Retrieved April 3, 2006, from: www.diam.unige.it/~idromet/avi080/jord1.htm

Salameh, E. and Bannayan, H. (1993). *Water Resources of Jordan: Present Status and Future Potentials*. Amman: Friedrich Ebert Stiftung.

Salman, S. and Boisson de Chazournes, L. (1998). International Watercourses: Enhancing Cooperation and Managing Conflict, Proceedings of the World Bank Seminar, Washington, DC: World Bank.

Sartori, G. (1991). Comparing and Miscomparing, *Journal of Theoretical Politics*, 3, pp. 243–257.

Scheumann, W. (1998). *Water in the Middle East: Potential for Conflicts and Prospects for Cooperation*. Berlin: Technical University of Berlin.

Schwartz, K. (2006). "Masters in Our Native Place": The Politics of Latvian National Parks on the Road from Communism to "Europe", *Political Geography*, 25, pp. 42–71

Scott, T. S. (1985). Lesotho: The Politics of Dependence. *Kurimoto Gakuen Souritsu Gojyushunen Kinen Nagoya Shouka Daigaku Ronshu*, 30, pp. 739–756.

Sechaba Consultants (2000). *Poverty and Livelihoods in Lesotho*. Maseru: Sechaba Consultants.

Selby, J. (2003). *Water Power, and Politics in the Middle East – the Other Israeli–Palestinian Conflict*. London: I.B. Tauris & Co Ltd.

Selby, J. (2005). *The Geopolitics of Water in the Middle East: Fantasies and Realities*. Brighton: Department of International Relations & Politics, University of Sussex.

Selby, J. (2013). Cooperation, Domination and Colonisation: The Israeli–Palestinian Joint Water Committee, *Water Alternatives*, 6 (1), pp. 1–24.

Shamir, U. (1998). "Water Agreements Between Israel and Its Neighbors." In J. Albert, M. Bernhardson, and R. Kenna (eds.), *Transformations of Middle Eastern Natural Environments: Legacies and Lessons*, Bulletin Series, Number 103. New Haven, CT: Yale School of Forestry and Environmental Studies.

Shamir, U. (2008). (Stephen and Nancy Grand Water Research Institute, Technion, Israel Institute of Technology, Haifa, Israel, and Member of the Israeli Negotiation Team in the Israel–Jordan Peace Talks and) personal phone interview by author, Austin, TX, September 18.

Shamir, U. and Haddadin, M. (2003). *Jordan Case Study: The Jordan River Basin, Part I and II*. Paris: UNESCO-IHP, PCCP Series Publication.

Shiva, V. (2002). *Water Wars: Privatization, Pollution and Profit*. Cambridge, MA: South End Press.

Shmueli, D. F. (1999). Approaches to Water Dispute Resolution: Applications to Arab–Israeli Negotiations, *International Negotiation*, 4 (2), pp. 295–325.

Shrestha, M. and Singh, L. M. (1996). "The Ganges–Brahmaputra System: A Nepalese Perspective in the Context of Regional Co-operation." In B. G. Verghese and T. Hashimoto (eds.), *Asian International Waters: From Ganges–Brahmputra to Mekong.* Oxford: Oxford University, pp. 81–94.

Shungur, S. A. (2005). *Cooperation Among Adversaries: Managing Transboundary Water Disputes In Conflict Settings.* Montreal: Department of Political Science, McGill University.

Shuval, H. (1992). Approaches To Resolving The Water Conflicts Between Israel and Her Neighbors – A Regional Water for Peace Plan, *Water International*, 17, pp. 133–143.

Shuval, H. (2000). A Proposal for an Equitable Resolution to the Conflicts Between the Israelis and the Palestinians Over the Shared Water Resources of the Mountain Aquifer, *Arab Studies Quarterly*, 22 (2), pp. 33–62.

Shuval, H. (2005). A Proposal for an Equitable Reallocation of the Shared Water Resources Between Israelis and Palestinians and the Other Riparians on the Jordan River Basin, First International Conference on Water Values and Rights, Ramallah, Palestine, May 2–4.

Shuval, H. (2007). "Meeting Vital Human Needs: Equitable Resolution of Conflicts over Shared Water Resources of Israelis and Palestinians." In H. Shuval and H. Dweik (eds.), *Water Resources in the Middle East.* Berlin: Springer.

Simpson, J. (2008). *Lesotho: It's all about the Pokotho* [electronic version]. Retrieved August 9, 2009, from: www.waterblogged.info/2008/09/19/lesotho-its-all-about-the-pokotho/

Slany W. Z. and Baehler, D. (1986). *Foreign Relations of the United States 1952–1954.* Washington, DC: United States Government Printing Office, Department of State.

Söderblom, J. D. (2003). *State of Inequity: The UN Partition Plan of 1947.* Canberra: The Terrorism Intelligence Centre.

Sosland, J. (2007). *Cooperating Rivals: The Riparian Politics of the Jordan River Basin.* Albany, NY: SUNY Press.

SPIDR (1997). *Best Practices for Government Agencies: Guidelines for Using Collaborative Agreement-Seeking Processes.* Albany, NY: State University of New York Press.

Spiegel, C. (2005). *International Water Law: the Contributions of Western United States Water Law to the United Nations Convention on the Law of the Non-navigable Uses of International Watercourses.* Durham, NC: Duke University School of Law.

Stake, R. (1995). *The Art of Case Study Research.* New York: Sage Publications.

Starr, J. R. (1991). Water Wars, *Foreign Policy*, 82 (Spring), pp. 17–36.

Stevens, R. P. (1967). *Lesotho, Botswana, and Swaziland: The Former High Commission Territories in Southern Africa.* New York: Fredrick A. Praeger Publishers.

Susskind, L. (1994). *Environmental Diplomacy: Negotiating More Effective Global Agreements.* New York: Oxford University Press.

Susskind, L. and Cruikshank, J. (1987). *Breaking The Impasse: Consensual Approaches To Resolving Public Disputes.* New York: Basic Books.

Swain, A. (2004). *Managing Water Conflict.* Oxford: Routledge.

TCTA–LHDA (Trans-Caledon Tunnel Authority–Lesotho Highlands Development Authority) (2003). *Sustainable Development: Lesotho Highlands Water Project* [Brochure]. Pretoria: TCTA–LHDA.

TFDD (*Transboundary Freshwater Dispute Database*) (2008). Oregon State University. [electronic version]. Retrieved April 15, 2008, from: www.transboundarywaters.orst. edu/.

Towfique, B. (2002). International Bilateral Water Treaties: An Economic and Institutional Analysis. PhD Dissertation, Clemson University.

Treaty of Peace (1994). Treaty of Peace between the State of Israel and the Hashemite Kingdom of Jordan, October 26, 1994.

Tromp, L. (2006). Lesotho Highlands: the Socio-Economics of Exporting Water, *Proceedings of the Institution of Civil Engineers – Civil Engineering*, 159, pp. 44–49.

Turner, S. D. (2003). *The Southern African Food Crisis: Lesotho Literature Review*. Care, Johannesburg. Retrieved June 8, 2013, from: www.sarpn.org/documents/d0000689/

Turton, A. (2003). "An Overview of the Hydropolitical Dynamics Of The Orange River Basin." In N. Mikiyasu (ed.), *International Waters in Southern Africa*. Tokyo: United Nations University Press.

Turton, A. and Henwood, R. (eds.) (2002). *Hydropolitics in the Developing World: A Southern African Perspective*. Pretoria: African Water Issues Research Unit (AWIRU), University of Pretoria.

Turton, A. R., Meissner, R., Mampane, P. M., and Seremo, O. (2004). *A Hydropolitical History Of South Africa's International River Basins*. Report No. 1220/1/04 to the Water Research Commission. Pretoria: Water Research Commission.

Tyler, T. R. and Huo, Y. J. (2002). *Trust in the Law: Encouraging Public Cooperation with the Police and Courts*. New York: Russell-Sage Foundation.

Uitto, J. (2008) (Academic Officer at the United Nations University in Tokyo, Japan) personal phone interview by author, Austin, TX, October 29.

UNESCO (United Nations Educational, Scientific and Cultural Organization) (2006). *Case Studies: South Africa*. World Water Assessment Programme (WWAP). Retrieved June 30, 2008, from: www.unesco.org/water/wwap/case_studies/index.shtml

UNFAO (United Nations Food and Agriculture Organization) (1978). *Systematic Index of International Water Resources Treaties, Declarations, Acts and Cases, by Basin: Volume I*. Legislative Study #15.

UNFAO (1984). *Systematic Index of International Water Resources Treaties, Declarations, Acts and Cases, by Basin: Volume II*. Legislative Study #34.

United Nations (1997). *Convention on the Law of the Non-navigational Uses of International Watercourses*. Retrieved May 20, 2008, from: www.thewaterpage.com/UN_Convention_97.html

United Nations (1999). *United Nations Treaty Collection: Treaty Reference Guide*. United Nations. Retrieved June 12, 2013, from: http://untreaty.un.org/ola-internet/assistance/guide.htm

United Nations HDR (Human Development Report) (2006). *Beyond Scarcity: Power, Poverty and the Global Water Crisis*. New York: United Nations Development Programme.

United Press International (2003). Turkish Prime Minister to Visit Syria. 3 January.

Upreti, T. (2006). *International Watercourses Law and Its Application in South Asia*. Kathmandu: Pairavi Prakashan.

Viessman, W. and Welty, C. (1985). *Water Management: Technology And Institutions*. New York: Harper & Row.

Villiers, G. and De, T. (1996). South Africa's Water Resources and the Lesotho Highlands Water Scheme: A Partial Solution to the Country's Water Problems, *International Journal of Water Resources Development*, 12 (1), pp. 65–78.

Vinogradov, S., Wouters, P., and Jones, P. (2003). *Transforming Potential Conflict into Cooperation Potential: The Role of International Water Law*. Dundee: University of Dundee.

Wade, J. H. (2004). Duelling Experts in Mediation and Negotiation: How to Respond When Eager Expensive Entrenched Expert Egos Escalate Enmity, *Conflict Resolution Quarterly*, 21 (4), pp. 419–436.

Walker, P. (2006). Political Ecology: Where is the Policy? *Progress in Human Geography*, 30 (3), pp. 382–295.

Wallis, S. (1996). *Lesotho Highlands Water Project, Volume 2.* Surrey: Laserline.

Wallis, S. (2000). *Lesotho Highlands Water Project.* Surrey: Laserline.

Waterbury, J. (1979). *Hydropolitics of the Nile Valley.* Syracuse, NY: Syracuse University Press.

Waterbury, J. (2002). *The Nile Basin: National Determinants of Collective Action.* New Haven, CT: Yale University Press.

Water Care (2004). Water. Multilateral Working Group on Water Resources, Middle East Peace Process. Retrieved June 3, 2009, from: www.watercare.org/WaterCare/index.htm

WBCSD (World Business Council for Sustainable Development) (2005). *Facts and Trends: Water.* Geneva: WBCSD.

Weisfelder, R. (1979). "Lesotho: Changing Patterns of Dependence". In G. Carter and P. O'Meara (eds.), *Southern Africa: The Continuing Crisis.* Bloomington, IN: Indiana University Press.

Whann, C. A. (1995). *The Revenue Imperative and States Management in Lesotho.* Madison, WI: The University of Wisconsin.

Whiteman, M. M. (1964). *Digest of International Law, Vol. 3.* Washington, DC: Department of State.

WHO (World Health Organization) (2003). *Right to Water.* Retrieved May 18, 2006, from: www.who.int/water_sanitation_health/en/

Willemse, N. E. (2007). Actual versus Predicted Transboundary Impact: A Case Study of Phase 1B of the Lesotho Highlands Water Project, *International Journal of Water Resources Development*, 23 (3), pp. 457–472.

Wolf, A. (1994). "A hydro-political history of the Nile, Jordan, and Euphrates River Basins." In A. K. Biswas (ed.), *International Water of the Middle East from Euphrates–Tigris to Nile.* New York: Oxford University Press.

Wolf, A. (1995). *Hydropolitics along the Jordan River – Scarce Water and its Impact on the Arab–Israeli Conflict.* Tokyo: United Nations University Press.

Wolf, A. (1996). *Middle East Water Conflicts and Directions for Conflict Resolution.* Washington, DC: International Food Policy Research Institute.

Wolf, A. (1997). International Water Conflict Resolution: Lessons From Comparative Analysis, *International Journal of Water Resources Development*, 13 (3), pp. 333–365.

Wolf, A. (1998). Conflict and Cooperation Along International Waterways, *Water Policy*, 1 (2), pp. 251–265.

Wolf, A. (1999). Criteria For Equitable Allocations: The Heart of International Water Conflict, *Natural Resources Forum*, 23 (1), pp. 3–30.

Wolf, A. (2000). Indigenous Approaches to Water Conflict Negotiations and Implications for International Waters, *International Negotiation. A Journal of Theory and Practice*, 5 (2), p. 17.

Wolf, A. (2001). Transboundary Waters: Sharing Benefits, Lessons Learned. International conference on Freshwater, Bonn, Germany, December 3–7.

Wolf, A. (2002). *Atlas of International Freshwater Agreements.* United Nations Environment Programme. Corvallis, OR: Oregon State University.

Wolf, A. (2003). "Southern Africa and the International Water Problematique." In A. R. Turton, P. Ashton, and E. Cloete (eds.), *Transboundary Rivers, Sovereignty and*

Development: Hydropolitical Drivers in the Okavango River Basin. Pretoria: University of Pretoria.

Wolf, A. (2004). *Regional Water Cooperation as Confidence Building: Water Management as a Strategy for Peace*. Berlin: Adelphi Research.

Wolf, A. (2006). Human Development Report: Conflict and Cooperation Over Transboundary Waters. Corvallis, OR: UNDP, Oregon State University.

Wolf, A. T. and Newton, J. T. (2008). "Case Studies of Transboundary Dispute Resolution." In J. Delli Priscoli and A. T. Wolf (eds.), *Managing and Transforming Water Conflicts*. Cambridge: Cambridge University Press.

Wolf, A. and Newton, J. (2010a). "The Indus Treaty: Conflicting Riparians and the Role of a Third Party." In A. Earle, A. Jägerskog, and J. Öjendal (eds.), *Transboundary Water Management: Principles and Practice*. Washington, DC: Earthscan.

Wolf, A. and Newton, J. (2010b). "Benefit sharing of the Lesotho Highlands Water Project." In A. Earle, A. Jägerskog, and J. Öjendal (eds.), *Transboundary Water Management: Principles and Practice*. Washington, DC: Earthscan.

Wolf, A., Yoff, S. B., and Giordano M. (2003). *International Waters: Indicators for Identifying Basins at Risk*. Corvallis, OR: Department of Geosciences, Oregon State University.

Wolf, A., Kramer, A., Carius, A., and Dabelko, G. (2006). *Water Can be a Pathway to Peace, Not War, Navigating Peace*. No. 1, July. Washington, DC: Woodrow Wilson International Center for Scholars.

Wolford, W. (2004). This Land Is Ours Now: Spatial Imaginaries and the Struggle for Land in Brazil, *Annals of the Association of American Geographers*, 94 (2), pp. 409–424.

Woodhouse, M. (2008). (implementation advisor on the ORACOM and the SADC Protocol at an operational level) personal phone interview by author, Austin, TX, December 13.

World Bank (2000). *2000 World Development Indicators*. Washington, DC: The International Bank for Reconstruction and Development.

van Wyk, J. A. (1998). Towards Water Security in Southern Africa, *African Security Review*, 7(2). Retrieved July 20, 2009, from: www.iss.co.za/PUBS/ASR/7No2/VanWyk.html

Yin, R. (1994). *Case Study Research: Design and Methods*, 2nd edn. Beverly Hills, CA: Sage Publishing.

Yin, R. (2003). *Case Study Research*, 3rd edn. Thousand Oaks, CA: Sage Publications.

Zartman, I. W. (1991). "The Structure of Negotiation." In V. Kremenyuk (ed.), *International Negotiation*. San Francisco, CA: Jossey-Bass.

Zawahri, N. A. (2004). *The Water Weapon: Havoc and Harmony over International Rivers*. Charlottesville, VA: Department of Politics, University of Virginia.

Zawahri, N. A. (2006). Stabilizing Iraq's Water Supply: What the Euphrates and Tigris Rivers can Learn from the Indus, *Third World Quarterly*, 27 (6), pp. 1041–1058.

Zawahri, N. (2009). Attempts at Governing the Jordan River System, Transboundary Water Crises conference, New Mexico State University, January 22.

Zeitoun, M. (2008a). *Power and Water in the Middle East: The Hidden Politics of the Palestinian–Israeli Water Conflict*. London: I. B. Tauris.

Zeitoun, M. (2008b). (Senior Lecturer in Development Studies, Centre for Environmental Policy and Governance, Department of Geography & Environment London School of Economics and Political Science), personal phone interview by author, Austin, TX, November 26.

Zeitoun, M. and Allan, J. (2008). Applying Hegemony and Power Theory to Transboundary Water Analysis, *Water Policy*, 10 (S2), pp. 3–12.

Index